Getting Biodiversity Projects to Work

Biology and Resource Management Series

Biology and Resource Management Series
Edited by Michael J. Balick, Anthony B. Anderson, Charles M. Peters, and Kent H. Redford

Alternatives to Deforestation: Steps Toward Sustainable Use of the Amazon Rain Forest,
edited by Anthony B. Anderson

Useful Palms of the World: A Synoptic Bibliography,
compiled and edited by Michael J. Balick and Hans T. Beck

The Subsidy from Nature: Palm Forests, Peasantry, and Development on an Amazon Frontier,
by Anthony B. Anderson, Peter H. May, and Michael J. Balick

Contested Frontiers in Amazonia,
by Marianne Schmink and Charles H. Wood

Conservation of Neotropical Forests: Working from Traditional Resource Use,
edited by Kent H. Redford and Christine Padoch

The African Leopard: Ecology and Behavior of a Solitary Felid,
by Theodore N. Bailey

Footprints of the Forest: Ka'apor Ethnobotany—The Historical Ecology of Plant Utilization by an Amazonian People,
by William Balée

Medicinal Resources of the Tropical Forest: Biodiversity and Its Importance to Human Health,
edited by Michael J. Balick, Elaine Elisabetsky, and Sarah A. Laird

Tropical Deforestation: The Human Dimension,
edited by Leslie K. Sponsel, Thomas N. Headland, and Robert C. Bailey

The Catfish Connection: Ecology, Migration, and Conservation of Amazon Predators,
by Ronaldo Barthem and Michael Goulding

So Fruitful a Fish: Ecology, Conservation, and Aquaculture of the Amazon's Tambaqui,
by Carlos Araujo-Lima and Michael Goulding

GIS Methodologies for Developing Conservation Strategies: Tropical Forest Recovery and Wildlife Management in Costa Rica,
edited by Basil G. Savitsky and Thomas E. Lacher Jr.

Hunting for Sustainability in Tropical Forests,
edited by John G. Robinson and Elizabeth L. Bennett

The Cutting Edge: Conserving Wildlife in Logged Tropical Forests,
edited by Robert A. Fimbel, Alejandro Grajal, and John G. Robinson

The Return of the Unicorns: The Natural History and Conservation of the Greater One-Horned Rhinoceros,
by Eric Dinerstein

Getting Biodiversity Projects to Work

Towards More Effective Conservation and Development

Edited by Thomas O. McShane and Michael P. Wells

Columbia University Press
New York

Columbia University Press
Publishers Since 1893
New York Chichester, West Sussex

Library of Congress Cataloging-in-Publication Data

Getting biodiversity projects to work : towards more effective conservation and development / edited by Thomas O. McShane and Michael P. Wells.
p. cm. — (Biology and resource management series)
Includes bibliographical references (p.).
ISBN 0–231–12764–2 (cloth : alk. paper) — ISBN 0–231–12765–0 (pbk. : alk. paper)
1. Biological diversity conservation. 2. Sustainable development. 3. Conservation projects (Natural resources). I. McShane, Thomas O. II. Wells, Michael P. III. Biology and resource management in the tropics series.

QH75.G475 2004
333.95'16—dc22 2003068811

Columiba University Press books are printed on permanent and durable acid-free paper.

Printed in the United States of America
c 10 9 8 7 6 5 4 3 2 1
p 10 9 8 7 6 5 4 3 2 1

Contents

Foreword

It is a simple truism that humans cannot exist without nature. We are all part of it and will forever depend on the natural environment for food, water, air, and innumerable goods. Planet Earth's biosphere is essential for the survival of humanity, not the other way round. However, what is an incontestable fact at the planetary level may not be quite the same on a smaller scale. Local conditions are often influenced by economic and environmental factors emanating from afar. The livelihood, opportunities, and prospects of local communities are increasingly affected by global influences, whether through adverse trade effects or global environmental impacts such as climate change or toxic pollutants.

The tension between global and local is one of the basic difficulties in balancing the interests of conservation and development. This was reflected at the UN Conference on Environment and Development (UNCED) in Rio de Janeiro in 1992 when global aspects were addressed (e.g., through the Conventions on Biological Diversity and Climate Change) and local considerations were represented through Agenda 21. Ten years later at the World Summit on Sustainable Development (WSSD) in Johannesburg, these issues were still alive. The integrated conservation and development project (ICDP) as a defined concept dates back roughly to the same period as the Rio de Janeiro conference in 1992. Although the practice of integrating biodiversity conservation aims with community development objectives was not new, it became a more clearly defined and more deliberate approach, transposing the conference objectives to local contexts.

The other element of tension ingrained in the integrated conservation and development project concept relates to the fact that "conservation" strives for long-term stability—the preservation of biodiversity and maintenance of ecological processes. "Development," on the other hand, signals a dynamic process—a response to the short-term economic and social needs of local communities as well as their long-term aspirations. Reconciling such fundamentally different objectives is clearly a challenging task. Idealizing the interdependence of conservation and development—or worse, giving it a romantic touch, as some northern groups have tended to do—is not helpful or realistic. As much as the harmony between humans and nature is a global prerequisite for the future of the planet, forcing this philosophy into a microcosmos without considering the policy context, the macroeconomic root causes of poverty, and global effects is to do the ICDP concept a fatal disservice.

One might ask the perfectly legitimate question: Is a concept loaded with such obvious tensions still a valid one? But what are the alternatives? Development at the expense of the environment goes against long-term sustainability and thus the livelihood of local communities. On the other hand, biodiversity conservation without respect for the needs of local communities, as some radical nature protectionists advocate, requires policing, leads to alienation and land-use conflicts, and ultimately defeats itself. Reconciling conservation with development at the local level is clearly not going to be easy, and the fact that ICDPs have proven difficult to translate into lasting success stories does not justify a conclusion that the approach is flawed. Rather, it signals the need to take a long, hard look at the experience gained so far, identify the lessons, and ensure that they are reflected in our future efforts.

Thomas McShane and Michael Wells have closely followed the spectacular rise of ICDPs, as well as the subsequent period of doubt and uncertainty. They are both convinced that demonstrating constructive ways of involving local stakeholders in the conservation and sustainable use of biodiversity in and around the most significant protected areas remains one of the most important challenges and priorities for nature conservation at the beginning of the twenty-first century. For this book they have been joined by an impressive group of colleagues to revisit key aspects of the ICDP experience, reinterpret these in the light of today's conservation priorities, identify the most promising approaches, and map out a way forward. Their message is an important one.

If we are serious about the future of the planet, then we must conserve biodiversity. To do this, we have to make protected areas more effective. Protected areas can have little future without the cooperation and support of local populations. How to achieve this cooperation and support without jeopardizing conservation is the subject of this book.

Dr. Claude Martin
Director General
WWF International
Gland, Switzerland

Preface

THERE once was a sage resting quietly in a garden outside a monastery built into evergreens in a ravine on a high hill. At the entrance gate at the mouth of the ravine stood a young man with searching eyes. "Oh wise one, may I enter your refuge?" the young man called. "Yes, my son," replied the sage. The young man ascended the steep hill by leafy walks that passed beneath the delicate light-filled red-bronze of Japanese maple mixed among pale bamboo. The young man knelt in front of the sage.

"Oh sage, they speak of you everywhere I go. Your knowledge is known far and wide. Tell me, how have you become so wise?" Bronze pigeons crossed the trees where the ravine vanished into the forest. In a pond beneath a moss-green wall, the sun glinted on the raised red and yellow head of an old turtle resting on an ancient rock, drawing its slow eye closed as it stared and listened, winking out the world, then letting in the world slowly, slowly once again.

"My son," the sage replied, "the decisions I have made in my life have been resolute and firm, correct in most every way. These have guided me through life and in how I live." Sunlight descended through a canopy of delicate maple leaves, alighting here and there on pink sprays of wild azaleas. A bird called in the distance.

"Oh wise one, tell me, how were you able to make such decisions, so resolute and firm?" the young man asked. As the sun peered down over the eastern wall of the ravine, lighting silver fires in the stream above the monastery, two wagtails teetered on river stones of shining black, stuck with pink cherry blossoms, and a dipper fluttered on the water sparkle, rounding a bend.

The sage's eyes closed. "Ah, my son, such decisions are based on experience. I am an old man. I have wandered far and wide. I have seen many things, I have done many things, and I have not forgotten." In the distance kelp and oyster beds were being tended by small boats, while solitary old women of another epoch, bent by big baskets, up to their hips in water, foraged for shellfish in the tidal rocks. Far out in the bay, a small, still boat on the silver sea, against the islands, evoked that immemorial solitude that draws painters and poets.

"Oh sage, how does one get such experience to make such sound decisions to become so wise? I want to learn." A small fishing village sat perched at the mouth of the river, while gulls, terns, herons, and an osprey convened around the calm waters of an estuary pool behind the beach. The beach was disfigured by the wrack of plastic and industrial flotsam, which, even on this remote coast, so far from cities, was left stranded on the sand after each tide. The eager materialism of humanity.

The sage smiled as he gazed down on the scene. "Ah, my son. Very bad decisions. Lots of very bad decisions." In the distance could be heard the lovely song of an unknown bird.

For many conservation organizations, government departments, and development agencies, the need to integrate conservation and development is obvious. As fragmentation of natural habitat reduces protected areas to islands in a sea of humanity, only a broad-scale conservation strategy that takes account of people's needs will suffice to conserve the world's biodiversity. As a result, conservationists increasingly have looked to work in communities that border protected areas to better address conservation goals and objectives.

Most of the financial resources available for biodiversity conservation in developing countries have been spent on the most popular approach, the integrated conservation and development project, or ICDP. The ICDP concept includes a variety of initiatives that aim to link the conservation of biodiversity, usually in protected areas, with the social and economic development of neighboring communities. The underlying assumption is that providing alternative sources of livelihood to communities in or near protected areas will ultimately reduce pressure on the exploitation of resources in these areas. Although ICDPs were originally promoted and supported mainly by non-governmental organizations (NGOs), they have also attracted significant amounts of mainstream funding from international development agencies and other official donors.

In practice, the ICDP record has been weak. There are still important unanswered questions concerning the effectiveness, and even appropriateness, of the ICDP approach. Many of the key underlying assumptions remain untested, while others have proven false. Some conservationists have argued that the ICDP focus on development and poverty alleviation effectively dilutes biodiversity conservation goals.

Over the last five years the World Wildlife Fund (WWF) has been implementing a program of ICDPs in collaboration with the Royal Netherlands Development Agency (DGIS). The DGIS-WWF Tropical Forest Portfolio is composed of seven field projects in six countries in Africa (Gabon [2], Ethiopia), Latin America (Ecuador, Honduras), and Asia (Pakistan, Philippines). Since its inception in 1996, the DGIS-WWF Tropical Forest Portfolio has aimed to draw lessons on the effectiveness of the ICDP approach. In bringing the Portfolio to a close, the WWF and DGIS held a workshop October 10–12, 2000, on integrated conservation and development.

The aim of the workshop was to address the question: "What is the appropriate role for ICDPs in tropical forest conservation and management?" During the workshop the question was addressed more generally as "What is the appropriate role for ICDPs?" The objectives were the following:

- Provide a forum to review existing ICDP approaches and assess their effectiveness in conserving and managing tropical forests and other biomes
- Explore cross-cutting and thematic issues (e.g., scales of effectiveness, biodiversity conservation, poverty alleviation, corresponding policy issues, institutional constraints, capacity requirements, etc.)
- Synthesize the existing experiences to better inform conservationists and decision makers about ICDP implementation
- Inform audiences of the role ICDPs should play in biodiversity conservation and development

Rather than issuing a general call for papers, we identified a number of critical thinkers and current ICDP implementers and commissioned papers from our distinguished colleagues on a variety of subjects. The result is presented in this book.

As conservationists, development agencies, government departments, local organizations, and resource users struggle to link biodiversity conservation with development, there continues to be a pressing need to better assess the experiences gained thus far and learn from the difficulties that others have experienced. This book is designed to help consolidate our knowledge thus far, and to influence the primary supporters and implementers of such approaches.

Clearly much remains to be learned about the ICDP experience. ICDPs cannot be all things to all people, as they have so often been marketed to raise funds. We have to acknowledge trade-offs between biodiversity, economics, and societies. The promotion of "win-win" outcomes is politically correct at best and naïve at worst. If anything, the one clear message that has come out of the ICDP experience is that we have to learn from our mistakes.

This book was made possible by the contributions and hard work of a number of people. We would like to thank the Royal Netherlands Government's Directorate General for International Cooperation (DGIS) and the minister for development cooperation for their support for the DGIS-WWF Tropical Forest Portfolio over the last five years. Their boldness and commitment to the portfolio as a learning experience rather than a series of quick fixes has allowed us to be creative and flexible while remaining focused on the issues at hand. Perhaps most importantly, they have been open to examining their own policies in light of our findings.

The WWF has been the institutional home of the Tropical Forest Portfolio. Its continued commitment to examining conservation and development issues has greatly facilitated the implementation of the program. Timothy Geer, Chris Elliott, and Jean-Paul Jeanrenaud have been most supportive. Astrid-Tine Bjorvik and Carole Hurlimann have kept the portfolio on a sound administrative and financial track, while handling much of the communications work.

Special thanks are due to the DGIS-WWF Tropical Forest Portfolio field implementers and partners who continue to struggle with many of the issues discussed in this book: Ejaz Ahmad, Vicente Alvarez, Hervé Ndong Allogho, Ermais Bekele, Soledad Cruz, Steve Gretzinger, Fetene Hailu, Syed Ali Hasnain, Sergio Herrera, Bas Huijbregts, Tesfaye Hundessa, Syed Najam Khurshid, Octave Mboumba, Aurélien Mofouma, Oswaldo Munguia, Gezahegn Negussie, Sosthène Ndong Obiang, Noorum Nisa, Jean-Marie Nkombé, Adalberto Padilla, Matthew Perl, Fayyaz Rasool, Jorge Rivas, Harlan Rivas, Ruth Elena Ruiz, Chrisma Salao, Rolando Tan, Marc Thibault, Edgardo Tongson, Romy Trono, Shahzadi Tunio, Carlos Vijil Moreno, Carmen Villaseñor, and Pauwel de Wachter.

Robin Smith, senior executive editor, sciences, at Columbia University Press has been enthusiastic and supportive of the publication from the start. His confidence in this work is most appreciated.

Guidance, debate, and discussion on ICDPs, the workshop, and this book have come from many people. We are grateful to Nigel Dudley, Arthur Ebregt, Tom Hammond, Erica McShane-Caluzi, Gonzalo Oviedo, John Ramsay, Kent Redford, Alan Rodgers, Hans van de Veen, Ton van der Zon, Marion van Schaik, Aneke Wevers, Sejal Worah, and three reviewers who wish to remain anonymous. All have generously provided their time and expertise. Finally, thanks are due to the contributors for bearing with us during the review period and publication process. We hope that the product lives up to their expectations.

Thomas O. McShane
Michael P. Wells

Contributors

Thomas Blomley, CARE Uganda, P.O. Box 7280, Kampala, Uganda, *blomley@careug.org*

Katrina Brandon, Center for Applied Biodiversity Science, Conservation International, 1919 M Street, N.W., Suite 600, Washington, DC 20036, U.S.A., *k.brandon@conservation.org*

Katrina Brown, School of Development Studies and CSERGE (Centre for Social and Economic Research on the Global Environment), University of East Anglia, Norwich NR4 7TJ, U.K., *k.brown@uea.ac.uk*

Brian Child, Development Services and Initiatives, P.O. Box 320207, 30G Sable Road, Kabulonga, Lusaka, Zambia, *adichild@zamnet.zm*

Barry Dalal-Clayton, International Institute for Environment and Development, 3 Endsleigh Street, London WC1H ODD, U.K., *Barry.dalal-clayton@iied.org*

Marisel Dino, Kabang-Kalikasan ng Pilipinas, LBI Building, 57 Kalayan Avenue, Diliman, Quezon City 1101, Philippines, *mdino@wwf.org.ph*

Holly T. Dublin, African Elephant Specialist Group, IUCN–The World Conservation Union, P.O. Box 68200, Nairobi, Kenya, *holly.dublin@ssc.iucn.org*

Nigel Dudley, Equilibrium, 47 The Quays, Cumberland Road, Spike Island, Bristol BS1 6UQ, U.K., *equilibrium@compuserve.com*

Phil Franks, CARE International, P.O. Box 43864–00100, Nairobi, Kenya, *phil@care.or.ke*

Curtis Freese, World Wildlife Fund–U.S., P.O. Box 7276, Bozeman, MT 59771, U.S.A., *cfreese@mcn.net*

Steve Gartlan, Projet Transfrontalière, Cameroun-Congo-Gabon, B.P. 9144, Libreville, Gabon, *sgartlan@mweb.co.za*

Dennis Glick, Sonoran Institute, 201 S. Wallace, Bozeman, MT 59715, U.S.A., *dennis@sonoran.org*

William Jackson, IUCN–The World Conservation Union, rue Mauverney 28, CH-1196, Gland, Switzerland, *wjj@iucn.org*

Agnes Kiss, Africa Environment and Social Development Group, J-6–604, The World Bank, 1818 H Street, N.W., Washington, DC 20433, U.S.A., *akiss@worldbank.org*

Stewart Maginnis, IUCN–The World Conservation Union, rue Mauverney 28, CH-1196, Gland, Switzerland, *Stewart.maginnis@iucn.org*

Richard Margoluis, Foundations of Success, 147 Willow Street, New Haven, CT 06511, U.S.A., *Richard@FOSonline.org*

Thomas O. McShane, WWF International, Avenue du Mont-Blanc, CH-1196, Gland, Switzerland, *tmcshane@wwfint.org*

Suad A. Newby, 40 rue Marius Jacotot, 92800 Puteaux, France, *suad .newby@wanadoo.fr*

Sheila O'Connor, WWF International, 39 Stoke Gabriel Road, Galmpton, Nr Brixham, Devon TQ5 0NQ, U.K., *soconnor@wwfint.org*

Michelle O'Herron, Graduate Program in Sustainable Development and Conservation Biology, Biology-Psychology Building, University of Maryland, College Park, MD 20742, U.S.A., *oherron@hotmail.com*

Kent H. Redford, Wildlife Conservation Society, 2300 Southern Blvd., Bronx, NY 10460, U.S.A., *kredford@wcs.org*

John G. Robinson, Wildlife Conservation Society, 2300 Southern Blvd., Bronx, NY 10460, U.S.A., *jrobinson@wcs.org*

Nick Salafsky, Foundations of Success, 4109 Maryland Avenue, Bethesda, MD 20816, U.S.A., *Nick@FOSonline.org*

Jeffrey Sayer, WWF International, Avenue du Mont-Blanc, CH-1196, Gland, Switzerland, *jsayer@wwfint.org*, and CIRAD–Forêt, Montpellier, France

Arpan Sharma, Indian Institute of Public Administration, IP Estate, New Delhi 110 002, India, *arpan@samrakshan.org*

Gill Shepherd, Forest Policy and Environment Group, Overseas Development Institute, Portland House, Stag Place, London, SW1E 5DP, U.K., *g.shepherd@odi.org.uk*

Shekhar Singh, Samya–Centre for Equity Studies, C17A Munirka, New Delhi 110 067, India, *shekharsingh@vsnl.com*

Edgardo Tongson, Kabang-Kalikasan ng Pilipinas, LBI Building, 57 Kalayan Avenue, Diliman, Quezon City 1101, Philippines, *etongson@wwf.org.ph*

Michael P. Wells, Tunnelveien 3, 3400 Lier, Norway, *wells@online.no*

Part One

The Challenge of Linking Conservation and Development

1

Integrated Conservation and Development?

Thomas O. McShane and Michael P. Wells

The ICDP Approach

Integrated conservation and development projects (ICDPs) have become one of the most widely implemented and yet controversial approaches to biodiversity conservation. The term itself emerged as a collective label for a new generation of projects that started to go outside park and reserve boundaries and pay particular attention to the welfare of local people (Wells and Brandon 1992). Most of these efforts took place in developing countries with international financial support. During the last two decades, ICDPs and their equivalents (comparable initiatives have used a variety of other labels) have exploded in popularity, rapidly metamorphosing from an untested idea attracting seed money for pilot projects to become widespread "best practice" for biodiversity conservation (Larson, Freudenberger, and Wyckoff-Baird 1998; Wells et al. 1999). ICDPs have attracted the lion's share of the relatively large investments in conservation projects by bilateral development agencies and the Global Environment Facility (GEF), the interim financing mechanism for the Convention on Biological Diversity, during the 1990s. A key factor behind the growth in popularity of ICDPs seems to have been the prospect of delivering working models of "sustainable development," which had become an overarching priority since the 1992 UN Conference on Environment and Development in Rio de Janeiro.

Even as the ICDP momentum was building, the initial experiences of some of the early field projects were disappointing. There was considerable uncertainty and debate over how much emphasis to put on biological versus social or economic goals, and the combining of conservation and development objectives was proving more difficult than had been anticipated (Wells and Brandon 1992). These possible indicators of the need for caution were generally overlooked, however, or were attributed to start-up problems that would be surmountable over time. Analysts began to question the contribution of ICDPs to biodiversity conservation, both from ecological perspectives (Redclift 1987; Sachs 1991; Stocking and Perkin 1992; Robinson 1993; Barrett and Arcese 1995) and social perspectives (West and Brechin 1991; Murphree 1993; Colchester 1994; Ghimire 1994; Ghimire and Pimbert 1997). Waves of new ICDPs continued to be launched, however, with few signs that lessons from the existing projects had been taken into account.

The seeds of doubt grew as more time passed, and convincing cases where ICDPs had effectively helped reconcile local people's development needs with protected area management remained difficult to find. It became increasingly clear that many of the success stories reported earlier had been based more on overoptimistic ICDP goals and objectives than on a calm analysis of actual experience. By the millennium, concern among the organizations both implementing and financing ICDPs had become widespread, fueled by an expanding barrage of mostly critical literature (Agrawal 1997; Sanjayan, Shen, and Jansen 1997; Larson, Freudenberger, and Wyckoff-Baird 1998; Wells et al. 1999; McShane 1999; Few 2001; Hughes and Flintan 2001).

A backlash against the ICDP approach has begun to gain momentum among the organizations actually carrying out conservation activities in developing countries, as well as among the agencies providing the funding. The staff of some conservation organizations have begun questioning whether projects that emphasize sustainable development—a term that remains frustratingly elusive to define—can in practice be compatible with biodiversity conservation. Meanwhile, as the twenty-first century begins, the international development agencies show few signs of prioritizing biodiversity conservation or ICDPs in their efforts to help achieve the Millennium Development Goals by 2015.

The Aim of This Book

The editors of and contributors to this book all have extensive experience in various aspects of biodiversity conservation, from ecological, economic, and social perspectives, and include both practitioners and researchers. All have seen and experienced the changing priorities, issues, and trends in the practice of biodiversity conservation during the last decade or so. The rapid rise of ICDPs and now the uncertainty over the future of this approach have motivated these contributors to take this in-depth look at experience to date, to work out what has happened and why, and to map out a way forward.

It seems clear that the ICDP approach to biodiversity conservation should not and will not continue forward in the headlong rush seen in the past. But it seems equally clear that the approach should not be abandoned. This book tries to carefully analyze what has been learned and to point toward the implications of these lessons for future interventions. The challenge of establishing or maintaining protected areas while addressing the needs of local populations remains. Funding priorities may change over time, as they always do, but any long-term solution is going to need to draw on key aspects of the ICDP approach. Hence this book.

Learning from Experience

The papers that follow take a hard look at the ICDP approach. The authors examine the ecological, social, economic, and historical forces that have shaped the relationship between conservation and development. We have selected studies that cover a broad geographic range and delve into the key issues that influence the conservation initiatives linked to protected areas. Throughout the book we have tried to cross-reference papers to facilitate the readers' exploration of these topics.

John Robinson and Kent Redford (ch. 2) argue that conservation and development objectives are often based on unrealistic assumptions, with profound implications for ICDPs. Thomas McShane and Suad Newby (ch. 4) examine some of these key assumptions underlying ICDPs, finding that many of the constraints being experienced involve capacity constraints related to social organization, governance, advocacy, and competing economic factors. Capacity weaknesses are also highlighted by Shekhar Singh and Arpan Sharma (ch. 13) in their analysis of ecodevelopment in India. All of these cases raise issues that conservationists are sometimes uncomfortable dealing with. Phil Franks and Thomas Blomley (ch. 5) and Katrina Brown (ch. 11) address the tug of opposing forces within ICDPs—where biodiversity goals can marginalize the interests of local stakeholders, while economic development goals can marginalize legitimate national and international interests in biodiversity. These authors identify the key challenge of distinguishing between conservation and development goals while reconciling the different interests of multiple stakeholders.

While some ICDPs have tried—and failed—to satisfy all interest groups, a more realistic approach must recognize the need for trade-offs. Practical approaches to trade-off analysis are explored by Katrina Brown (ch. 11), and identifying and negotiating trade-offs is a recurring theme in case studies involving Yellowstone National Park (Dennis Glick and Curtis Freese, ch. 7), indigenous peoples in the Philippines (Edgardo Tongson and Marisel Dino, ch. 9), and community-based natural resource management in Zambia (Brian Child and Barry Dalal-Clayton, ch. 12). Steve Gartlan (ch. 10) compares and contrasts the Korup and Kilum ICDPs in Cameroon and documents how tenurial rights have influenced project implementation.

The institutional characteristics of ICDPs are usually critical. These include legal and institutional frameworks, formal and informal property rights and resource management rules, and the norms and traditions of the various stakeholders and actors. Brian Child and Barry Dalal-Clayton, analyzing the Luangwa Valley of Zambia (ch. 12), as well as Shekhar Singh and Arpan Sharma, examining ecodevelopment in India (ch. 13), argue that ICDPs require institutional forms with the capacity to deal with ecological, social, economic, and even political change, necessitating adaptive approaches to management, a topic explored in depth by Nick Salafsky and Richard Margoluis (ch. 16). Gill Shepherd (ch. 15) argues that poor monitoring of ICDP biodiversity and livelihood impacts has prevented these projects from learning from their own efforts.

Jeffrey Sayer and Michael Wells (ch. 3) explore the inherent contradictions in the conventional project framework, where local actors are expected to achieve "ownership" and project activities are expected to achieve "sustainability" even though the objectives, design, time frame, and budget are largely determined by outsiders. Similar concerns are voiced by Phil Franks and Thomas Blomley (ch. 5), who review CARE's experience of fitting integrated conservation and development into a project framework and argue for a vision-driven planning approach.

ICDPs have tended to concentrate their efforts on mitigating the threats to protected areas from the activities of local communities. Broader threats are often of greater significance, however, and can be addressed through improved policy formulation and implementation (Katrina Brandon and Michelle O'Herron in Costa Rica, ch. 8), greater influence over public and private investment (Dennis Glick and Curtis Freese in Yellowstone National Park, ch. 7), and spatial planning (Stewart Maginnis, William Jackson, and Nigel Dudley, ch. 14). As an alternative approach to the conventional project framework, Agnes Kiss (ch. 6) examines in detail the potential and limitations of direct payments to stakeholders and argues that this approach could be more cost-effective than many ICDPs.

The predominant paradigm of most of the major international conservation organizations has shifted from ICDPs to conservation at larger scales, whether ecoregions, bioregions, hotspots, or landscapes. Most of these efforts have focused on defining large-scale conservation priorities and conservation-friendly policies across diverse sectors. Stewart Maginnis, William Jackson, and Nigel Dudley (ch. 14) examine ICDPs within a larger, landscape perspective, proposing that this perspective holds the potential for reconciling conflicting stakeholder needs much more effectively than at the local level. There is general acknowledgment that ICDPs must address issues at multiple scales, though the form this might take varies. Phil Franks and Thomas Blomley (ch. 5) question whether these large-scale approaches are the new paradigm for integrated conservation and development.

To conclude, a group of authors attempt an overall synthesis of all of the contributors' findings (ch. 17). Despite the discouraging record of ICDPs, they argue that no other approach has been more effective. Linking protected area management with the interests of local stakeholders remains one of the few widely applicable approaches to site-based biodiversity conservation that offers a realistic prospect of success. Likewise, integrated conservation and development is one of the pillars of sustainable development and is well embedded in international aid policy (cf. IUCN 1980; WCED 1987; Munro 1991; OECD 1996). These authors conclude that learning more lessons is less important than applying the ones that are already available. Some of the key characteristics and ingredients needed for successful ICDPs and similar initiatives are identified.

References

Agrawal, A. 1997. *Community in Conservation: Beyond Enchantment and Disenchantment.* Gainesville, Fla.: Conservation and Development Forum.

Barrett, C. S. and P. Arcese. 1995. Are integrated conservation and development projects sustainable? On the conservation of large mammals in sub-Saharan Africa. *World Development* 23:1073–1084.

Colchester, M. 1994. *Salvaging Nature: Indigenous Peoples, Protected Areas, and Biodiversity Conservation.* Discussion Paper no. 55. Geneva: UN Research Institute for Social Development.

Few, R. 2001. Containment and counter-containment: Planner/community relations in conservation planning. *The Geographical Journal* 167:111–124.

Ghimire, K. B. 1994. Parks and people: Livelihood issues in national parks management in Thailand and Madagascar. *Development and Change* 25:195–229.

Ghimire, K. B. and M. P. Pimbert, eds. 1997. *Social Change and Conservation: Environmental Politics and Impacts of National Parks and Protected Areas.* London: Earthscan.

Hughes, R. and F. Flintan. 2001. *Integrated Conservation and Development Experience: A Review and Bibliography of the ICDP Literature.* London: International Institute for Environment and Development (IIED).

IUCN (International Union for Conservation of Nature and Natural Resources). 1980. *World Conservation Strategy: Living Resource Conservation for Sustainable Development.* Gland, Switzerland: International Union for Conservation of Nature and Natural Resources.

Larson, P. S., M. Freudenberger, and B. Wyckoff-Baird. 1998. *WWF Integrated Conservation and Development Projects: Ten Lessons from the Field 1985–1996.* Washington, D.C.: World Wildlife Fund.

McShane, T. O. 1999. Voyages of discovery: Four lessons from the DGIS-WWF Tropical Forest Portfolio. *Arborvitae* suppl.: 1–6.

Munro, David, dir. 1991. *Caring for the Earth: A Strategy for Sustainable Living.* Gland, Switzerland: International Union for Conservation of Nature and Natural Resources, UN Environment Program, World Wildlife Fund.

Murphree, M. W. 1993. *Communities as Resource Management Institutions.* Gatekeeper Series no. 36. London: International Institute for Environment and Development (IIED).

OECD (Organisation for Economic Co-operation and Development). 1996. *Shaping the 21st Century: The Contribution of Development Coöperation.* Paris: Organisation for Economic Co-operation and Development.

Redclift, M. 1987. *Sustainable Development: Exploring the Contradictions.* London: Methuen.

Robinson, J. G. 1993. The limits to caring: Sustainable living and the loss of biodiversity. *Conservation Biology* 7:20–28.

Sachs, W. 1991. Environment and development: The story of a dangerous liaison. *The Ecologist* 21:252–257.

Sanjayan, M. A., S. Shen, and M. Jansen. 1997. *Experiences with Integrated Conservation and Development Projects in Asia.* World Bank Technical Paper no. 38. Washington, D.C.: The World Bank.

Stocking, M. and S. Perkin. 1992. Conservation-with-development: An application of the concept in the Usambara Mountains, Tanzania. *Transactions of the Institute of British Geographers* 17:337–349.

WCED (World Commission on Environment and Development). 1987. *Our Common Future.* Oxford: Oxford University Press.

Wells, M. and K. Brandon. 1992. *People and Parks: Linking Protected Area Management with Local Communities.* Washington, D.C.: The World Bank, World Wildlife Fund, and U.S. Agency for International Development.

Wells, M., S. Guggenheim, A. Khan, W. Wardojo, and P. Jepson. 1999. *Investing in Biodiversity: A Review of Indonesia's Integrated Conservation and Development Projects.* Washington, D.C.: The World Bank.

West, P. C. and S. R. Brechin, eds. 1991. *Resident Peoples and National Parks: Social Dilemmas and Strategies in International Conservation.* Tucson: University of Arizona Press.

2

Jack of All Trades, Master of None: Inherent Contradictions Among ICD Approaches

John G. Robinson and Kent H. Redford

Introduction

Integrated conservation and development (ICD) initiatives offer the opportunity to simultaneously address two major societal goals: the promotion of socioeconomic development and the conservation of nature. This paper will explore *(a)* the intellectual roots of ICD projects (ICDPs) and the consequent approaches adopted by these projects, *(b)* the relationship between these approaches and the outcomes of projects—and the ensuing need to decide between development and conservation goals—and *(c)* the conditions under which ICD initiatives are likely to be successful.

The philosophy of ICD initiatives cannot be understood without an appreciation of the historical events that constrain and create that philosophy. Ever since the Marshall Plan was launched in 1948, the key ingredient for economic development has been considered to be economic growth (Redclift 1987; Sachs 1991). After World War II, development assistance, first to Europe, then to the recently independent countries of Asia and Africa, followed the Keynesian model of encouraging growth in the national gross domestic product (GDP) through increases in the consumption of natural resources and in the export and import of goods. By the early 1970s, however, it was clear that this strategy was resulting in resource depletion and environmental degradation (Club of Rome 1972). Economic results were no better: per capita incomes had not risen in many countries, and distribution of wealth between the North and South was becoming more skewed (Brandt Commission 1983). There was a recognition that development purely through economic growth was not sustainable and that natural resources were finite.

Added to this recognition was concern over the survival of the world's species and ecosystems, arguably inspired by Rachel Carson's 1962 *Silent Spring.* Untrammeled economic growth was increasingly seen as responsible for species extinction and ecosystem degradation. The conservation community argued forcefully for the integration of ecological considerations into economic development plans (see Redclift 1987; Holdgate 1999; Oates 1999). These ideas were first articulated in a fully formed way in the World Conservation Strategy (WCS), published in 1980, which linked conservation to development by defining one in terms of the other (IUCN 1980). Conservation was defined as the management of human use of the biosphere so that it may "yield the greatest sustainable development to present generations while maintaining its potential to meet the needs and aspirations of future generations." Conservation of biological diversity and ecological processes was seen as underlying sustained development through the sustainable use of natural resources. *Sustainable development,* a term first used by the Cocoyoc Declaration on environment and development in the early 1970s (Redclift 1987), was posited as fundamentally different from earlier economic development.

The concept of sustainable development was elaborated more fully in the 1987 report of the World Commission on Environment and Development (WCED), an independent body established by the United Nations. In contrast to the WCS, the WCED defined sustainable development, without explicitly mentioning conservation, as development that "meets the needs of the present without compromising the ability of future generations to meet their own needs." The WCED incorporated the concept of inter- and intragenerational equity (Solow 1974:43). While the phrase "without compromising" acknowledged the need to change social and economic reactions to natural resources, the WCED did not formally recognize that resources can be limited and natural systems fragile. Instead, the mechanism posited to achieve sustainable development was social and economic change. Sustainable development could only happen, it was posited, if human society changed fundamentally so as to allow continued social and economic development. How was this to happen? The WCED suggested that this required meeting three objectives:

- Poverty Alleviation. "Sustainable development requires meeting the basic needs of all and extending to all the opportunity to fulfill their aspirations for a better life" (WCED 1987:8). The argument was that the poor countries and poor people degrade their natural environment through overuse of their natural resources. There is a strong correlation between resource degradation and deteriorating incomes among the poorest (Sanderson 1992), and the WCED noted this correlation and assumed causality. The argument was that poverty leads to environmental destruction, and thus that raising people's incomes above some level will decrease environmental degradation and protect the environment (Lélé 1991; Broad 1994).
- Improved Social Organization. The WCED made the argument that development efforts should help poor nations and poor people to manage their own natural resources more effectively through improved social organization. "Technology and social organization can be both managed and improved to make way for a new era of economic growth" (WCED 1987:8). Developmental assistance should encourage national self-reliance, cost-effectiveness, and the use of appropriate technologies. Indigenous and local institutions should be strengthened. The WCED rejected the notion that socioeconomic development deterministically required economic growth (Verburgh and Wiegel 1997) or that economic growth by itself would lead to development. It promoted the notion that social change can lead to socioeconomic development without environmental degradation.
- Social Equity and Justice. To raise people's incomes, to develop the institutions to prevent environmental degradation, and to allow poor people to take control of their own lives and resources, the argument went, social equity and justice must be promoted. Sustainable development requires an "assurance that these poor get a fair share of the resources required to sustain that growth. Such equity would be aided by political systems that secure effective citizen participation in decision making and by greater democracy in international decision making" (WCED 1987:8). Social justice and equity require devolution and decentralization of authority and broader participation in decision making (Lélé 1991). Greater equity in resource consumption must be fostered. In some ways, it is exactly this redistribution of wealth, power, and responsibility that distinguishes sustainable development from traditional development through economic growth.

These social and economic changes, it was posited, would mitigate increases in human populations and average per capita consumption and seamlessly allow both the conservation of the natural environment and continued human social and economic development.

The WCED report was followed, in 1991, by *Caring for the Earth (CE)*, which, like WCS, was published by IUCN–The World Conservation Union, the World Wildlife Fund (WWF), and the UN Environment Program (UNEP) (Munro 1991). This manifesto attempted to graft the ideas of sustainable development back onto a conservation agenda. *CE* embraced the need to conserve "all species of plants, animals, and other organisms" (not just natural resources for people) and maintain "ecological processes that keep the planet fit for life." But the all-challenging need to improve "the quality of life of several billion people" meant that *CE* repeated key objectives of the WCED report: "improving the quality of life" (= poverty alleviation); "provide a national framework for integrating development and conservation" (= improving social organization); and "enable communities to care for their own environment" (= social equity and justice). The resulting document is thus an uneasy hybrid—subsuming conservation under the development agenda and confusing the distinct goals of conservation and development (see Robinson 1993; Redford and Richter 1999). It is this confusion that forms the intellectual antecedents of ICD efforts.

WCS, WCED, and *CE* together provide the intellectual framework for sustainable development. More recent efforts (such as the 1992 UN Conference on Environment and Development) have refined but not fundamentally changed that framework. While the internal logical contradictions of sustainable development have been richly discussed (e.g., Redclift 1987; Lélé 1991; Dovers and Handmer 1993; Verburgh and Wiegel 1997; Oates 1999), we will not repeat these arguments nor attempt to resolve them here. Instead we will explore how, in practice, these definitions of sustainable development have influenced ICD initiatives. In particular, we will focus on the confusion of the goals of conservation and socioeconomic development.

In addition we will examine the other legacy that *CE* bequeathed to ICDPs: the multiple project approaches that are used to achieve the objectives. On the one hand are the traditional approaches of conservation: park management and management of natural resources. On the other hand are the approaches that derive from the approaches originally proposed in the WCED report and *CE* to achieve sustainable development. Where the WCED and *CE* noted the need to alleviate the poverty of poor nations, ICDPs adopted poverty alleviation as a core method for rural peoples. Where the WCED and *CE* noted the need to change social organizations, ICDPs focused on institution and capacity building. And where the WCED and *CE* promoted social justice and equity, ICDPs urged the empowerment and participation in decision making of rural peoples. The broad concepts of sustainable development were operationalized by ICDPs as project methods at the project level. The challenge is that these methods have different, often contradictory, outcomes.

A Confusion of Goals and Objectives

Officially christened in 1992, ICDPs are seen by many people as purely the expression of the ideas of sustainable development at the project level. Gartlan (1997:2), for instance, formally links "sustainable development and its principal tool, the integrated conservation and development project (ICDP)" and states that these projects are "focused on the development needs of people, [while] the conservation goals tend to be neglected." To other observers, ICDPs aim to conserve natural resources for local people. Brown and Wyckoff-Baird (1992:xiii) state that the ICDP has "a dual goal of improving the management of natural resources and the quality of life of people," and they see the "sustainable management of the resources [as] the ultimate goal of the [ICDP]." To still others, the goal of ICDPs is to preserve wild species and wild areas. Brandon and Wells (1992:560), the midwives of the term *ICDP*, assert that "the major objective of ICDPs is to reduce the pressure on a protected area." Finally, there are those who believe that ICDPs can reliably meet both the goals of conservation and development: "Their main goal is to link conservation and development such that each fosters the other" (Alpert 1996:845). Or as Sanjayan, Shen, and Jansen (1997:10) state: "It is not conservation *through* development, or conservation *with* development, or even conservation *adjoined* with development....[I]t is the achievement of conservation goals and development needs together."

This confusion persists when one examines specific objectives of projects, which can usefully be placed in three categories (Robinson 2001): projects that seek to "conserve species," those that seek to maintain "ecosystem health," and those that promote "human livelihoods," an inclusive category that contains multiple development objectives. Even in Wells and Brandon's (1992) original review of ICDPs, where all projects were located in and around protected areas, the specific objectives varied. Some projects were clearly to conserve specific wild species. For instance, the Parc des Volcans project in Rwanda had as its primary goal the protection of the mountain gorilla. The Monarch

Table 2.1 *Indicators of success with different objectives*

Objectives	*Indicators of success*
Species conservation	1. Populations show no consistent decline.
	2. Populations are not vulnerable to extinction.
	3. Populations maintain ecological role.
Ecosystem functioning	1. Species richness and diversity maintained.
	2. Primary productivity maintained.
	3. Nutrient cycling maintained.
	4. Landscape patterns maintained.
Human livelihoods	1. Resource availability maintained.
	2. Poverty alleviated.
	3. Per capita income increased.
	4. Local management institutions strengthened.
	5. Participation by local people in governance increased.

Butterfly Overwintering Reserves project in Mexico was clearly focused on that species. In contrast, the Chitwan National Park project in Nepal, or the ADMADE program in Zambia, had as its primary objective the maintenance of the ecological health of natural systems. Finally, there were projects, such as those in the Annapurna Conservation Area in Nepal or the Sian Ka'an Biosphere Reserve in Mexico, whose stated objectives were to improve human livelihoods.

If conservation and development goals were universally compatible with one another, this confusion of goals would not be a problem. If the conservation of overwintering sites for monarchs also conserves the ecosystem where the animals are found, and in addition improves the livelihoods of people in and around the reserves, then there is a synergy and complementarity. Unfortunately, this is not always the case. Table 2.1 distinguishes the objectives of species conservation, ecosystem functioning, and human livelihoods and lists some indicators of project success. There is no a priori reason to assume that these indicators will reflect identical conservation and development outcomes, or even that they need be compatible with one another.

The realization that outcomes vary has spawned a whole taxonomy of ICDPs. A recent internal review by CARE of the organization's ongoing projects (Blomley 1997; see also Franks and Blomley, this volume) found that seven (50 percent) aimed to accomplish development using conservation methods, two aimed to accomplish conservation using development methods, one was simply a conservation project with no development involvement, and only three had the dual objective of conservation and development. One begins to question whether "integrated" in the phrase "integrated conservation and development" is more frequently than not a misnomer.

A Multitude of Approaches

Paralleling this confusion of objectives is the multitude of approaches used by ICDPs. Before the advent of ICDPs, if the objective of a project was to foster conservation (for instance, see McNeely 1988; Wells and Brandon 1992; Caldecott 1996), the two approaches most frequently used were to strengthen park protection and management and to manage renewable natural resources. The insight provided by the ICD perspective was that the greatest threat to many protected areas is resource extraction by the rural poor living in the park or in adjacent areas. The prescribed response was to use the approaches of rural development: poverty alleviation, promotion of local institutions, and empowerment and involvement of local people in decisions affecting the area. As a result, most efforts and the great proportion of program budgets in most ICDPs focus on the sustainable development of local human populations. "ICDPs target human populations as primary beneficiaries so that biodiversity can survive and flourish" (Brown and Wyckoff-Baird 1992).

In the logic of sustainable development, in well-designed projects there should be no conflict between "conservation" approaches and "development" approaches. This logic indicates that local changes in social organization and increases in participation will inevitably alleviate poverty and thus take pressure off the protected area. Therefore, resources in protected areas need not be placed "off limits" because the development of new economic activities provides new alternatives ("substitution"), which are likely to be less dependent on natural resource extraction. In practice, recognizing that this panacea is not automatic, many ICDPs have acknowledged the need for additional "compensation." The aim is to create a direct linkage between material benefit to rural people and a desired conservation outcome (Brown and Wyckoff-Baird 1992). "Conservation policies will work only if local communities receive sufficient benefits to change their behavior from taking wildlife to conserving it" (Gibson and Marks 1995:944). The

Table 2.2 *Suggested applicability of different approaches to different objectives*

	Approaches				
Objectives	*Park management*	*Management of natural resources*	*Poverty alleviation*	*Capacity building*	*Participation and empowerment*
Species conservation	+	+/0	0/–	0/–	0/–
Ecosystem health	+/0	+/0	+/0/–	+/0	0/–
Human livelihoods	–	+/0/–	+	+	+

+ = positive
0 = neutral
– = negative

assumption is that compensation will not be required forever, because economic development will create new and better ways of making a living and reduce dependence on natural resource extraction (see also McShane and Newby, this volume).

Although different ICDPs have emphasized different approaches, there has evolved over the last ten years a distinct ICDP menu of approaches. Three approaches derive from the imperative of sustainable development (poverty alleviation, institution and capacity building, and empowerment and participation). Two approaches derive from conservation (park protection and management, management of natural resources). Of course, these different approaches are most applicable to different project objectives (table 2.2). Where the objective is species conservation, park management and resource management are clearly the methods of choice. Where the objective is improving human livelihoods, strategies that focus on poverty alleviation, socioeconomic development, and the governance by local people are generally more efficacious. Where the objective is ecosystem health, a number of methods are useful, depending on the specific context.

We now ask two questions: How have these approaches been put into practice in ICDPs? How have each of these approaches worked in attaining the different objectives of species conservation, ecosystem health, and human livelihoods?

POVERTY ALLEVIATION

"The only hope for breaking the destructive patterns of resource use is to reduce rural poverty, and improve income levels, nutrition, health care and education" (Brandon and Wells 1992:561). The logic of sustainable development indicates that poverty alleviation is essential for successful project implementation, and ICDPs generally address this issue directly. Most ICDPs either assume that management of natural resources (see the following discussion) will allow the capture of the commercial value of biodiversity (Caldecott 1996), leading to increased income levels for local people, or promote local enterprises that are considered to be environmentally neutral or positive (Alpert 1996).

How successful have ICDPs been at alleviating poverty? To date there has not been a systematic review across projects, but case studies are instructive. In their review of the Communal Areas Management Programme for Indigenous Resources (CAMPFIRE) program in Zimbabwe, Getz et al. (1999) argue that the wildlife management raised income levels of poor rural communities (though Campbell, Sithole, and Frost 2000 worry that emerging problems could derail gains). A recent review of fishermen income in the Mamiraua Sustainable Development Reserve in Central Amazonia reported that average income rose from R$320 in 1999 to R$845 in 2001 in two management sectors (Ayres, personal communication 2002), based largely on an increase in fish production from management lakes from 6.2 to 15 tons.

Does this poverty alleviation contribute to conservation gains? In both the CAMPFIRE program (Getz et al. 1999) and in Mamiraua, rises in local income co-occurred with increasing wildlife and fish populations. In the management lakes at Mamiraua, populations of pirarucu (*Arapaima*), the most important fisheries species, tripled in density (Ayres, personal communication 2002). It is less clear that poverty alleviation *caused* these conservation gains by taking the pressure off natural resources, as predicted by the original theory of sustainable development. In both of these cases, conservation gains were posted because local people, through the project, were able to exclude other potential stakeholders from access to natural resources, not because of increases in income per se. Indeed, the income needs and expectations of people are not likely to be fixed at a certain level, and increased income derived from ICDP activities is frequently accepted by people in addition to, rather than in lieu of, income derived from access to protected areas (Sanjayan, Shen, and Jansen 1997).

This absence of a documented causality means that the specific conditions of a project override any simple relationship between poverty alleviation and conservation gains (see table 2.3). In some cases, as noted previously, poverty alleviation is associated with conservation.

Table 2.3 *Suggested compatibility among ICDP methods*

	Park management	*Management of natural resources*	*Poverty alleviation*	*Community development*	*Participation and empowerment*
Park management	+	+	0/–	0/–	–
Management of natural resources		+	+/0	+/0	+/0
Poverty alleviation			+	+	+/0
Community development				+	+
Participation and empowerment					+

+ = compatible
0 = neutral
– = incompatible

In other cases, poverty alleviation can work against conservation. For instance, in the Xishuangbanna Nature Reserve in China, a project encouraging the cultivation by local people of *Amomum villosum*, a medicinal herb, has converted most of the riparian land within the reserve to shaded miniplantations (Albers and Grinspoon 1997). ICDPs generate employment, provide infrastructure and services, and can act as a pole for in-migration (Brandon and Wells 1992). Broad comparisons, as in Salafsky, Cauley, et al. (2001), generally find no correlation between the success of economic enterprises and conservation outcomes. Establishing and enforcing a linkage between community and personal benefits on one hand, and reduction in the impact on natural systems on the other, remains a major challenge for ICD initiatives.

Institution and Capacity Building

The focus of ICDPs is often on strengthening the management institutions, such as for harvested species (cf. Alpert 1996), for tourism (Caldecott 1996), and managing coasts and seas (Caldecott 1996). The strengthening of tenure and usufruct rights over land and resources has been a major push of many projects (see Tongson and Dino, this volume). The core argument is that if local people have rights and control over land and resources, they will act as stewards of those entities. And it is clear that institutional strengthening, and the establishment and protection of tenure and usufruct rights, has been a powerful way of stabilizing land and resource relationships and building a constituency for conservation (Brandon 1998).

Nevertheless, it is less clear that institution and capacity building support both livelihood and conservation goals. While these forms of institutional strengthening do limit the flexibility and range of options of people, the increased predictability of action does not a priori favor either development or conservation outcomes (Sanderson 1998). Much depends on the choices taken by projects as to which people are to be the beneficiaries of development assistance. Projects "privilege certain actors and marginalize others" (Brosius 1999:38), and most ICDPs have an impact on only a small percentage of local people (Albers and Grinspoon 1997). Agrawal (1997) points out that "the local community" is not an easily defined entity and that there are multiple actors with divergent interests within and outside any specified community. Nor is tenure a unitary phenomenon. Tenure over land is generally different from that of resources. The tenures of different groups overlap and are in conflict, and "strengthening tenure" works for some people and against others (Naughton-Treves and Sanderson 1995). In the final analysis, the choice of what institutions to strengthen, and what people to favor, affects what goals are achieved.

The requirement to make choices on which institutions to strengthen leaves ICDPs open to criticism on all fronts. Those primarily concerned with the conservation of biodiversity, for example, see the focus on strengthening community-based management systems as frequently short-sighted and ineffective (Barrett et al. 2001). For those concerned with local empowerment and social change, institutional building can be counterproductive: "institutions in fact often obstruct meaningful change through endless negotiations, legalistic evasion, compromise among 'stakeholders,' and the creation of unwieldy projects aimed at top-down environmental management" (Brosius 1999:38). For those concerned with poverty alleviation, institution and capacity building are a poor substitute for economic growth, and trap people into a rural agricultural and extractive existence.

Empowerment and Participation

The logic of sustainable development indicates that social justice and equity are essential to successful project implementation, and this logic is followed in some ICD initiatives. Caldecott (1996:164), for instance, notes that projects need to address problems where "concentrations of wealth and power are combined with a lack of public accountability," creating "an elite whose self-interested decisions can over-rule those of environmentally aware civil servants and local people." Yet most ICD initiatives shy away from direct political action. Most assume that social equity, decentralization, and local participation are interchangeable terms, and project documents tend to focus on local participation even though it is "clear that some form of participation is necessary but not sufficient for achieving equity and social justice" (Lélé 1991:165).

While achieving social equity and justice is considered to be beyond the remit of most projects, many project managers promote the notion that local people need to participate at some level in the project. While local participation is sometimes justified solely on the basis that it increases project effectiveness (Paul 1987), most projects consider empowerment and participation as necessary because *(a)* local people possess the knowledge and resource management institutions that are essential for conservation, or *(b)* ignoring local resource rights or failing to compensate people for their exclusion from resources will lead to hostility to conservation goals (Kjersgård 1997).

The extent of that participation is highly variable. Cohen and Uphoff (1980) note that local people can participate in making decisions, implementing projects, distributing the benefits, and evaluating the results.

Many projects adopt some but not all of these levels of participation. Overlapping political jurisdictions, retention of control by central governments, and lack of technical knowledge, political capacity, and skills at the local level all can confound project efforts (Wyckoff-Baird et al. 2001). This is not to say that the effort to increase participation of local entities in natural resource management is not a necessary step in the development of good governance (Hyden 1998), but it is very difficult to achieve. Indeed, most analyses conclude that projects have not been very successful at achieving even the more limited objectives of participation (e.g., Agrawal and Yadama 1997).

How does greater participation contribute to other conservation and human livelihood goals? While greater participation is a sine qua non of sustainable development, there is little evidence that greater participation is directly linked to poverty alleviation. Neither is there much evidence that greater participation leads to desired conservation outcomes (table 2.3). Sanjayan, Shen, and Jansen (1997:15) are categorical: "More participation does not necessarily lead to a better [conservation] project."

Park Protection and Management

Managing parks and protected areas remains the preeminent method of protecting wildlife and wildlands (McNeely and Miller 1984; Lucas 1992), and ICDPs frequently invest significant effort in strengthening park management, developing regulatory mechanisms, establishing buffer zones around the park, and encouraging boundary patrols.

However, by their very nature, most parks exclude local people from natural resources to which previously they might have had access or to which they wish access. Where people live in parks (a situation more frequent than conservation policy admits), park management generally seeks to minimize natural resource harvests. ICDPs invariably attempt to reconcile this exclusion of people from resources with project activities that compensate people. For instance, projects frequently seek to return a portion of the benefits of the park to local people. Involving people in ecotourism or distributing a portion of gate receipts from protected areas (Alpert 1996) is a common approach. A common problem, however, is that activities such as park management and ecotourism do not normally generate a lot of jobs or revenue that might flow to the local community.

Park management by itself generally does not contribute a lot to human livelihood goals, and should not be expected to (though important exceptions might involve ecosystems services). Brandon (1998:418) cautions that we should not expect parks to cure "structural problems such as poverty, unequal land distribution and resource allocation, corruption, economic injustice and market failures." In fact, we should expect park management to frequently conflict with other project objectives (see table 2.3). The same point is made by Wells and Brandon (1992) in their review of ICDPs, which notes that ICDP projects often moved to halt environmental degradation and encroachment, and in so doing abrogated the social contract with local people.

Management of Natural Resources

A core premise of ICDPs is that the appropriate management of natural resources can allow both conservation of biological diversity and an increased flow of resources to local communities. The key assumption is that management will allow greater efficiency of harvest—minimizing the impact on biodiversity while maximizing the harvest (Mace and Hudson 1999; Robinson 2001). Many ICDP projects shift extraction away from protected areas and compensate local people by providing and seeking to increase resource production in buffer zones or multiple-use areas.

Resource management does not have to involve local communities in management decisions, but most ICDPs have assumed that effective management requires community control and have delegated responsibility for management of wildlife (cf. Kiss 1990; IIED 1994), forests (cf. IES 1995), and fisheries (cf. Pomeroy 1994). The efficacy of community-based management of natural resources has been widely debated, and clearly it varies (Western and Wright 1994). The variables of institutional capability at the community level, consumer demand for goods from that resource, and the limits to resource availability interact to determine whether resources can be managed in a sustainable way (Freese 1998). A couple of examples will suffice to illustrate the challenges. Community management of wildlife in tropical forests for meat is unlikely to be successful once human population densities rise above 1 person/square kilometer because of limits of wildlife productivity (Robinson and Bennett 2000). Community management of resources producing high-value goods like elephant ivory is exceedingly difficult (Freese 1998), and efforts like CAMPFIRE in Zimbabwe and Administration Management Design (ADMADE) in Zambia have had to incorporate significant central government involvement. One can draw the general conclusion that community management of natural resources is possible in some but not all settings.

Management of natural resources as a project method is widely adopted in ICDPs because the approach is generally compatible with other project methods (table 2.3). If resources can be managed and harvests made more efficient, then revenues are potentially available to address human livelihood goals. It is instructive that in a review of WWF projects, Larson, Freudenberger, and Wyckoff-Baird (1998) concluded that more successful ICDP projects focus on land outside of protected areas, treating people as resource managers and emphasizing their rights and responsibilities for resource management.

Table 2.3 summarizes our suggested compatibility among the different ICD approaches. The traditional conservation approaches of park and natural resource management are generally compatible with one another, but less so with the approaches of socioeconomic development. Similarly, the development approaches are broadly compatible with one another, though not in all cases, and the link between poverty alleviation, on the one hand, and participation and empowerment, on the other, is weak.

Future Guiding Principles

Perhaps the fairest thing to say about the current practice of ICDPs is that their failings are not of their own making but should be seen in the light of the innocence of their conceptual antecedents. Born in the cacophony of inexplicit assumptions, confusion of objectives, and naïve expectations of win-win solutions, ICDPs, not surprisingly, have been criticized by conservationists, by social advocates, and by developmental economists alike.

Those, such as the originators of the ICDP concept, who viewed ICDPs as being mostly about improved biodiversity conservation have been disappointed. Reviewing ICDPs in Indonesia, Wells et al. (1999:2) concluded that "very few ICDPs in Indonesia can realistically claim that biodiversity conservation has been or is likely to be significantly enhanced as a result of current or planned project activities." Oates (1999:230) seconds this conclusion, stating that the "strategy of integrating conservation with economic development has failed to protect wildlife in the forests of West Africa."

Those who expected ICDPs to achieve myriad social goals have concluded that the approach has also been unsuccessful in that. In a broad comparison of ICDPs in Asia, for example, Salafsky, Cauley, et al. (2001) concluded that there was no general relationship between successful economic enterprises and management of natural resources. Successful economic enterprises were frequently independent of traditional resource extraction and depended on a suite of other determinants. Other social objectives have proved equally elusive (Brandon 2001).

Is this lack of success inherent to ICDPs, or does the next generation of ICD approaches promise to address the inadequacies of the past? It is instructive that the ICDP approach is not disappearing. If anything, it seems to be strengthening. A recent report on the status of ICDP activities in Vietnam is symptomatic, with the document concluding, "ICDPs in Viet Nam are important because they are the country's primary initiative to conserve Viet Nam's extraordinary biodiversity. The failure of ICDPs in Viet Nam would mean the continued rapid decline of the country's biodiversity" (*Proceedings of Integrated Conservation and Development Projects Lessons Learned Workshop* 2000).

For projects faced with the necessity of combining conservation and development, what is the way ahead? We would suggest that five principles must all be addressed before the next generation of integrated projects will be successful.

Principle 1: Specify Goals

Many, if not all, ICDPs that have been implemented to date have had multiple and general goals. But it is impossible to develop informed collaboration and develop innovative implementation methods, let alone measure success, without specific goals. In the absence of this specificity, it has been all too easy for different stakeholders to advocate and influence the project's goals. With a frequently changing and discordant set of expectations, disappointment and even anger are natural outcomes. Projects must have clear, measurable, achievable, and appropriate goals (Margoluis and Salafsky 1998; Salafsky, Margoluis, and Redford 2001; see also Salafsky and Margoluis, this volume).

Most importantly, there needs to be an informed separation of these two significant societal goals: socioeconomic development and the conservation of nature. ICD approaches incorporate both, but within each ICDP one goal must be recognized as preeminent. Conservation projects must include a strategic development component, and development projects must include strategic conservation components. In recognition of the "codependency," we suggest distinguishing between development projects with conservation (DPC) and conservation projects with development (CPD). Table 2.4 distinguishes these, if in too dichotomous a fashion. This differentiation will ease the tension that is generated by the simplistic desire to seek win-win projects when experience shows that they are few and far between (Redford and Sanderson 1992; Brandon 2001; Lee, Ferraro, and Barrett 2001). Others (e.g., Stocking and Perkins 1992; Brannstrom 2001; Franks and Blomley this volume) argue for a similar distinction.

Table 2.4 *Comparison of selected parameters of CPD and DPC*

	Conservation projects with development (CPD)	*Development projects with conservation (DPC)*
Target	Species, ecosystems	Local peoples
Desired condition	Viable populations and areas	Socioeconomic development; equitable social conditions
Role of humans	Threat	Target
Principal activities	Protection; restoration; threat alleviation; stakeholder education	Enterprise development; institution building; improved livelihoods; empowerment

CPD projects will tend to focus on protected areas and resource management, with parks and reserves serving as the cornerstone of biodiversity conservation. Improved park management should not be equated with sustainable development (Brandon 1998; Redford and Richter 1999; Brandon 2001). By contrast, DPC projects will tend to focus on the complicated task of alleviating poverty, building the necessary capacity and institutions, and addressing social justice and equity. In this case, the conservation of all the attributes and components of biodiversity conservation is neither a necessary nor sufficient condition for project success (Veit, Mascarenhas, and Ampadu-Agyei 1995; Wells et al. 1999; Salafsky and Wollenberg 2000; Lee, Ferraro, and Barrett 2001).

PRINCIPLE 2: ACKNOWLEDGE TRADE-OFFS

Difficulties in implementing ICDPs are often due, in major part, to the fact that project designers and implementers have failed to properly evaluate the cost of human activities on the conservation of components and attributes of the biodiversity. Almost all human use and exploitation of biological systems reduces biodiversity (Robinson 1993), with different human uses affecting different attributes and components of that biodiversity (Redford and Richter 1999). There are trade-offs between use and conservation. These trade-offs are the variables that must be solved in the calculus of compatibility and sustainability. Putz et al. (2000, see esp. figure 2.4) illustrates how this calculus might be applied in a tropical forest context when assessing the compatibility of a given conservation target with different forestry practices. This sort of analysis allows developing a specific set of trade-offs that can be used in calculating how and where to achieve conservation of the specific targets (see also Brown this volume).

Principle 3: Respect Context

Conservation and development projects are tools for achieving certain socially defined goals. The choice of which tool to use at a given location should always be based on an assessment of the ecological, social, and political context at that site. Yet rarely has this assessment been done; instead, approaches are usually applied because of funding or institutional conviction, irrespective of context. This is nowhere more true than with the ICDP approach, which historically has been applied in a wide variety of contexts—where threats are local and extralocal and where solutions are local and extralocal. One of the lessons of the ICDP experience is that one approach does not fit all situations (Newmark and Hough 2000; McShane and Newby this volume). Another is that ICDPs are tools that probably should be applied only in settings where the threat and the solution are both local—where locally applied solutions alleviate threats originating locally.

Several authors have detailed other lessons relevant to this principle. ICD initiatives have been most successful in areas where *(a)* a primary threat to biodiversity is from local people living in the immediate vicinity of the protected area, *(b)* the types and scales of pressures are relatively limited, *(c)* realistic opportunity exists to generate income from limited local development activities, *(d)* policies exist that are conducive for dialogue among stakeholders (Sanjayan, Shen, and Jansen 1997), *(e)* communities are strong and intact, *(f)* immigration (or emigration) is controlled, and *(g)* resources of economic value (but not too great a value, Freese 1998) to the outside world can be exploited sustainably (Gartlan 1997; see also Brandon 1998).

Principle 4: Respect Scale and Heterogeneity

ICD efforts must answer the question, "integration at what scale?" There are two dimensions to scale, extent and grain. *Extent* refers to the total area under consideration, while *grain* is determined by the size of the unit of analysis (Weins 1989). Both grain and extent contribute to the heterogeneity of all areas. Most projects to date have not recognized the inherent heterogeneity of the areas in which they operate, seeking to integrate irrespective of scale.

It is clear that integrating conservation and development is easier at larger scales. There is increased area for parks, sustainable use zones, and regions of development. Grain must also be considered. The grain, or size of the units of analysis, within the project context could be considered as land-use types. Land-use management in ICDPs is in its infancy. While ICD efforts frequently incorporate different land uses—protected areas, buffer zones, support zones, agricultural areas—there has been little analysis of the relative area that should be allocated to each of these land uses. What watershed area needs to be protected to supply agriculture areas? What amount of forest is necessary to supply the resource extraction needs of local people? How will these vary with the nature of the natural resource extraction, the socioeconomic development of local people, or the nature and extent of outside subsidies? Grain size needs to be examined carefully both within individual projects and between projects. Much has been written about the need for big parks, but little about the size of buffer areas, sustainable use zones, or agricultural development areas.

Finally, it is critical to appreciate the scaling of values. In a path-breaking study, Kremen et al. (2000) showed that the value of the forest in Masoala National Park, Madagascar, was different at local, national, and global scales. In assessing the sustainability and success of ICDPs, careful consideration must be given to the scaling of these questions.

PRINCIPLE 5: LEARN FROM DOING

In order for practitioners and participants to successfully integrate conservation and development, it is vital to structure the learning that should take place from project implementation. Learning has not been part of the ICD past. ICD efforts must be designed and implemented using adaptive management. Adaptive management—the integration of design, management, and monitoring to systematically test assumptions in order to adapt and learn (Salafsky, Margoluis, and Redford 2001)—creates a framework to learn systematically from an implementer's successes and failures as well as the successes and failures of other project implementers.

The following facts about implementing conservation and development projects indicate the need to adopt adaptive management: projects take place in complex systems; the world is a constantly and unpredictably changing place; immediate action is required; there is no such thing as complete information; and learning and improvement are possible (from Salafsky, Margoluis, and Redford 2001).

Within an adaptive management context, the project must be based on a conceptual model incorporating goals, cause, and effect. Making assumptions explicit is an integral part of this context (see also McShane and Newby this volume). This is particularly important in projects that integrate conservation and development, for as

Brandon (2001:421) points out: "Flawed assumptions frequently serve as the intellectual foundation for the vast majority of conservation activities being developed." Brandon continues: "Evidence is increasingly showing that many of these assumptions do not hold much of the time and that fewer 'win-win' outcomes are possible than might be imagined. The fact that conservation agendas are actively being designed around these assumptions is likely to lead to numerous and significant failures in biodiversity conservation." Making explicit these assumptions, testing them, and learning from doing are vital parts of this principle.

If all ICDPs were to be based on an adaptive management model, there would be enormously increased power from project implementation that could be harnessed. In the current practice this power is being squandered.

Looking to the Future

Have the conceptual underpinnings of ICDPs changed since the WCED report was penned some fifteen years ago? While the policy framework underlying ICD has continued to evolve, the fundamental bifurcation of the goals of nature conservation and socioeconomic development remains. The Convention on Biological Diversity, adopted at the UN Conference on Environment and Development in Rio de Janeiro, is the most recent example. The preamble begins, "Conscious of the intrinsic value of biological diversity," a statement that recognizes the "inherent right of all components of biodiversity to exist independent of their value to humankind" (Glowka, Burhenne-Guilmin, and Synge 1994:9). This is at variance with the last paragraph of the preamble, which says that the parties are "determined to conserve and sustainably use biological diversity for the benefit of present and future generations," a statement that explicitly acknowledges that "conservation of biodiversity and sustainable use of its components should be accomplished for the benefit of people" (Glowka, Burhenne-Guilmin, and Synge 1994:14). These are both worthy goals, but the first stresses conservation while the second promotes socioeconomic development.

ICD initiatives continue to grapple with this dual mandate. One approach to resolve the contradictions has been to recognize that in a large enough area, there is an opportunity to accomplish conservation in some areas and development in others. Larson, Freudenberger, and Wyckoff-Baird (1998), in their review of ICDPs undertaken by the WWF, suggest that the most effective efforts have been those that address conservation and development at the landscape level. At that scale, the heterogeneity of different land uses allows variation in outcomes and allows the attainment of different goals. This approach is perhaps best illustrated in the Living Landscapes program of the Wildlife Conservation Society (Sanderson et al. 2002; see also Maginnis, Jackson, and Dudley this volume).

The other approach is the one we have advocated in this paper: recognize that certain ICD efforts have conservation as the primary goal (CPD), while others have primarily development objectives (DPC). This avoids the danger that we subsume one goal under the other and pretend that both goals are being fully attained. This danger is evident in the current shift in multilateral and bilateral funding toward poverty alleviation (cf. Biodiversity in Development Project 2001), in which the developing policy is to subsume the goal of conservation into oblivion.

The question is not whether conservation and development should be integrated, but how. Decades of experience with both development and conservation projects have shown that the former must include a component of conservation and the latter a component of development. However, the answers provided by the present generation of ICD efforts are not compelling. Future efforts must distinguish conservation and development, acknowledge trade-offs, respect context, respect scale and heterogeneity, and learn from past experience.

References

Agrawal, A. 1997. *Community in Conservation: Beyond Enchantment and Disenchantment*. Gainesville, Fla.: Conservation and Development Forum.

Agrawal, A. and G. Yadama. 1997. How do local institutions mediate the impact of market and population pressures on resource use? *Development and Change* 28:435–465.

Albers, H. J. and E. Grinspoon. 1997. A comparison of the enforcement of access restrictions between Xishuangbanna Nature Reserve (China) and Khao Yai National Park (Thailand). *Environmental Conservation* 24:331–362.

Alpert, P. 1996. Integrated conservation and development projects. *BioScience* 46:845–855.

Barrett, C. B., K. Brandon, C. Gibson, and H. Gjertsen. 2001. Conserving tropical biodiversity amid weak institutions. *BioScience* 51:497–502.

Biodiversity in Development Project. 2001. *Strategic Approach for Integrating Biodiversity in Development Cooperation*. Brussels: European Commission; Gland, Switzerland: IUCN–The World Conservation Union.

Blomley, T. 1997. Overview of CARE International's ICDP portfolio. In T. Blomley and P. Mundy, eds., *Integrated Conservation and Development Projects in CARE International*, 54–55. Copenhagen: CARE–Denmark.

Brandon, K. E. 1998. Perils to parks: The social context of threats. In K. Brandon, K. H. Redford, and S. E. Sanderson, eds., *Parks in Peril: People, Politics, and Protected Areas,* 415–439. Washington, D.C.: Island Press.

Brandon, K. E. 2001. Moving beyond integrated conservation and development projects (ICDPs) to achieve biodiversity conservation. In D. R. Lee and C. B. Barrett, eds., *Tradeoffs or Synergies?* 417–432. Wallingford, Oxon, U.K.: CABI.

Brandon, K. E. and M. Wells. 1992. Planning for people and parks: Design dilemmas. *World Development* 20:557–570.

Brandt Commission. 1983. *Common Crisis.* London: Pan Books.

Brannstrom, C. 2001. Conservation-with-Development models in Brazil's agro-pastoral landscapes. *World Development* 29:1345–1359.

Broad, R. 1994. The poor and the environment: Friends or foes? *World Development* 22:881–893.

Brosius, J. P. 1999. Green dots, pink hearts: Displacing politics from the Malaysian rain forest. *American Anthropologist* 101:36–57.

Brown, M. and B. Wyckoff-Baird. 1992. *Designing Integrated Conservation and Development Projects.* Washington, D.C.: Biodiversity Support Program.

Caldecott, J. 1996. *Designing Conservation Projects.* Cambridge: Cambridge University Press.

Campbell, B. M., B. Sithole, and P. Frost. 2000. CAMPFIRE experiences in Zimbabwe. *Science* 287:42–43.

Carson, R. 1962. *Silent Spring.* Boston: Houghton Mifflin.

Club of Rome. 1972. *The Limits to Growth.* New York: Universe Books.

Cohen, J. and N. Uphoff. 1980. Participation's place in rural development: Seeking to clarify through specificity. *World Development* 8:213–235.

Dovers, S. R. and J. W. Handmer. 1993. Contradictions in sustainability. *Environmental Conservation* 20:217–222.

Freese, C. H. 1998. *Wild Species as Commodities: Managing Markets and Ecosystems for Sustainability.* Washington, D.C.: Island Press.

Gartlan, S. 1997. Falling between two stools: The false premises of sustainable development. Gland, Switzerland: WWF Annual Conference: People and Conservation.

Getz, W. M., L. Fortmann, D. Cumming, J. du Toit, J. Hilty, R. Martin, M. Murphree, N. Owen-Smith, A. M. Starfield, and M. I. Westphal. 1999. Sustaining natural and human capital: Villagers and scientists. *Science* 283:1855–1856.

Gibson, C. C. and S. A. Marks. 1995. Transforming rural hunters into conservationists: An assessment of community-based wildlife management programs in Africa. *World Development* 23:941–957.

Glowka, L., F. Burhenne-Guilmin, and H. Synge (with J. A. McNeely and L. Gündling). 1994. *A Guide to the Convention on Biological Diversity.* Environmental Policy and Law Paper no. 30. Gland, Switzerland: IUCN–The World Conservation Union.

Holdgate, M. 1999. *The Green Web.* Gland, Switzerland: IUCN–The World Conservation Union.

Hyden, G. 1998. Governance in conservation and development. In G. Hyden, ed., *Governance Issues in Conservation and Development,* 1–9. Gainesville, Fla.: Conservation and Development Forum.

IES (Institute for Environmental Studies). 1995. *Case Studies of Community-Based Forestry Enterprises in the Americas*. Madison: University of Wisconsin.

IIED (International Institute for Environment and Development). 1994. *Whose Eden? An Overview of Community Approaches to Wildlife Management*. London: Russell Press.

IUCN (International Union for Conservation of Nature and Natural Resources). 1980. *World Conservation Strategy: Living Resource Conservation for Sustainable Development*. Gland, Switzerland: International Union for Conservation of Nature and Natural Resources.

Kiss, A., ed. 1990. *Living with Wildlife: Wildlife Resource Management with Local Participation in Africa*. Washington, D.C.: The World Bank.

Kjersgård, L. 1997. Developing participatory systems of protected area management. In T. Blomley and P. Mundy, eds., *Integrated Conservation and Development Projects in CARE International*, 54–55. Copenhagen: CARE–Denmark.

Kremen, C., J. O. Niles, M. G. Dalton, G. C. Daily, P. R. Ehrlich, J. P. Jay, D. Grewal, and R. P. Guillery. 2000. Economic incentives for rain forest conservation across scales. *Science* 288:1828–1832.

Larson, P. S., M. Freudenberger, and B. Wyckoff-Baird. 1998. *WWF Integrated Conservation and Development Projects: Ten Lessons from the Field 1985–1996*. Washington, D.C.: World Wildlife Fund.

Lee, D. R., P. J. Ferraro, and C. B. Barrett. 2001. Introduction: Changing perspectives on agricultural intensification, economic development, and the environment. In D. R. Lee and C. B. Barrett, eds., *Tradeoffs or Synergies?* 1–15. Wallingford, Oxon, U.K.: CABI.

Lélé, S. M. 1991. Sustainable development: A critical review. *World Development* 19:607–621.

Lucas, P. H. C. 1992. *Protected Landscapes*. London: Chapman and Hall.

Mace, G. M. and E. J. Hudson. 1999. Attitudes towards sustainability and extinction. *Conservation Biology* 13:242–246.

Margoluis, R. and N. Salafsky. 1998. *Measures of Success*. Washington, D.C.: Island Press.

McNeely, J. A. 1988. *Economics and Biological Diversity: Developing and Using Economic Incentives to Conserve Biological Resources*. Gland, Switzerland: International Union for Conservation of Nature and Natural Resources.

McNeely, J. A. and K. Miller, eds. 1984. *National Parks, Conservation, and Development: The Role of Protected Areas in Sustaining Societies*. Washington, D.C.: Smithsonian Institution Press.

Munro, David, dir. 1991. *Caring for the Earth: A Strategy for Sustainable Living*. Gland, Switzerland: IUCN–The World Conservation Union, UN Environment Program, World Wildlife Fund.

Naughton-Treves, L. and S. Sanderson. 1995. Property, politics, and wildlife conservation. *World Development* 23:1265–1275.

Newmark, W. D. and J. L. Hough. 2000. Conserving wildlife in Africa: Integrated conservation and development projects and beyond. *BioScience* 50:585–592.

Oates, J. F. 1999. *Myth and Reality in the Rain Forest. How Conservation Strategies Are Failing in West Africa*. Berkeley and Los Angeles: University of California Press.

Paul, S. 1987. *Community Participation in Development Projects: The World Bank Experience.* Discussion Paper no. 6. Washington, D.C.: The World Bank.

Pomeroy, R. S. 1994. *Community Management and Common Property of Coastal Fisheries in Asia and the Pacific: Concepts, Methods, and Experiences.* Manila, Philippines: International Center for Living Aquatic Resources Management (ICLARM).

Proceedings of Integrated Conservation and Development Projects Lessons Learned Workshop. 2000. Hanoi, Vietnam.

Putz, F. E., K. H. Redford, J. G. Robinson, R. Fimbel, and G. M. Blate. 2000. *Biodiversity Conservation in the Context of Tropical Forest Management.* World Bank Environmental Department Papers no. 75. Washington, D.C.: The World Bank.

Redclift, M. 1987. *Sustainable Development: Exploring the Contradictions.* London: Methuen.

Redford, K. H. and B. Richter. 1999. Conservation of biodiversity in a world of use. *Conservation Biology* 13:1246–1256.

Redford, K. H. and S. E. Sanderson. 1992. The brief, barren marriage of biodiversity and sustainability. *Bulletin of the Ecological Society of America* 73:36–39.

Robinson, J. G. 1993. The limits to caring: Sustainable living and the loss of biodiversity. *Conservation Biology* 7:20–28.

Robinson, J. G. 2001. Using "sustainable use" approaches to conserve exploited populations. In J. D. Reynolds, G. M. Mace, K. H. Redford, and J. G. Robinson, eds., *Conservation of Exploited Populations,* 485–498. Cambridge: Cambridge University Press.

Robinson, J. G. and E. L. Bennett, eds. 2000. *Hunting for Sustainability in Tropical Forests.* New York: Columbia University Press.

Sachs, W. 1991. Environment and development: The story of a dangerous liaison. *The Ecologist* 21:252–257.

Salafsky, N., H. Cauley, G. Balachander, B. Cordes, J. Parks, C. Margoluis, S. Bhatt, C. Encarnacion, D. Russell, and R. Margoluis. 2001. A systematic test of an enterprise strategy for community-based biodiversity conservation. *Conservation Biology* 15:1585–1595.

Salafsky, N., R. Margoluis, and K. Redford. 2001. *Adaptive Management: A Tool for Conservation Practitioners.* Washington, D.C.: Biodiversity Support Program.

Salafsky, N., R. Margoluis, K. H. Redford, and J. G. Robinson. 2002. Improving the practice of conservation: A conceptual framework and agenda for conservation science. *Conservation Biology* 16:1469–1479.

Salafsky, N. and L. Wollenberg. 2000. Linking livelihoods and conservation: A conceptual framework for assessing the integration of human needs and biodiversity. *World Development* 28:1421–1438.

Sanderson, E. W., K. H. Redford, A. Vedder, S. E. Ward, and P. B. Coppolillo. 2002. A conceptual model for conservation planning based on landscape species requirements. *Landscape and Urban Planning* 58:41–56.

Sanderson, S. E. 1992. *The Politics of Trade in Latin American Development.* Stanford, Calif.: Stanford University Press.

Sanderson, S. E. 1998. The new politics of protected areas. In K. Brandon, K. H. Redford, and S. E. Sanderson, eds., *Parks in Peril: People, Politics, and Protected Areas*, 441–454. Washington, D.C.: Island Press.

Sanjayan, M. A., S. Shen, and M. Jansen. 1997. *Experiences with Integrated-Conservation Development Projects in Asia.* World Bank Technical Paper no. 388. Washington, D.C.: The World Bank.

Solow, R. M. 1974. Intergenerational equity and exhaustible resources. *Review of Economic Studies* 41:29–46.

Stocking, M. and S. Perkins. 1992. Conservation-with-development: An application of the concept in the Usambara Mountains, Tanzania. *Transactions of the Institute of British Geographers* 17:337–349.

Veit, P. G., A. Mascarenhas, and O. Ampadu-Agyei. 1995. *Lessons from the Ground Up: African Development That Works*. Washington, D.C.: World Resources Institute (WRI).

Verburgh, R. M. and V. Wiegel. 1997. On the compatibility of sustainability and economic growth. *Environmental Ethics* 19:247–265.

Weins, J. 1989. Spatial scale in ecology. *Functional Ecology* 3:385–397.

Wells, M. and K. E. Brandon. 1992. *People and Parks: Linking Protected Area Management with Local Communities*. Washington, D.C.: The World Bank.

Wells, M., S. Guggenheim, A. Khan, W. Wardojo, and P. Jepson. 1999. *Investing in Biodiversity: A Review of Indonesia's Integrated Conservation and Development Projects*. Washington, D.C.: The World Bank.

Western, D. and R. M. Wright, eds. 1994. *Natural Connections: Perspectives in Community-Based Conservation*. Washington, D.C.: Island Press.

WCED (World Commission on Environment and Development). 1987. *Our Common Future*. Oxford: Oxford University Press.

Wyckoff-Baird, B., A. Kaus, C. A. Christen, and K. Keck. 2001. *Shifting the Power: Decentralization and Biodiversity Conservation*. Washington, D.C.: Biodiversity Support Program.

3

The Pathology of Projects

Jeffrey Sayer and Michael P. Wells

Introduction

In the early years of development assistance it was common for aid funds to be invested in "institutional support" for government agencies in developing countries. International advisers were sent to work within the host institutions. Later the drive for accountability and the need for international donors to be able to target their support more precisely led to the emergence of the "development project" as the main delivery mechanism. This meant that donors worked with their national counterparts to define discrete, time-bound packages of development assistance. These packages have allowed donors to apply their own accountability mechanisms and allowed development to be reduced to bite-sized components for which donors can assume responsibility and take credit.

This trend away from institutional support and toward projects has been reflected in international development agency support for biodiversity conservation. Earlier aid programs supported game rangers, wardens, researchers, and others working within national protected area programs. As international donor support for biodiversity conservation has expanded significantly during the last two decades, however, these institutionally based programs have been displaced by projects targeting a subset of the problems facing biologically rich sites, mainly protected areas. Most international development agency support for biodiversity conservation since the 1980s has been in the form of projects, mainly integrated conservation and development projects (ICDPs).

Alternative, nonproject financing models have recently grown in importance within at least some international development agencies. For example, more than half of the World Bank's lending is now in the form of programmatic support linked to structural adjustment, sector-wide investment programs, and social action funds. However, projects continue to be the dominant method of financing in the environmental sector, especially in biodiversity conservation, and there is no sign that they are about to decline in importance.

This paper discusses how and why the project model has come to dominate biodiversity conservation activities supported by development agencies. The constraints imposed by this model are reviewed and some recommendations made for improving effectiveness, while recognizing that projects, in some shape or form, are undoubtedly here to stay.

Why Are Projects So Ubiquitous?

The popularity of projects in development assistance, and particularly in biodiversity conservation, seems to be due less to their inherent advantages than to the often-fatal weaknesses associated with any alternatives proposed so far.

Government agencies responsible for delivering public services in many developing countries are relatively weak. While such agencies have been targeted for institutional strengthening through donor assistance programs, the results have often been disappointing. This is why projects are so popular. If official institutions in developing countries were stronger and more effective—with better technical capacities, human resources, financial controls, and service delivery mechanisms—there would be no need for projects. Development agencies would simply provide funding for agreed-on priority programs and then supply specialized technical advice on demand. While this does happen in a few cases, it remains rare. Donors provide funding to government agencies through projects because they lack confidence in these agencies' normal operating procedures. To be fair, most developing country institutions do suffer from appalling human and financial resource limitations that make it almost impossible for them to carry out their mandate.

Project Pathologies

Ownership and Accountability

Projects do provide donors with more control over the use of funds, usually through planning and target-oriented management techniques that individual donor agencies have developed over time. Such approaches help to standardize the approach to projects, track the use of funds, reduce administrative costs, and measure and report progress to their own constituencies. For national counterparts to understand and become familiar with the project design and implementation methods of a single donor agency is extremely challenging, and the terminology and requirements that have evolved within each agency are rarely transferable to other agencies' projects. One result is that effective decision making and the real control over the project development process almost invariably lie with development agency staff and the consultants they hire. This external control discourages the recipient country authorities from taking responsibility, dilutes any real sense of genuine partnership, and often seems in conflict with the concept of national project ownership.

Donors do emphasize the importance of projects being "owned" by institutions in the recipient countries, recognizing that externally imposed sets of activities that are disconnected from existing national activities rarely succeed. However, the degree of ownership varies considerably between projects. Some projects are embedded in a genuine national process of development, while in others national ownership is minimal. Unfortunately, ICDPs have tended to be in the latter category. When the key national counterpart agencies are relatively weak, as is often the case for wildlife or protected area management departments and ministries of environment, the power imbalance between donor agencies and the recipients can be further exacerbated. When the intended beneficiaries are local communities, a genuine partnership to facilitate negotiations toward mutually acceptable outcomes is even more difficult to achieve.

Why Biodiversity Projects Are Particularly Problematic

The project model was originally developed and introduced to facilitate the building of roads and bridges and the establishment of large-scale agricultural projects. In the case of ICDPs, this same approach has been used for the subtle and difficult task of changing social and economic incentives to better support biodiversity conservation, a task for which projects were not designed and are often unsuited.

There are some important mismatches between what conservation requires and what projects offer (Kiss 1999). Projects are intrinsically limited in space, time, and numbers of beneficiaries, while the main cause of biodiversity decline is the loss, fragmentation, and degradation of habitat over large areas due to many different kinds of human activities. Halting or mitigating biodiversity loss requires changing the behavior of large numbers of people dispersed over large areas for long periods of time. Projects are inherently unsuited to this.

Projects also tend to focus on activities rather than impacts. Most ICDPs aim to conserve species and ecosystems. However, ICDPs have often emphasized community-level social and economic development activities as an indirect step toward more effective conservation in the long-term future. When the linkage between the eventual goal and the activities selected is vague or seems distant, as has often been the case with ICDPs, attention inevitably becomes focused on the project activities themselves and not the impacts of these activities. This leads to an excessive focus on getting activities completed or checked off (on the part of the donor) or getting as much benefit as possible out of the project (on the part of the beneficiaries). The frequent disconnect between activities and desired impacts within ICDPs has also led to a divergence between the benefits obtainable from biodiversity conservation and the benefits obtainable from the project. Many projects start by emphasizing the former but end up concentrating almost entirely on the latter.

Protected areas are usually major components of large, complex landscapes where land uses vary and where a variety of economic activities are being undertaken (see Maginnis, Jackson, and Dudley this volume). They are often under pressure by diverse interest groups. They are subject to unpredictable influences resulting from changes in local economies, access to markets, population movements, climate change, and a host of other exogenous forces. Most of these forces come to bear on the boundaries of protected areas where most ICDPs are operating. Many ICDPs are trying to shoehorn the complex and dynamic realities of the protected area frontier into the constraints of a time-bound, tightly planned, highly predictable project. This often does not work.

PREPARATION VERSUS IMPLEMENTATION

Frustrated at their inability to achieve successful examples of conservation and development, some donors have reacted by more rigorous application of the tools of the development assistance trade. They have planned their projects in more and more detail. They have commissioned more careful studies to reduce the likelihood of surprises. They have developed more sophisticated monitoring and evaluation systems to ensure that everything is staying on track. And the end result has been a generation of ICDPs that are so locked into a rigid donor-driven framework that they have little relevance to the changeable real world in which protected areas and their managers have to survive.

Within projects there tends to be a disproportionate emphasis on planning at the expense of implementation. Expatriate experts who will have no subsequent involvement in implementation often prepare detailed project plans. When reality turns out to deviate from the plans, there is rarely sufficient management capacity or budget flexibility to respond appropriately. These problems could be addressed by formally adopting an adaptive management approach, but adaptive management is so poorly understood by most organizations that it is often mistaken for trial-and-error management or simply poor planning (Salafsky and Margoluis this volume).

Projects tend to invest heavily in extensive background studies and elaborate plans made by outside experts, generating reports that too often are rarely used, culturally irrelevant, and quickly obsolete (Singh and Sharma this volume). The commissioning of studies by teams of experts in order to characterize "the problem" is a prevalent ICDP pathology. In practice, the informal knowledge of local people has to be the basis of most of the resource management decisions that will be taken by a project. It is the behavior of these local people that projects will strive to influence. This local knowledge is often the scarcest resource. One reason that projects often begin to become effective only after several years of operation is that it can take this long for international project advisers to become sufficiently attuned to local realities and begin to tap the informal local knowledge that is so important to success.

Repeatedly, one finds examples of preparation missions identifying and describing problems in ways that must seem quite bizarre to local people. The authors have frequently experienced cases where literate local people in areas targeted by major ICDPs describe the conservation and development problems they confront in their everyday lives in ways that are startlingly different from the assumptions underlying the projects. Even the more recent ICDPs developed using participatory techniques are still often predicated on fundamental and incorrect assumptions (see McShane and Newby this volume). Maps produced with local participation are still often subject to the overriding influence of the outside specialists with their own vision of what needs to be mapped. Such maps create their own realities (Scott 1998).

LACK OF FLEXIBILITY

A main feature of the project paradigm in development assistance is the attempt to reduce uncertainty. Projects attempt to reduce the level of complexity and to tease out a subset of actions with precise costs and indicators that can be verified. This is very different from the real-life task of a protected area manager. The job of the manager is not to attempt to reduce or eliminate complexity and uncertainty but rather to exercise professional judgment in dealing with these real-world situations. Good protected area managers have always been "adaptive managers"; their professionalism lies in their ability to make good judgments in response to the constant surprises that confront them in their day-to-day activities.

Logframes have become popular project preparation tools, and are now required by many donor agencies. Properly used, a logframe can indeed be a valuable basis for clarifying assumptions and facilitating a transparent process of negotiation toward desired outcomes (see also Franks and Blomley this volume). But too often logframes are used to limit the flexibility of projects. The logframe becomes the master rather than the tool. It ties participants into courses of action that were determined at the beginning of the project rather than being used to negotiate course changes and adaptability. While logframes can encourage implicit assumptions to be identified in advance and can work well as tools for adaptive management, more often they are used in ways that seem to suppress innovation and effective decision making during project implementation.

These problems are exacerbated by the extreme pressure that development agencies are under to reduce their management costs and cut back on staff. As a result of these pressures, there is an increasing reliance on standard, off-the-shelf project approaches. There is little incentive for managers within these organizations to work on complex conservation and development projects requiring unique approaches. Adaptive management, which is dependent on substantial high-level staff inputs from agencies over a long period, may be difficult to reconcile with the current trend of reductions in human and financial resources for project management.

DISBURSEMENT OF FUNDS

One particularly worrying element of the project paradigm is that relatively little money seems to reach the intended beneficiaries on the ground, especially in the early phases of projects. The surveys and planning required to get started inevitably seem to absorb a large part of the financial resources of projects. It is relatively common for successful efforts to raise local expectations to be followed by remarkably long periods of inactivity before projects begin to support local activities. A surprisingly large number of ICDPs never do provide significant benefits to local people. They may make new technologies available or temporarily improve some social facilities, schools, roads, etc. But often these benefits account for only a small proportion of the total project budget, and often the patience of local people is tried as they await the recompense for their investments of time and knowledge, and possibly practice restraint from livelihood activities they have been told are harmful to conservation and are unsustainable.

The relatively large amounts of money that some projects inject into limited spatial areas are often dramatically disproportionate to very modest local absorptive capacities. Projects are often like flash floods that generate massive amounts of money for several years and then abruptly end, with little to show afterward. Disbursing funds in smaller quantities over significantly longer periods in ways that are more compatible with local absorptive capacities seems almost impossible for most donor-financed projects to achieve. Once projects have been approved, there is often significant pressure to maintain the speed of disbursements, with lagging disbursements being blamed on "inefficient" management. One reason for haste is that most projects establish some form of project management unit that exists as a separate entity while the project is in existence. These units are often staffed at least partially by expatriates and are expensive to maintain. Extending the life of a project could mean that the costs of maintaining these management units would eat up an even larger share of the project budget.

Tangible "milestones" are the measure of success for many donors. ICDPs usually aim to provide alternative livelihood opportunities and/or to compensate for costs associated with reduced access to or use of biodiversity. Although rarely achieved in practice, these are at least measurable. If long-term change and sustainable development are the goal, however, learning and negotiation processes in most ICDPs are no less important than technical deliverables. The institutional and social learning that changes people's behavior is the ultimate outcome pursued by ICDPs. Change takes time, and many projects have suffered the long-term costs of imperfect processes in their excessive haste to disburse funds and achieve deadlines. The more successful examples of ICDPs have been those where small amounts of money were made available flexibly and sensitively over a long period. There have been many failures when large amounts of money have been thrown at problems too rapidly. The management challenge, of course, is how to distinguish projects that are progressing well toward long-term goals by building foundations in ways that are not easily measurable from projects that are going nowhere at great cost. This may require more innovative, and perhaps costly, approaches to project monitoring.

Local Participation

While increased local participation has become mandatory in projects, the commitment to participation is sometimes more rhetorical than real, with participation often degenerating into yet another requirement that project managers have to fulfill. In the worst cases, local participation seems little more than the process of explaining to local people what the project plans to do. Finding out who gets to participate in what, and how and when they get to participate, can often reveal the extent of genuine participation. Often, insufficient efforts are made to build genuine constituencies of support among local communities that will persevere after the relatively brief project period has ended (see Brown this volume). This project problem is by no means limited to ICDPs (Kiss 1999).

There are too many examples of local populations being completely unaware of why ICDPs were being undertaken—even in projects that have been operating for a few years. Many projects have made insufficient efforts to build genuine constituencies of local support for common objectives. Rather, the major emphasis seems to be on co-opting certain targeted groups to either participate in or at least not oppose the activities designed and supported by the project. We talk of ownership of projects by local people, but frequently we have to invest a lot of effort in trying to secure "their" ownership of "our" project. Projects can be a useful way of helping people do something they want to, but they are usually not a good way of getting them to want to do it.

Broader Contexts

Projects have a mixed record at being well linked with developments in other sectors that influence local outcomes. Failure of projects is often attributed to unpredicted changes in the macroeconomic or political context. Local political support is often essential to the success of projects, even though political changes may cause this support to evaporate rapidly. An international market for a newly introduced crop promoted by a project may collapse because of changes in exchange rates or trade policies. These are examples of the pathology of project bounding. The tendency for donors to circumscribe a project in a way that makes it a self-contained package seems to make it difficult for projects to be managed in ways that make them responsive to changes in their external environment. This aspect of ICDPs is explored further in several papers in this volume and will not be elaborated here.

Recognition and Success

Even some apparently trivial aspects of the project are inimical to success. Most donors and their executing agencies want their contributions recognized, not least because such recognition can go a long way toward justifying future budget allocations. The same donors who require local ownership of projects also want their logos on the vehicles and on the cover page of publications. They want their proposals and reports written in international languages and prepared in ways that only international experts can manage. Donors want to visit their projects, and preferably they want to bring politicians to see the good work. They want to see clear evidence of their own contributions, and they also have high expectations of success. ICDP interventions enjoy successes and failures—often quite a lot of the latter. But all the incentives favor the exaggeration of successes and the rationalization or downplaying of any failures. Yet it is these very failures that teach us the lessons from which long-term success may emerge (cf. Redford and Taber 2000).

There is sometimes reluctance among NGOs and private foundations, as well as development agencies, to recognize and learn from failure. Our culture is dominated by the search for success. We do not think that people unrealistically exaggerate their claims for proposed project benefits because they are uninformed or misguided; it is more often that the prevailing culture only rewards project proposals that exaggerate project anticipated benefits. While identifying and learning from success stories can be beneficial, success has been overemphasized to the point where it is the only outcome we are able to acknowledge. This puts enormous pressure on the internal dynamics of donors to avoid genuine innovation and risk taking.

Exit Strategies and Sustainability

Projects are generally conceived as short-term interventions to catalyze long-term change. In the case of ICDPs, the objectives are usually behavioral changes by key stakeholders toward activities that support rather then threaten biodiversity conservation. Since projects are usually for a fixed period, typically three to five years, a key question is, What happens afterward?

In practice the "What happens afterward?" question is rarely answered satisfactorily in advance. This is not because of uncertainties over the future situation; it is more related to the fact that funding for biodiversity conservation in developing countries is hard to find, that the problems facing key sites are often urgent, that three to five years seems a long time away, and that the forthcoming project is often the first substantial outside assistance the site has received. Key project promoters are often so pleased to finally get some support for conservation that what happens half a decade later is not an immediate concern.

Much has been written about the need for any kind of project based on substantial local participation to adopt a time frame of a decade or more, even though the opportunities for this in practice are limited. Three to five years is not only a very short period to achieve the often-ambitious goals of an ICDP, but many of these projects find themselves moving forward at a much slower pace than they had originally intended. As a result, the tangible progress made as the project approaches completion is often less than anticipated, even if there have been advances in critically important areas such as trust building among local communities, awareness building, and negotiations with key stakeholders.

Many ICDPs unfortunately make the rash assumption that the conservation benefits unleashed by the project will be sufficient to finance recurring management costs after project funding has been used up. Whether project designers really believe this is possible or they simply provide the language that the donor community has become conditioned to expect is difficult to say. Except in the rare cases of projects being able to establish a trust fund or endowment, such optimism usually proves totally unfounded. Irrespective of project design, in very few situations will biodiversity conservation in developing countries become self-financing as a result of a project or otherwise. In most cases, substantial external subsidies are clearly going to be required for as long as some biodiversity remains to be protected.

Lack of financial sustainability has meant that, without further injections of funds, many ICDP activities face the prospect of financial collapse and abrupt termination once projects are completed. Certainly, expensive project management units have proved completely unsustainable from a financial point of view. Few, if any, government conservation agencies are capable of absorbing such operations, even if they had been pressured to commit to do so before a project began.

Interestingly, many development projects do not actually collapse from a lack of funds—simply because of the political capital that has been invested in them. Flawed projects are sometimes not cancelled during their implementation period for the same reason. Experience shows that invested political capital often provides unstoppable momentum that seems relatively impervious to disappointing results. Donors and governments are extremely reluctant to stop a project once it has started. As a result, a surprisingly large number of dubious projects are not cancelled and do not die a natural death when their funding runs out, but reemerge in a second phase to "build on the successes" of the first phase. The larger the project, the more frequently this phenomenon seems to occur (see also Child and Dalal-Clayton this volume).

Approaches That Might Work

Donors have the right to expect accountability and to know what they are paying for. They will always need to know if and when the objectives of their interventions have been attained. Donor interventions will always have to be bounded in some way. So how can the present project approach be modified to increase the likelihood of achieving the dual goals of improving local livelihoods and conserving biodiversity?

Our main focus has been on projects themselves, as we expect these will continue to be the dominant biodiversity conservation vehicle. However, it is important to consider alternatives, including direct payments for conservation services and more flexible forms of funding (see Kiss this volume). There are many examples of innovative approaches that contain the elements of success. The World Bank has experience in other sectors in using approaches such as "adaptive program loans." These have similarities with programs of some bilateral donors who make long-term commitments to a general goal but manage the process continuously to ensure that the assistance program adapts to changing circumstances.

One key to success is to commit for the long term to a general conservation and development goal that is genuinely shared by the donor and the host nation. It is neither necessary nor desirable to be too precise about the exact nature of the final outcome nor the pathway to getting there. This is the concept of the "lighthouse" as formulated by Lee (1993). The lighthouse is the distant target toward which all efforts are directed. The compass and the gyroscope are metaphors for the management tools needed to get there. The lighthouse could be the goal of achieving an acceptable balance between local development benefits and global biodiversity values. The exact nature of the compromise that will eventually emerge cannot be known in advance—neither can the route that will be taken to arrive at the final outcome. Once an agreement has been reached on the ultimate goal, the donors and their agents must slowly engage with the local stakeholders. Instead of project interventions being front-loaded, with major investments at the beginning, they should be back-loaded. A long time should be invested in getting to know the local situation, developing sensitivity to local conditions, and simply getting accepted as another stakeholder—of being part of the system.

Inevitably the views of the donor or conservation organization on the desirable outcomes of a project will differ from those of many local stakeholders. Negotiations and trade-offs will be required. These differences in interests have to be made explicit from the beginning. Outside conservation agencies also have to bring something to the negotiating table. Local interest groups have to be given some plausible reason to settle for outcomes that are less than ideal from their local perspective. There has to be something in it for them, and it helps if some of these local benefits begin to flow reasonably early in the project process.

Under this scenario, donor or conservation organizations will cease to see themselves as the owners of a project, and begin to see themselves as interest groups who are participants in a process. They will no longer be dominant participants with superior knowledge and exclusive control of resources but participants with different kinds of knowledge and the ability, for a limited period, to bring financial and technical resources. Local interest groups bring to the negotiating table their traditional rights and knowledge and a large degree of ownership over the natural resources whose conservation is at stake (see also Tongson and Dino, Singh and Sharma, Child and Dalal-Clayton, all this volume). Measures of success will not be defined ex-ante by the donor but negotiated continually by the different interest groups.

It is difficult to escape from the model of the donor controlling project processes, which has been part of the development landscape for too long. But it is possible to see things differently. Government programs to improve the livelihoods and conserve the local environments of Aboriginal peoples in Australia have turned the project paradigm on its head. The government no longer contributes and manages projects; instead, it makes funds available to the Aboriginal peoples, who are then able to make their own choices about their conservation and development goals and to select and manage any experts that they might need to help them achieve their desired outcomes. The donor, in this situation the government, negotiates the limits of the type of activities that may be supported, but otherwise takes a backseat in the process.

The Way Forward

We have pointed out a lot of problems that have occurred in trying to achieve better biodiversity conservation in developing countries through projects. There are two sets of issues to bear in mind. The first set arises from the scale mismatch between the biodiversity "problem" as commonly manifested and the rather limited and rigid "solution" offered by the project model (Kiss 1999). These issues can be adequately addressed only by working on broad fronts that link site-specific project work with complementary activities designed to strengthen policies, institutions, and governance. The second set of issues arises from less-than-optimal project design and execution, where the remedies are easier to identify. To address this second set of issues we advocate the application of seven basic principles that could serve to modify classic projects in ways that could significantly improve the likelihood of achieving success:

1. *Set general goals and recognize that adaptability and learning will be required to reach them.* It is counterproductive to attempt to define outcomes too precisely ex-ante. Goals should be expressed in general terms, and negotiations and trade-offs should be accepted as part of the process leading to their attainment. All participants have to learn from the process, and adaptation will be required at all stages.
2. *There must be a fundamental commitment to an equitable relationship with local interest groups.* Project staff must show humility and respect in their dealings with local interest groups. They must genuinely value local knowledge and have a commitment to local values and judgments. The arrogance of Western-based science and financial clout must not be allowed to persist.
3. *Outside interest groups must bring something to the table.* The concept of a "donor" is itself outmoded. Aid agencies and conservation organizations are interested parties. They have an agenda, and they must negotiate like everyone else. Instead of adopting the stance of donors, they should be contributors who will provide money in order to achieve specified outcomes.
4. *All parties must commit to the process for the long term.* It is too easy for a donor to walk away after a fixed period and leave the local interest groups with "their" problem. If it is a genuinely shared problem, all parties must anticipate long-term involvement, even if this is conditional on other stakeholders living up to their commitments. Education, awareness, and communications must be integral. Periods of awareness raising as well as trust and constituency building must occur before more expensive "project" activities should be considered.
5. *All must move at the pace of the slowest.* The sorts of changes in resource management practices that ICDPs seek to bring about

will inevitably take time. One is seeking to change the behavior of cautious and conservative people. Some interest groups may be exposed to risks and will justifiably be cautious about accepting change. Externally determined deadlines should not be allowed to predominate. The artificial and counterproductive separation between the preparation and the implementation of projects needs to be reduced. More attention should be given to implementation by empowered managers and less to detailed advance planning.

6. *Everyone's expectations must be realistic.* The rate and extent of change that is possible in resource management systems is often much less than outside experts tend to imagine. Proponents of projects must avoid the exuberant hyperbole that often goes with effective fund-raising. Too many projects have totally unrealistic expectations.
7. *Funding must follow process.* More flexible funding vehicles should be made available so they can be adapted to the needs of the situation and especially to the local absorptive capacity.

While it may not be feasible to escape from the ubiquity of projects as vehicles for development assistance as well as for biodiversity conservation, the pathology of projects should always be borne in mind as we struggle to design and implement more effective interventions.

References

Kiss, A. 1999. Making community-based conservation work. Paper delivered at annual meeting of the Society for Conservation Biology, July, at Greenbelt, Md.

Lee, K. L. 1993. *Compass and Gyroscope: Integrating Science and Politics for the Environment.* Washington, D.C.: Island Press.

Redford, K. H. and A. Taber. 2000. Writing the wrongs: Developing a safe-fail culture in conservation. *Conservation Biology* 14:1567–1568.

Scott, J. C. 1998. *Seeing Like a State: How Certain Schemes to Improve the Human Condition Have Failed.* New Haven, Conn.: Yale University Press.

4

Expecting the Unattainable: The Assumptions Behind ICDPs

Thomas O. McShane and Suad A. Newby

Introduction

Almost two decades after the first integrated conservation and development projects (ICDPs) were implemented, results remain inconclusive and questions are being raised as to whether ICDPs have succeeded in achieving their joint conservation/development objectives and whether or not the approach is appropriate (Barrett and Arcese 1995; Wells 1995; Newby 1996; Wells et al. 1999). The ICDP experience has yielded few examples of protected areas generating adequate benefits to local communities to create sufficient incentives for conservation. Criticism of ICDPs has also focused on their lack of ability to address the underlying root causes of biodiversity loss, as well as the financial and technical sustainability of such initiatives (Barrett and Arcese 1995; McShane 1999a; Wood, Stedman-Edwards, and Mang 2000).

It has been postulated that many ICDP weaknesses are related to the assumptions made during their design and implementation, many of which are found to be false or unrealistic (Sanjayan, Shen, and Jansen 1997; Larsen, Freudenberger, and Wyckoff-Baird 1998; Worah 2000; Hughes and Flintan 2001). ICDPs may be categorized as being based on the following beliefs:

- Diversified local livelihood options will reduce human pressure on biodiversity, leading to its improved conservation.
- Local people and their livelihood practices, rather than "external factors," constitute the most important threat to biodiversity.
- ICDPs offer sustainable alternatives to traditional protectionist approaches to protected area management.

However, lessons from ICDP experiences point to the following realities:

- Equity issues are often poorly addressed (Colchester 1994; Gibson and Marks 1995; Wainwright and Wehrmeyer 1998).
- Conservation and development links are not clearly established (Kremen, Merenlender, and Murphy 1994; Wells et al. 1999; Mittleman 2000).
- Wider threats such as policy and market constraints are poorly accounted for (Barbier, Burgess, and Folke 1994; Clay 1996, 1997; McNeely 1998; Wood, Stedman-Edwards, and Mang 2000).

Testing assumptions is about systematically trying different interventions to achieve a desired outcome (cf. Bell and McShane-Caluzi 1984; Margoluis and Salafsky 1998). The making of assumptions during project design is meant to identify those issues that are likely to directly affect the project's goal (or target condition). Project activities are designed to address these causal factors. Underlying assumptions include the impacts that other, less directly linked conditions and factors might have on project activities (Margoluis and Salafsky 1998). While certain assumptions are clearly beyond an ICDP's remit or competence to address (e.g., climate conditions, political stability), others, if not addressed, may significantly constrain project progress (e.g., community participation, technical inputs, government support and policies, human and institutional capacity, funding, etc.).

This paper is a study of the seven ICDPs that make up the DGIS-WWF Tropical Forest Portfolio. In it, we review the impact of assumptions made by the Portfolio ICDPs on project effectiveness that identify constraints to project design and implementation. The underlying issues that the projects need to address, either directly or indirectly, are discussed. We believe that failure to address the issues embodied in the assumptions made will inevitably lead to a poor or failed ICDP.

Methodology

The DGIS-WWF Tropical Forest Portfolio is composed of seven field projects in six countries: Gabon (2), Ethiopia, Ecuador, Honduras,

Pakistan, and the Philippines. Since its implementation in 1996, the Portfolio has aimed to draw lessons on the effectiveness of ICDPs (cf. McShane 1999a). In putting the Portfolio together, the WWF recognized that the ICDP approach was constrained by various circumstances that collectively threaten projects such as these. To identify and better understand these constraints, resources were made available for rigorous monitoring, technical inputs, capacity development, and improved information exchange. As a result, an essentially loose collection of field projects was transformed into a coherent program to better understand the dynamics of conservation and development.

To collect information from as diverse a set of field projects as the Portfolio, a standardized monitoring system was developed with the aim of identifying common strengths and weaknesses in the implementation of these ICDPs (McShane 1997, 1999a). In going about this, four issues were addressed:

Framework: a common approach or methodology to report on experiences gained in trying to meet project objectives. The Portfolio required a tool that identified objectives, verifiable indicators, and assumptions as developed in the planning phase of the program. It also required an effective method of analyzing the results—targets achieved, problems experienced, lessons learned, and measures taken to address the problems.

Process: the development of a framework that is acceptable to all of the projects expected to use it. The challenge was to get a diverse group of people speaking a number of languages across three continents to buy into a common system. The Portfolio brought all of the field projects together in a facilitated meeting to have participants design a framework that would best fit their needs. In this way the Portfolio developed a common monitoring and reporting methodology that would be used across all the field projects.

Capacity: ensuring that the project implementers have the skills to undertake the monitoring so that information quality and accuracy remain high. As a result, the first two years of the five-year program were spent building capacity and developing the tools and methods that would allow learning and adaptation to take place. This proved to be a labor-intensive process.

Willingness: field staff sees the benefits of the framework in improving their ability to meet project goals and objectives. Reviews of the Portfolio have highlighted the effectiveness of the monitoring system, its adaptive management qualities, and the support it enjoys with the field projects (Hillegars and Conradi 2000). The framework was developed by the field project implementers and responds to the questions and issues they have to answer to do their work. The process was participative and collaborative, and project implementers have the skills and capacity to undertake the monitoring.

For this study we reviewed all of the Portfolio's field project proposals from 1996 and the twice-yearly project monitoring reports from 1997 to 2001 (a total of nine monitoring reports for each field project). The monitoring reports are presented as double matrices with a minimum amount of narrative (McShane 1997). The first matrix describes project objectives, verifiable indicators, and assumptions. The second matrix undertakes a results analysis of targets achieved for the previous six months, problems encountered, lessons learned, and measures taken against project objectives and activities. These results were then assessed against the assumptions made during the project design phase to determine the possible impact of the assumptions on project performance and whether or not they were taken into consideration during project implementation. We specifically looked for constraining factors common to all of the field projects for which assumptions had been made. Examples from the Portfolio field projects are used to illustrate how certain assumptions have either proven false or have had such an impact on project progress that they need to be recognized and addressed more directly and more effectively.

The Assumptions

Four principal themes, embodied in many of the assumptions made by the Portfolio field projects, were identified as acting to constrain progress in implementation:

- Livelihood development
- Capacity development
- Institutional support
- Policy factors

Although we have lumped the assumptions into these four themes, we acknowledge that they are interrelated and influence one another.

LIVELIHOOD DEVELOPMENT

Table 4.1 summarizes the assumptions concerning livelihood development made across the DGIS-WWF Tropical Forest Portfolio projects, along with the most common resulting constraints.

Table 4.1 *Livelihood development assumptions and constraints*

Assumptions	*Constraints*
1. Livelihood development through the promotion of income-generating and ecologically sustainable activities will lead to biodiversity conservation.	1.1 High expectations by local communities for project livelihood activities. 1.2 Activities proposed do not always generate sufficient revenue. 1.3 Lack of capacity of implementing organizations to effectively carry out chosen activities.
2. Communities are homogeneous and static.	2.1 Marginal groups. 2.2 In-migration from other areas.
3. Main threats to biodiversity arise from communities adjacent to protected areas.	3.1 The root causes of biodiversity loss are not well identified.

Assumption 1: Livelihood development through the promotion of income-generating and ecologically sustainable activities will lead to biodiversity conservation.

The concern for the economic and social development of communities living near protected areas is what distinguishes ICDPs from other conservation projects. Rural development becomes a means of achieving conservation goals, and income-generating activities are used to improve livelihoods. In some cases the development of income-generating activities is seen as a compensation for the loss of access to, and use of, protected areas (IIED 1994; Newby 1996). However, the linkages between the conservation and the development components are difficult to establish. There are few good examples illustrating that such linkages exist and that biodiversity conservation is enhanced by development activities when there is no effective enforcement of sanctions and penalties (Wells and Brandon 1992; Kremen, Merenlender, and Murphy 1994; Scott 1998). It is also held that increasing the living standards of local communities may stimulate the demand for meat and other wildlife products and/or attract in-migration, which will lead to increasing pressure on biodiversity (Robinson 1993; Gartlan 1998; Brandon, Redford, and Sanderson 1998; Oates 1999).

In order to offer opportunities and influence changes in resource-use patterns, knowledge of local socioeconomic, biological, and cultural factors is critical. For the Portfolio project in Gamba (Gabon), for instance, considerable time and effort were spent in setting up baselines with information on the state of the environment, local use of natural resources, community requirements, and ways to reconcile these with biodiversity conservation (Blaney and Thibault 2000). However, identifying needs and developing the participation of local communities is a long process. The delay between the first consultations with communities and the actual implementation of livelihood activities resulted in a loss of confidence and involvement by communities, as they perceived little or no development benefits and their expectations of project benefits were not fulfilled (Marc Thibault, personal communication 1998).

In contrast, the effectiveness of the Honduras livelihoods program stems from the project identifying and implementing income-generating activities that are compatible with the cultural, environmental, and economic realities of the local indigenous communities. Small-scale land-use alternatives have included an ecotourism program and agroforestry activities that include training, technical assistance, forest certification, and marketing of cacao products through the local producers organization. The ecotourism initiative at Las Marías has seen visitor numbers rise from 155 in 1996 to 535 in 2001. Consequently, ecotourism income to the community was estimated to be U.S.$96,000 in 2000 (Erik Hansen, personal communication 2001). This new source of income has served to reduce hunting activities and limit the spread of agricultural clearing around Las Marías within the Rio Platano Biosphere Reserve. It has also helped the community to adapt and develop new organizational structures. These positive results, however, are further complicated by increasing population movements and land tenure problems, important threats to biodiversity that cannot be addressed via community development activities alone (Herlihy 1997). Law enforcement remains a key aspect of the management of the reserve.

Assumption 2: Communities are homogenous and static.

Conservation organizations are slowly beginning to realize how complex conservation can be at the local level. Local communities consist of different groups, defined by age, gender, ethnicity, class, and religion, and contain a range of internal imbalances of interest and power (Doornbos, Saith, and White 2000). Outsiders have rarely recognized or taken into account these differences, basing their policies and views on the assumption that local "communities" are a homogeneous group, easily defined and recognizable, and that social cohesion allows the community to become allied as a whole (Flintan 1999). This is particularly the case for conservation NGOs where a vision exists of a situation where "outside the community conflicts prevail; within, harmony reigns" (Agrawal and Gibson 1999:631). The assumption is that those who come forward and participate are a fair and true representation of the community as a whole, when in fact communities contain groups that harbor different aspirations for leadership, wealth, and degrees of resource use. For example, women still bear proportionately greater degrees of poverty than men, especially in experiencing excessive workloads and a lack of decision-making authority (Shaffer 1998; Tunio 2000).

Another important issue for biodiversity conservation relates to movements of large numbers of people and their impact on natural resources. Local population growth directly affects the use of resources and their degradation, as encroachment on protected areas and illegal off-take of resources increase, particularly in areas where land tenure systems are unclear and law enforcement is weak (Wood, Stedman-Edwards, and Mang 2000).

In Ethiopia, the Portfolio has assessed the linkages between gender and integrated conservation and development. The work has focused on gender roles and differences in mobility, social organization, current livelihood practices, and perceptions of the protected area and "conservation." The program has carried out semistructured interviews with village/town inhabitants and/or key informants; conducted a survey of women traders in the marketplace; and employed Rapid Rural Appraisal techniques such as mobility/resource mapping and transect walks (Fiona Flintan, personal communication 2000). It was found that women have fewer environmental entitlements than men and less access to resources and decision-making processes. Women and girls are marginalized groups in the society. Large gender inequities exist in schooling, health care, and with institutional support. Women are mainly responsible for the household and men for agriculture. However, women often work on the land as well, most notably in collecting firewood. This has significant local impacts on the resources of Bale Mountains National Park.

Despite the fact that under Philippine law the indigenous people of Sibuyan Island qualify for special status and are entitled to resource use and property titles, their rights over large parts of the island are contested by other inhabitants (see Tongson and Dino this volume). In support of indigenous peoples, the Portfolio project in collaboration with Anthropology Watch, the Legal Assistance Center for Indigenous Filipinos (PANLIPI), and the Philippine Association for Intercultural Development (PAFID) have assisted the Sibuyan Mangyan Tagabukid (the indigenous people of the island) in their application for a Certificate of Ancestral Domain Title and the formulation of their Ancestral Domain Sustainable Development and Protection Plan. At present, the project is focusing its efforts on empowering this group and building their capacity in order for them to enforce their newly obtained rights over the land and natural resources.

Assumption 3: Main threats to biodiversity arise from communities adjacent to protected areas.

Until the 1980s, the identification of threats to biodiversity focused mainly on direct and proximate causes, which often included habitat alteration and loss, overharvesting of species and disease introduction, and pollution and climate change. However, the failure of local initiatives in conserving biodiversity in a meaningful and long-term way has led to looking at other factors that contribute to the loss of biodiversity. Root causes analysis consists of an assessment of underlying biological, social, political, economic, or cultural factors that drive resource degradation and/or biodiversity loss (Wood, Stedman-Edwards, and Mang 2000). In order to analyze and address the underlying causes of such loss, the social, economic, political, and cultural factors at various levels that influence resource-use decisions at the local level need to be identified. The following are seen as driving forces to environmental degradation and biodiversity loss: demographic change, poverty and inequality, markets and politics, and macroeconomic policies and structures (McNeely 1998; Wood, Stedman-Edwards, and Mang 2000).

The Pakistan project, Conservation of Mangrove Forest at Coastal Areas of Sindh and Balochistan, was developed to address three issues: community use of mangroves, public awareness of the importance of mangroves and wetlands, and the restoration of degraded mangroves. To do this, a wetland center was built, communities were organized to sustainably manage the natural resources, and mangroves were replanted in areas where they had been degraded or lost. An analysis of the root causes of biodiversity loss in Pakistan's mangrove ecosystems, undertaken in 1998, revealed that the greatest impacts were from activities "upstream" from the project site (Wood, Stedman-Edwards, and Mang 2000, ch. 12). Pollution, the diminishing supply of freshwater due to agriculture, and overexploitation of fish stocks were identified as the constraining factors. As a result, the project has begun to reorient itself to specifically address the upstream factors (McShane 1999b). It is now working to change agricultural policies that contribute to reduced water flows and is involved with cotton growers in the development of more environmentally friendly production of cotton, thereby reducing water extraction from rivers critical to maintenance of mangrove ecosystems. The project has also engaged with the Pakistan navy to monitor illegal fishing and enforce catch limits, and with the Karachi Port Authority to reduce pollution.

CAPACITY DEVELOPMENT

The combination of capacity shortfalls and lack of skills is perhaps the greatest constraint to the sustainability of ICDPs. Table 4.2 summarizes

Table 4.2 *Capacity development assumptions and constraints*

Assumptions	*Constraints*
1. Skills are already in place.	1.1 Lack of skilled labor delays and constrains the implementation of activities. 1.2 Local capacity is underestimated.
2. By strengthening the capacity of government organizations to implement community-based approaches to natural resources management, benefits will accrue to communities.	2.1 Government bureaucracies are resistant to the devolution of natural resource management responsibilities to local communities. 2.2 Mistrust of government institutions by communities due to past and present law enforcement roles.
3. Local communities are best positioned to manage natural resources.	3.1 Capacity development is too focused on technical activities. 3.2 Local organizational capacity requirements are given low priority.

the capacity development assumptions made across the DGIS-WWF Tropical Forest Portfolio projects, along with the most common resulting constraints.

Assumption 1: Skills are already in place.

In many countries, government priorities for development in other sectors has resulted in understaffed and underskilled conservation departments. For greater and more sustainable impact, conservation efforts have to better institutionalize capacity by providing government agencies, NGOs, and community-based organizations with skills to make conservation-based decisions (Brandon and Wells 1992; IIED 1994; Barrett et al. 2001). The lack of skilled personnel was identified as a significant constraint on all Portfolio projects.

The experiences of Pakistan, the Philippines, and Ecuador demonstrate that prolonged collaboration with national institutions (governmental and academic) helps support a broader institutional vision for biodiversity conservation, ensuring better continuity and commitment toward the achievement of long-term objectives. In Minkébé (Gabon), the project aims to strengthen capacity by involving students from the National School of Waters and Forests. Once graduated, most of these students end up working for the Ministry of Waters and Forests and can put to use their experience gained from working on the project (Pauwel de Wachter, personal communication 2001).

Despite the current decentralization process taking place in Ethiopia, which aims to devolve natural resources management responsibilities to regional government authorities, the federal Ethiopian Wildlife Conservation Organisation (EWCO) continues to play an important national policy-making role. The Portfolio project is working to strengthen both EWCO's and the regional authorities' institutional capacities. The project has established a small capacity-building unit within EWCO. The purpose of this unit is to provide training and other support to build EWCO's capacity for protected area and wildlife management and, perhaps more importantly, to facilitate collaboration between EWCO and the Oromya (regional) natural resource management agency. By facilitating collaboration between these institutions, the project is providing a model for decentralized regional protected area management in Ethiopia.

Capacity development investments tend to be carried out on an ad hoc basis and are concentrated at the level of implementing partners and government organizations, while the importance of developing the skills of local groups is often undervalued. In Gamba (Gabon), the outcomes of training sessions on ecological surveys and participatory rural appraisal (PRA) involving government agents, environmental NGOs, and community partners showed that, respectively, 76.2 percent and 60 percent of the local community members, as opposed to 7.7 percent and 0 percent of government employees and 66.7 percent and 0 percent of the NGOs' staff, were still carrying out ecological surveys and PRA two years after the training (Thibault and Blaney 2001). This is not to say that government organizations and implementing partners should not see their skills and capacities strengthened. Rather, capacity is required at multiple levels, and may be more effective where least expected.

Assumption 2: By strengthening the capacity of government organizations to implement community-based approaches to natural resources management, benefits will accrue to communities.

This assumption is based on the view that governments, primarily through their national resource management agencies, are supportive of community-based approaches to natural resource management. Unfortunately, the reality is often different. Central government agencies have often shown themselves unwilling to cede control of resources, financial as well as physical, to local authorities (cf. Gamaledinn 1987; Murphree 1994). Likewise, communities remain distrustful of government authorities that have traditionally suppressed resource use, at times arbitrarily, under the guise of law enforcement (McShane 1989; Adams and McShane 1996). The Portfolio found that these issues often get in the way of developing capacity that promotes the integration of conservation and development.

AFE-COHDEFOR, the government agency charged with administering all state forest lands and protected areas in Honduras, is the Portfolio's government counterpart agency and represents a considerable challenge to the implementation of project activities. Despite the project's objective to strengthen the technical and administrative capacity of AFE-COHDEFOR, the institution has been unable to overcome bottlenecks, mostly of an administrative nature (e.g., poor internal communications, delays in procurement and hiring, no decentralization of decision making). To get around some of these constraints, the Portfolio is working with a number of communities around the Rio Platano Biosphere Reserve to transfer forest management responsibility from AFE-COHDEFOR to the community. The negotiation process to effect such a transfer of responsibility has been long and complicated by the resistance of the government agency to such a change. Many of the conditions AFE-COHDEFOR required to be in place for the transfer were of the type they themselves could not meet.

Collaboration between government agents and local communities is sometimes very difficult to achieve, in part because of the history of law enforcement. In places like Ecuador and the Philippines, where there is a history of conflict between the local communities and government agents, and where the latter are still seen as policemen, this represents an important obstacle in winning people's trust (see Tongson and Dino this volume). In Ecuador the communities' resentment of the national parks authority, INEFAN, has a lot to do with the establishment of Sangay National Park and the expropriation of their land without consultation or compensation (Rivas and Ruiz 2001). Being deprived of land that is legally theirs (obtained during the land reform in 1970) has not gone down well with the communities, especially when their property rights were again ignored during the construction of the Guamote-Macas road that bisects the park (Jorge Rivas, personal communications 1999). Now the Portfolio is attempting to bring together INEFAN and these same communities to negotiate a co-management agreement.

Assumption 3: Local communities are best positioned to manage natural resources.

Capacity building is about putting in place the human, financial, and technical resources necessary to make institutions and individuals self-sufficient and sustainable (cf. McShane 1999a). In theory, training is geared to foster knowledge and skills, and it aims to increase the self-sustaining capacity of community institutions so they can make decisions and solve their own problems. However, the capacity of communities to manage their natural resources cannot be assumed, though this is difficult when community-based approaches have become so popular that they are little questioned (Barrett et al. 2001). Although

Portfolio project training and capacity-building activities involve community members, they appear to be, first, geared toward implementing agencies and, second, of a too technical nature (e.g., GPS reading and mapping, data collection for ecological monitoring). Many of these activities do not address local needs and aspirations.

In Ecuador, Honduras, Pakistan, and the Philippines, capacity development efforts have focused on strengthening the organizational capacities of local communities. In these countries there are good examples of organizational training that involve issues such as conflict management, advocacy, negotiation, and management. However, this type of capacity development requires considerable time and understanding of community dynamics, and is not always supported by partner governments. Currently, there are few compelling examples where the empowerment of partner groups through the development of planning and management skills has been applied by communities to improve their decision making on resource use, or enabled them to more effectively negotiate with and lobby government authorities (cf. Bebbington 1996). The approach in the Philippines to create a local organization to take over responsibilities in guiding resource management on Sibuyan Island has been delayed because of organizational problems, even following skills training (Edgardo Tongson, personal communication 2002).

INSTITUTIONAL SUPPORT

Institutional support from government agencies, local communities, and implementing organizations and donors was key to a number of assumptions made by Portfolio field projects. Table 4.3 summarizes these assumptions along with the common constraints concerning institutional support. Such assumptions mainly concern the preconditions necessary for a productive working environment and include technical, financial, and logistical support, active partner participation, and long-term commitment to project objectives.

Governments and Their Agencies

Assumption 1: The government will support community-based approaches to natural resource management.

Ensuring that communities have direct responsibility for natural resource management practices grew out of the recognition that centralized forms of control over resources have failed to halt resource degradation and that local control may be more effective where there is greater vested interest (cf. Brandon and Wells 1992; Murphree 1993, 1994). Moreover, the financial constraints facing many governments in

Table 4.3 *Institutional support assumptions and constraints*

Actors	*Assumptions*	*Constraints*
Governments and their agencies	1. Government will support community-based approaches to natural resource management.	1.1 Governments have other priorities and are not willing to turn responsibility for natural resource management over to communities.
	2. Government will provide support in terms of human, technical, financial, and logistical resources.	2.1 Government staff allocated to a project are often insufficient in number and lack the required skills. 2.2 Commitments are not always respected, especially financial ones.
Local communities	3. Local communities will support protected areas if they are involved in management process.	3.1 Local expectations are not always met. 3.2 Community trust takes time to build. 3.3 Community participation is impeded by slow process of implementation.
Implementing organizations and donor agencies	4. Implementing organizations and donor agencies will provide the necessary funds.	4.1 Projects are too dependent on donor funding, and exit strategies are rarely planned ahead of time. 4.2 Funding limited to a period of 3 to 5 years with little room for impact.
	5. Implementing organizations have all the skills necessary to integrate conservation and development.	5.1 Implementation is focused on task-in-time results rather than adaptive processes and learning.

developing countries and the low political priority accorded to conservation in the face of other priorities, such as health, education, and food security, undermine the ability of governments to allocate adequate resources for biodiversity conservation (Barrett et al. 2001). The lack of commitment to conservation objectives is often demonstrated by delays in providing resources, in passing laws and approving policy, in releasing funds, in allocating sufficient and appropriate personnel, and in respecting the legal status of protected areas. Often governments allow resource extraction activities to take place in areas already under protected area status.

In Gabon, the government provided special authority to the Shell Oil Company in the Gamba region to explore for and extract petroleum resources in violation of national legislation. The same allowances have been made by the government in other protected areas—for example, allowing forest exploitation (cf. White 1992). In such cases, the agencies responsible for the environment rarely possess the leverage to influence the government when up against the Finance and Mining Ministries that generate revenue for the country.

Minkébé (Gabon) is an interesting example that emphasizes the importance of working at the local and regional levels to build institutional support to achieve positive results. The Minkébé experience shows how discussion with stakeholders and participatory establishment of protected area boundaries can offset many constraints. Prior to the delineation process of the protected area, the project established a dialogue with the provincial authorities, the local communities, and their traditional leaders. The project specifically targeted these stakeholders to mitigate possible conflicts (Pauwel de Wachter, personal communication 2000). These efforts not only resulted in 6,000 square kilometers of protected area being gazetted but also contributed to mobilizing local, regional, and national support for the conservation of Minkébé.

Assumption 2: The government will provide support in terms of human, technical, financial, and logistical resources.

Many governments are poorly organized to undertake conservation and development activities and rarely possess the required resources to implement projects despite their best intentions (cf. Barbier, Burgess, and Folke 1994; Wells 1998). Likewise, governments make commitments to provide support to ensure that outside funding is forthcoming. The promises made, but not delivered on, by government partners have served to constrain in some way all seven projects.

While many of the countries studied here suffer from real limits to their ability to provide human, technical, financial, and logistical resources, Gabon provides a somewhat different example. A recent

Human Development Report issued by the UN Development Program is telling (UNDP 1999). The gap between Gabon's per capita gross domestic product (U.S.$4,230 in 1997) and investment in human development (as measured by income, life expectancy, adult literacy, and educational enrollment) is substantially greater than in the 173 other countries surveyed. This indicates that, despite significant income generated by natural resource exploitation, little is being invested in Gabon's own development (McShane and McShane-Caluzi 2003). This raises different questions about the country's commitment to conservation.

Local Communities

Assumption 3: Local communities will support protected areas if they are involved in the management process.

The devolution of decision-making power to local communities, it is believed, will not only motivate them to take action, but also alleviate the burden of management responsibilities on the state (Berkes 1989; IIED 1994; Little 1994; Barrett et al. 2001). However, concurrent with decision-making power is the transfer of ownership and access rights to local people. In reality, the decentralization of power needed to achieve this is moving very slowly, both on the part of governments and implementing agencies.

The lack of secure tenure is an issue that affects community support for project objectives. In the Rio Platano Biosphere Reserve, Honduras, for example, the government itself is largely to blame for the colonization problem (cf. Herlihy 1997). At the end of the last century it proposed removing some 50,000 landless farmers from other parts of Honduras to the western bank of the Río Sico. Officially, this area is just outside the boundary of the Rio Platano Biosphere Reserve, but it lies very close to its core zone. Following widespread protests, the plan was canceled, but it still had its effect. Thousands of poor Hondurans, whose expectations had been raised, migrated to the zone, and many of them crossed the river into the protected area. Given the weakness of AFE-COHDEFOR, the government agency charged with administering all state forestlands and protected areas in Honduras, there is now little control of resource use in this part of the reserve. Lack of land tenure coupled with weak law enforcement has left communities with little choice but to get what they can from the natural resource base before it is all gone.

Five years after its establishment, Mount Guiting-Guiting Natural Park in the Philippines is faced with challenges as well as opportunities for collaboration between the park and the indigenous community. A co-management framework looks to be a workable strategy for the overlap areas of the protected area and the indigenous people's land as evidenced by the common forest management objectives of the park and the indigenous people (see Tongson and Dino this volume). Co-management agreements, according to the International Union for Conservation of Nature and Natural Resources (IUCN) World Commission on Protected Areas, should establish common objectives and commitments to the conservation of protected areas; define responsibilities for conservation and sustainable use of biodiversity and natural resources contained in them; and be the basis for management objectives, standards, regulations, etc. Although involvement of indigenous communities, in theory, is a more cost-effective biodiversity conservation tool, the institutional changes, methodology development, capacity building, and consultations necessary to introduce co-management require considerable effort (cf. Barrett et al. 2001). Currently, there is limited enforcement by the government in the upland areas. It suffers from a shortage of staff to patrol the hinterlands primarily because of a lack of resources. Devolving the daily surveillance and patrol functions to indigenous tribal councils would save the park enforcement costs. While de facto enforcement by the Sibuyan Mangyan Tagabukid is ongoing, official recognition of their role through a co-management agreement could strengthen the basis of their enforcement actions. However, there remains a need to elaborate the products that may be extracted, harvestable levels, and monitoring systems. The project has been working with these people to strengthen their resource management capability and local institutional structures.

Implementing Organizations and Donor Agencies

Assumption 4: Implementing organizations and donor agencies will provide the necessary funds.

Many donors have become increasingly interested in funding ICDPs as part of their expanding environmental mandates and growing interest in links between conservation and development (Wells and Brandon 1992; Wells 1995; McShane and Wells this volume). However, they adopt time schedules that seem unrealistically short and that predict achieving financial self-sustainability within a few years. Given the ambitious goals aimed at, it is important that support is guaranteed over a sufficiently long period and that projects operate using flexible and experimental processes.

All of the Portfolio projects rely on donor funding as their primary source of support. None of the projects will be self-sustaining following the end of the current funding cycle. Thus all the field projects have expressed a need for new sources of funding to continue the work they started under the Portfolio. Nonetheless, despite the fact that exit strategies have not been developed, the issue of financial sustainability has often been raised, and several projects are exploring and developing alternative funding sources—trust funds and debt-swaps in Gabon and Ethiopia (Barry Spergel, personal communication 2002), a livelihood fund in the Philippines, and a biosphere fund in Honduras. Given the need for significant capital start-up costs, these efforts remain limited in scope.

Assumption 5: Implementing organizations have all the skills necessary to integrate conservation and development.

The realities of implementing ICDPs are not always straightforward (cf. Wells et al. 1999; McShane 1999a). Conservation and development are based on a set of disciplines—ecology, sociology, and economics—that are exceedingly complex and poorly understood. Therefore, implementers characteristically operate in situations where the outcome of their actions is uncertain. One cannot be certain that the results will be as expected, and one can be confident that there will be some unexpected side effects. As a result, the process of implementing ICDPs must be consciously structured to allow for these uncertainties, as well as for changes in value systems, policies, and technical capabilities. To address this, the Portfolio has attempted to organize itself as a self-testing and self-evaluating system operating by negative feedback in relation to clearly defined project objectives. Such a system of adaptive management is the continuous integration of design, management, and monitoring to systematically test assumptions in order to adapt and learn (Bell and McShane-Caluzi 1984; Salafsky and Margoluis this volume). It is idealistic in that it requires rethinking how projects go about the business of implementation, but it is realistic in that it assumes the need to correct mistakes, modify judgements, and learn directly from doing. This process has resulted in two projects, those in the Philippines and Pakistan, reviewing and revising their project goals.

Policy Factors

In most cases the Portfolio field projects assumed that either the necessary legal frameworks were in place to meet the stated goals and objectives, or some other group was advocating these issues. Table 4.4 summarizes these assumptions along with the common constraints concerning policy factors.

Governments and Their Agencies

Assumption 1: The government will pass laws and regulations to facilitate devolution of management responsibilities to decentralized levels and ensure more sustainable utilization of natural resources.

Table 4.4 *Policy factor assumptions and constraints*

Actors	*Assumptions*	*Constraints*
Governments and their agencies	1. The government will pass laws and regulations to facilitate devolution of management responsibilities to decentralized levels and ensure more sustainable utilization of natural resources.	1.1 Weak systems of advocacy. 1.2 Poor identification of supportive policies and regulations required to meet biodiversity conservation goals.
Local communities	2. Local communities are actively involved in the governance process.	2.1 Communities do not have the organizational capacity to participate in natural resource advocacy.

The role of governments and their agencies in conservation and development should be enabling and supportive. In practice this is not always the case. Governments have tended to assert power and resist the devolution of authority even when they lack the resources to fulfill their responsibilities (Feldmann 1994). Despite the fact that Portfolio projects recognized supportive policies as crucial to the long-term success of their objectives, policy advocacy was a weak point for all of the projects. It was often assumed that governments would support projects by ensuring that laws and regulations would be enacted and enforced. However, governments demonstrated little commitment toward undertaking such actions.

In the case of Gabon, the impact of commercial logging is relatively limited at present, but pressure on the forest will greatly increase as the government implements its policy of tripling exports and expanding the range of timber species currently harvested (McShane and McShane-Caluzi 2003). Hunting and the bushmeat trade are known to accompany logging (Wilkie, Sidle, and Boundzanga 1992; Lahm 2001). So, in the Minkébé region, the Portfolio, and government agencies responsible for monitoring logging, entered into negotiations with a Malaysian logging company and developed a protocol document making the logging company responsible for policing its concession against bushmeat hunters. These enforcement activities are further supported by the government agency responsible for forest management. The result is that hunting within the concession has been almost completely stopped (Pauwel de Wachter, personal communication 2001). Protocols are now being negotiated between the government, the project, and three other logging companies around the reserve. The expansion of this approach to other regions of Gabon, or elsewhere in central Africa, demonstrates how policy experience at a particular site can contribute to broader policies at the national, and perhaps international, levels.

In Pakistan, cotton irrigation and the consequent reduction of water flow in rivers critical to mangroves have led to policy dialogue between the WWF, international policy institutions, and the relevant national authorities. The aim of this dialogue is to find ways to reduce the adverse impacts of external factors such as agrochemicals, water supply, and land-use and trade investment policies and to initiate the adoption of actions that actively promote more sustainable agricultural practices (Ali Hasnain, personal communication 2000).

There is enough evidence from the Portfolio to emphasize the need for ICDPs to address the policy environment in which they operate and to link field activities with policy development. Methodologies to better understand and identify linkages between field activities and policy have been tested in Pakistan (McShane 1999b; Wood, Stedman, and Mang 2000, ch. 12) and Ecuador (Rivas and Ruiz 2001). These analyses have provided each project with relevant information highlighting the critical factors affecting progress as well as enabling projects to better target their activities and stakeholders.

Local Communities

Assumption 2: Local communities are actively involved in the governance process.

To actively advocate for representative government and supportive policies, communities need organizational structures and the legitimacy to defend their rights and responsibilities (cf. Murphree 1993; Bebbington 1996; Barrett et al. 2001). ICDPs need to examine issues such as decision making, representation, information flows, and the role and importance of traditional authorities. The DGIS-WWF Tropical Forest Portfolio is addressing these issues in several countries (e.g., the Philippines, where the project is supporting the rights of indigenous peoples to control and manage their resources, and in Ecuador and Honduras, where the projects have elaborated co-management approaches to expand local responsibility for natural resources management in cooperation with national management agencies).

In Honduras, the Portfolio is working in partnership with MOPAWI, which has significant outreach, credibility, and experience in the Mosquitia, to address the institutional and structural issues that impinge on the long-term effectiveness of the Rio Platano Biosphere Reserve.

MOPAWI (from the Miskito Indian words "Mosquitia Pawisa") was founded in 1985 as a private, nonprofit organization dedicated to the sustainable development of the Honduran Mosquitia and its indigenous inhabitants. MOPAWI's main objective is to improve the quality of life of the local populations in the Mosquitia through income-generating community development activities that are compatible with the local environmental and cultural realities. Specifically targeted in this respect is greater coordination and leadership among local and national NGOs and community organizations and strengthened links to government and other decision makers.

Conclusions

The experience of the DGIS-WWF Tropical Forest Portfolio shows that many of the assumptions ICDPs make about conservation and development revolve around social, organizational, and development skills, policies, governance, and advocacy, and in understanding competing economic factors. These are issues not usually found at the forefront of conservation organization expertise, but are crucial to functioning in today's world.

Linkages between national and international policies and local actions must be recognized early in ICDP development. Supportive laws, policies, and regulations must be in place if ICDPs are to be successful and sustainable. Projects cannot simply address field-based actions; they must also address policy issues in partnership with other efforts or through actions of their own. They must take a vertically integrated view toward implementation, meaning that policy advocacy and change is as integral to ICDP success as site-based action.

Trade-offs between conservation and development must be acknowledged. "Win-win" scenarios rarely, if ever, take place, and conflict is often the norm. Moreover, by pursuing development objectives, ICDPs may inevitably result in some losses to biodiversity. The challenge for ICDPs is to find ways to balance the trade-offs between improving livelihoods and conserving biodiversity. The main issues are how to negotiate these trade-offs, what level of biodiversity loss is acceptable, and who takes part in these decisions (see Brown this volume). To help address conflicts related to natural resource management, ICDPs must (1) build on shared interests and points of agreement; (2) foster a sense of ownership in the solution process of implementation; (3) utilize processes that resemble those already in existence to manage resource use conflicts; and (4) emphasize building capacity among affected interest groups at all levels (government, private industry, communities,

NGOs, etc.) so they become more effective facilitators, communicators, planners, and mediators of natural resource conflict.

ICDPs must learn from mistakes and adapt to change. The approach has been caught in a web of having to demonstrate its successes when, in fact, most learning comes from what goes wrong rather than right—and what may be right today can quickly go wrong tomorrow. Institutions require the skills, understanding, and flexibility to cope with the inevitable changes that will occur in the condition of natural resources and their users. A "safe-fail" approach to conservation and development implementation must be promoted by donors and practitioners (Redford and Taber 2000; Salafsky and Margoluis this volume).

ICDPs must find ways to complement their skills. Partnership with other disciplines is critical. This applies sectorally and institutionally. Sectorally, ICDPs have shown themselves effective at recognizing the need to take social development and economic well-being into account in planning and implementing biodiversity conservation programs. They have also shown themselves ineffective in applying socioeconomic principles, often to the detriment of what they are trying to accomplish. Institutionally, strong local systems of social cohesion and control support community-based approaches, while a competent bureaucracy is needed for effective government-run systems (Barrett et al. 2001). Support for both is required for ICDPs to be effective; rarely are both in place.

ICDP approaches will have to address issues simultaneously at different scales: for example, field programs demonstrating what works and what does not, policy initiatives influencing and changing factors across broader constituencies, and campaigns advocating action to achieve change. All of these actions may not be appropriate in a single ICDP, but may be necessary to achieve desired goals and objectives. Understanding the scale (ecoregion, landscape, country, project site), and the relationships between scales, is critical to effective implementation.

For example, in 1997 the government of Gabon set aside 6,000 square kilometers of the Minkébé region of northeast Gabon as a forest reserve, prohibiting logging. However, to better conserve the ecological processes throughout this relatively intact ecosystem, the protected area (or site) has to be seen in a larger context. The following are examples of this approach:

- An ecoregion vision has been developed, connecting landscapes in three countries (Gabon, Cameroon, and Congo) to a common future.
- At the landscape level, the WWF and its partners (the European Union (EU), World Bank, and GEF), in collaboration with the

various government agencies, are currently creating a complex of protected areas and management zones in the Gabon-Congo-Cameroon border region, where Minkébé is one of the cornerstones. In this way, large intact ecosystem-scale processes can be conserved in a transborder region with high defensibility.

- At the country level, knowing that key policy decisions are made in the national and provincial capitals, partners are actively engaged in these strategic places.
- At the site level, an effective reconnaissance service has been created for the Minkébé Reserve. Its actions are critical to the enforcement of national conservation laws. Teams are led by highly qualified people and are able to knowledgeably discuss issues with resource users and authorities, as well as collect biological and socioeconomic data. Resource-use plans are developed and negotiated with users. For example, protocols are being developed with logging companies to control hunting and ensure that the protected core area will have a high defensibility and can conserve its biodiversity in the long term.

Does changing the scale of different interventions change what we have been calling an ICDP? Perhaps. Many ICDP efforts have been constrained by their inability to take these different scales of intervention into account in both design and implementation. The challenge for practitioners is not to decide at which scale to operate, but rather which combination of actions at different scales is required (see Maginnis, Jackson, and Dudley this volume; Robinson and Redford this volume).

Many of the assumptions identified in this study are about what the field projects expected from others. In some cases these expectations were not realized, and in others they were unattainable from the beginning without significant social and political change. Are addressing these issues important for conservation? Yes, for without strong social and political foundations, biodiversity conservation is unlikely to be successful in the long term. Conservation organizations and their partners must rise to the challenge of conserving biodiversity in this complicated world. Conservation cannot ignore the needs of human beings, but development that runs roughshod over the environment is doomed.

References

Adams, J. S. and T. O. McShane. 1996. *The Myth of Wild Africa: Conservation Without Illusion.* Berkeley and Los Angeles: University of California Press.

Agrawal, B. and C. Gibson. 1999. Enchantment and disenchantment: The role of community in natural resource conservation. *World Development* 27:629–649.

Barbier, E. B., J. C. Burgess, and C. Folke. 1994. *Paradise Lost? The Ecological Economics of Biodiversity.* London: Earthscan.

Barrett, C. S. and P. Arcese. 1995. Are integrated conservation-development projects (ICDPs) sustainable? On the conservation of large mammals in sub-Saharan Africa. *World Development* 23:1073–1084.

Barrett, C. S., K. Brandon, C. Gibson, and H. Gjertsen. 2001. Conserving tropical biodiversity amid weak institutions. *Bioscience* 51:497–502.

Bebbington, A. 1996. Movements, modernizations, and markets: Indigenous organisations and agrarian strategies in Ecuador. In R. Peet and M. Watts, eds., *Liberation Ecologies: Environment, Development, Social Movements,* 86–109. London: Routledge.

Bell, R. H. V. and E. McShane-Caluzi, eds. 1984. *Conservation and Wildlife Management in Africa.* Washington, D.C.: U.S. Peace Corps.

Berkes, F., ed. 1989. *Common Property Resources.* London: Belhaven Press.

Blaney, S. and M. Thibault. 2000. *Une méthode pour l'étude et le suivi des caractéristiques socio-économiques des communautés rurales.* Libreville, Gabon: DGIS-WWF Tropical Forest Portfolio.

Brandon, K., K. Redford, and S. Sanderson, eds. 1998. *Parks in Peril: People, Politics, and Protected Areas.* Washington, D.C.: Island Press.

Brandon, K. and M. Wells. 1992. Planning for people and parks: Design dilemmas. *World Development* 20:557–570.

Clay, J. W. 1996. *Generating Income and Conserving Resources: Twenty Lessons from the Field.* Washington, D.C.: World Wildlife Fund.

Clay, J. W. 1997. Brazil nuts. In C. H. Freese, ed., *Harvesting Wild Species: Implications for Biodiversity Conservation,* 246–282. Baltimore: Johns Hopkins University Press.

Colchester, M. 1994. Sustaining the forests: The community-based approach in South and South-East Asia. *Development and Change* 25:69–100.

Doornbos, M., A. Saith, and B. White. 2000. Forest lives and struggles: An introduction. *Development and Change* 31:1–10.

Feldmann, F. 1994. Community environmental action: The national policy context. In D. Western, M. R. Wright, and S. C. Strum, eds., *Natural Connections: Perspectives in Community-Based Conservation,* 393–402. Washington, D.C.: Island Press.

Flintan, F. 1999. Unheard voices in a white man's wilderness? Women in conservation and development. Master's thesis, School of Oriental and African Studies, University of London.

Gamaledinn, M. 1987. State policy and famine in the Awash Valley of Ethiopia: The lessons for conservation. In D. Anderson and R. Grove, eds., *Conservation in Africa: People, Policies, and Practice,* 327–344. Cambridge: Cambridge University Press.

Gartlan, S. 1998. Falling between two stools: The false promise of sustainable development. In Nils Christoffersen et al., eds., *Communities and Sustainable*

Use: Pan-African Perspectives, 72–78. Harare, Zimbabwe: IUCN–The World Conservation Union.

Gibson, C. and S. A. Marks. 1995. Transforming rural hunters into conservationists: An assessment of community-based wildlife management programmes in Africa. *World Development* 23:941–957.

Herlihy, P. 1997. Indigenous peoples and biosphere reserve conservation in the Mosquitia rain forest corridor, Honduras. In S. Steven, ed., *Conservation Through Cultural Survival: Indigenous Peoples and Protected Areas,* 99–130. Washington, D.C.: Island Press.

Hillegars, P. and M. K. Conradi. 2000. *DGIS-WWF Tropical Forest Portfolio 1996–2001: Midterm Evaluation—Interregional Component.* Wageningen, The Netherlands: Alterra Green World Research and Manidis Roberts.

Hughes, R. and F. Flintan. 2001. *Integrated Conservation and Development Experience: A Review and Bibliography of the ICDP Literature.* London: International Institute for Environment and Development (IIED).

IIED (International Institute for Environment and Development). 1994. *Whose Eden? An Overview of Community Approaches to Wildlife Management.* London: International Institute for Environment and Development.

Kremen, C., A. M. Merenlender, and D. D. Murphy. 1994. Ecological monitoring: A vital need for integrated conservation and development programs in the tropics. *Conservation Biology* 8:388–397.

Lahm, S. A. 2001. Hunting and wildlife in northeastern Gabon. In W. Weber, L. J. T. White, A. Vedder, and L. Naughton-Trevers, eds., *African Rainforest Ecology and Conservation,* 344–354. New Haven, Conn.: Yale University Press.

Larsen, P. S., M. Freudenberger, and B. Wyckoff-Baird. 1998. *WWF Integrated Conservation and Development Projects: Ten Lessons from the Field 1985–1996.* Washington, D.C.: World Wildlife Fund.

Little, P. D. 1994. The link between local participation and improved conservation: A review of issues and experiences. In D. Western, M. R. Wright, and S. C. Strum, eds., *Natural Connections: Perspectives in Community-Based Conservation,* 347–372. Washington, D.C.: Island Press.

Margoluis, R. and N. Salafsky. 1998. *Measures of Success: Designing, Managing, and Monitoring Conservation and Development Projects.* Washington, D.C.: Island Press.

McNeely, J. A. 1998. *Economics and Biological Diversity: Developing and Using Economic Incentives to Conserve Biological Resources.* Gland, Switzerland: IUCN–The World Conservation Union.

McShane, T. O. 1989. Wildlands and human needs: Resource use in an African protected area. *Landscape and Urban Planning* 19:145–158.

McShane, T. O. 1997. *Adaptive Management, Monitoring, Evaluation, and Communications: Workshop Report. Cap Estarias, Gabon, 7–11 October 1997.* Gland, Switzerland: DGIS-WWF Tropical Forest Portfolio.

McShane, T. O. 1999a. Voyages of discovery: Four lessons from the DGIS-WWF Tropical Forest Portfolio. *Arborvitae* suppl.: 1–6.

McShane, T. O. 1999b. *Linking Field and Policy in Pakistan's Mangroves: Workshop Report. Karachi, Pakistan, 10–12 November 1999.* Gland, Switzerland: DGIS-WWF Tropical Forest Portfolio.

McShane, T. O. and E. McShane-Caluzi. 2003. Resource use in Gabon: Sustainability or biotic impoverishment? In M. C. Reed and J. F. Barnes, eds., *Culture, Ecology, and Politics in Gabon's Rainforest*, 7–36. Lewiston, N.Y.: Edwin Mellen Press.

Mittelman, A. 2000. Conservation and development linkages: Lessons learned from 15 years on ICDP experience in Thailand. Paper presented at an International Seminar on Integrated Conservation and Development: Contradiction of Terms? 4–5 May, at Copenhagen. CARE-Denmark, DANCED, DANIDA.

Murphree, M. W. 1993. *Communities as Resource Management Institutions.* Gatekeeper Series no. 36. London: International Institute for Environment and Development (IIED).

Murphree, M. W. 1994. The role of institutions in community-based conservation. In D. Western, M. R. Wright, and S. C. Strum, eds., *Natural Connections: Perspectives in Community-Based Conservation*, 403–427. Washington, D.C.: Island Press.

Newby, S. A. 1996. *The Strengths and Weaknesses of Integrated Conservation and Development Projects: Lessons Learnt from Cameroon.* Master's thesis, University of East Anglia.

Oates, J. F. 1999. *Myth and Reality in the Rain Forest. How Conservation Strategies Are Failing in West Africa*. Berkeley and Los Angeles: University of California Press.

Redford, K. H. and A. Taber. 2000. Writing the wrongs: Developing a safe-fail culture in conservation. *Conservation Biology* 14:1567–1568.

Rivas, J. and R. Ruiz. 2001. *National Policy and the Repercussions Thereof on Conservation and ICDPs in Ecuador.* Quito, Ecuador: Fundación Natura and DGIS-WWF Tropical Forest Portfolio.

Robinson, J. G. 1993. The limits to caring: Sustainable living and the loss of biodiversity. *Conservation Biology* 7:20–28.

Sanjayan, M. A., S. Shen, and M. Jansen. 1997. *Experiences with Integrated Conservation and Development Projects in Asia.* World Bank Technical Paper no. 38. Washington, D.C.: The World Bank.

Scott, P. 1998. *From Conflict to Collaboration: People and Forests at Mount Elgon, Uganda.* Gland, Switzerland: IUCN–The World Conservation Union.

Shaffer, P. 1998. Gender, poverty, and deprivation: Evidence from the Republic of Guinea. *World Development* 26:2119–2135.

Thibault, M. and S. Blaney. 2001. Sustainable human resources in a protected area in southwest Gabon. *Conservation Biology* 15:591–595.

Tunio, S. 2000. Gender issues of natural resource management in coastal communities of Pakistan. *DGIS-WWF Tropical Forest Portfolio Newsletter* 5:1–2.

UNDP (UN Development Program). 1999. *Human Development Report 1999.* New York: Oxford University Press.

Wainwright, C. and W. Wehrmeyer. 1998. Success in integrating conservation and development. A study from Zambia. *World Development* 26:933–944.

Wells, M. 1995. *Biodiversity Conservation and Local People's Development Aspirations: New Priorities for the 1990s.* Rural Development Forestry Network. Paper 18a. Winter 1994–Spring 1995.

Wells, M. 1998. Institutions and incentives for biodiversity conservation. *Biodiversity and Conservation* 7:815–835.

Wells, M. and K. Brandon. 1992. *People and Parks: Linking Protected Area Management with Local Communities.* Washington, D.C.: The World Bank.

Wells, M., S. Guggenheim, A. Khan, W. Wardojo, and P. Jepson. 1999. *Investing in Biodiversity: A Review of Indonesia's Integrated Conservation and Development Projects*. Washington, D.C.: The World Bank.

White, L. J. T. 1992. *Vegetation History and Logging Damage: Effects on Rain Forest Mammals in the Lopé Reserve, Gabon.* Ph.D. diss., University of Edinburgh.

Wilkie, D. S., J. G. Sidle, and G. C. Boundzanga. 1992. Mechanized logging, market hunting, and a bank loan in Congo. *Conservation Biology* 6:570–581.

Wood, A., P. Stedman-Edwards, and J. Mang, eds. 2000. *The Root Causes of Biodiversity Loss*. London: Earthscan.

Worah, S. 2000. International history of ICDPs. In *Proceedings of Integrated Conservation and Development Projects Lessons Learned Workshop, 12–13 June 2000,* 5–6. Hanoi: UN Development Program, The World Bank, World Wildlife Fund.

Part Two

Applications and Issues

5

Fitting ICD Into a Project Framework: A CARE Perspective

Phil Franks and Thomas Blomley

Introduction

CARE International is one of the world's largest relief and development non-governmental organizations (NGOs), with field programs in over seventy developing countries. Principal sectors of development programming include agriculture and natural resources (ANR), children's health, reproductive health, small economic activities, basic and girls' education, and water, sanitation, and environmental health.

Within CARE's ANR sector there are currently twenty-two site-based projects that in one way or another seek to promote a link between biodiversity conservation and the socioeconomic development of local communities, and thus conform to the classic definition of integrated conservation and development (ICD) projects (Wells and Brandon 1992). CARE is currently the only international development NGO with a substantial involvement in ICD.

CARE's involvement in ICD programming started in 1988, and as with other agencies the experience has been mixed. For CARE the challenge has been compounded by the fact that CARE's mission as a development agency is very different from that of the conservation agencies that pioneered the concept. That said, with the increasing emphasis on community-based approaches, there is a sense of need and opportunity for greater involvement of agencies such as CARE in ICD programming.

Over the last ten years there has been much criticism of the ICD concept, particularly focusing on the apparent lack of conservation impact from efforts to support local livelihoods, and on the failure of projects to identify and address the true underlying causes of environmental degradation and loss of biodiversity (Kremen, Merenlender, and Murphy 1994; Wells 1995; Wells et al. 1999; Newmark and Hough 1999; Wood, Stedman-Edwards, and Mang 2000; Hughes and Flintan 2001; Wunder 2001). On a more positive note, it is clear that lessons learned have resulted in a substantial evolution in the approach. In a major review of its ICD experience, the World Wildlife Fund (WWF) defined three generations of ICDP: a first generation emphasizing mitigation and substitution, a second generation emphasizing community participation in management and utilization of biodiversity resources, and a new generation based on so-called landscape approaches (Larson, Freudenberger, and Wyckoff-Baird 1998). This paper will refer to this simple typology.

At a theoretical level, the ICD concept, linking biodiversity conservation with socioeconomic development, is very attractive. Analyzed from an economic perspective, biodiversity generates a range of benefits that accrue to a wide range of stakeholders from the local to global level (Pearce and Moran 1994). Building on this, a political ecology framework can be used to examine the political and social dimensions of the various interests at stake, and the power relations through which these are expressed (Blaikie and Jeanrenaud 1997; Brown 1998). The result is a conceptual basis for ICD that appears very credible.

However, the underlying complexity and contradictions are revealed once we start to design an ICDP and define the roles of the various partners in implementation. At the heart of this exercise lies development of a logical framework, a widely used project planning format that summarizes a project's goals, strategies, and activities, critical assumptions in the project design, and the means of assessing achievement (European Commission 2001). Drawing on CARE's experience, this paper examines the outcome of the project design process—how a range of projects has arrived at different interpretations of the ICD concept, the goals of the project, and key strategies to achieve these goals.

This is followed by a discussion of how the outcome of project design appears conditioned by the nature of the process used, concluding that the weaknesses of ICDPs have much to do with the classic problem-driven planning methodologies employed in their design. Finally the paper considers CARE's experience in applying landscape approaches and how such approaches, coupled with an improved design process, may form the basis of a more successful ICD paradigm.

Confusing Goals

Table 5.1 presents statements of the overall goal (i.e., the long-term, ultimate objective) for eight different ICDPs supported by CARE International, which illustrate different interpretations of the ICD concept. Three of these projects conform to the typical ICDP model: an overall goal framed in terms of biodiversity conservation, and development at a local level appearing as a means to this end, i.e., "conservation through development" (CTD). In contrast, the Bardia project in Nepal has an overall goal framed in terms of livelihoods, with conservation of biodiversity expressed as a condition: "in a manner which safeguards ... existing biodiversity"—i.e., essentially "development through conservation" (DTC).

Two projects attempt to combine conservation and development objectives (i.e., "conservation and development," C&D), avoiding any sense of

Table 5.1 *Overall goal statements for a cross section of CARE-supported ICDPs*

Project	*Overall goal*	*Approach*
Awesome (Philippines)	To protect the resources of Mt. Malindang National Park (MMNP) by increasing awareness of the value of biodiversity and environmental conservation and increasing household livelihood security and income of buffer zone communities.	CTD
SUBIR (Ecuador)	Protect the Ecuadorian Choco and Amazon's unique biological diversity through sustainable natural resource management in native forests in and around the Cotocachi-Cayapas Ecological Reserve and the Huaorani Ethnic Territory Reserve, respectively.	CTD
U Minh Thuong (Vietnam)	The existing natural resources and biodiversity of U Minh Thuong Nature Reserve conserved through the socioeconomic development of participating buffer zone communities and the strengthening of institutional reserve management capacity.	CTD
Bardia (Nepal)	To improve the socioeconomic security of participating communities living in the buffer zone in a manner that safeguards the existing natural resources and biodiversity of the Royal Bardia National Park.	DTC

(continued)

Table 5.1. *Overall goal statements for a cross section of CARE-supported ICDPs (continued)*

Project	*Overall goal*	*Appraoch*
DTC (Uganda)	The environmental values of natural resources within Bwindi and Mgahinga National Parks and the livelihood security of resource-poor households in the twenty-four surrounding parishes are sustainably increased and the social and economic benefits are equitably shared within local communities and between local communities and the Ugandan people as a whole.	C&D
Madidi (Bolivia)	The natural resources and biodiversity of Madidi National Park and its area of influence, and associated social, economic, environmental, and cultural values are conserved and enhanced, for the benefit of local communities and the people of Bolivia as a whole.	C&D
Awash (Ethiopia)	Household livelihood security of pastoralist communities living in and around Awash National Park is improved, while conserving the park's unique and significant biodiversity and habitats.	DTC or C&D
Msitu Yetu (Tanzania)	The livelihood security of households in communities adjacent to Eastern Arc/Coastal Forests is improved while the globally important biodiversity of these areas continues to flourish.	DTC or C&D

hierarchy: Madidi in Bolivia and (despite its name) Development Through Conservation in Uganda. Finally there are two projects, Awash in Ethiopia and Msitu Yetu in Tanzania, where the overall goal statement could imply either C&D or DTC, depending on whether the word "while" is taken to mean "at the same time as" or implies a condition. Such differing interpretation of simple words further confuses the picture.

Conservation for Whom?

Applying a political ecology perspective to a study of grasslands within the Royal Bardia National Park in Nepal, Brown (1998) identifies a whole range of stakeholder interests in grassland resources, ranging from consumptive values to local people to existence values (of tigers)

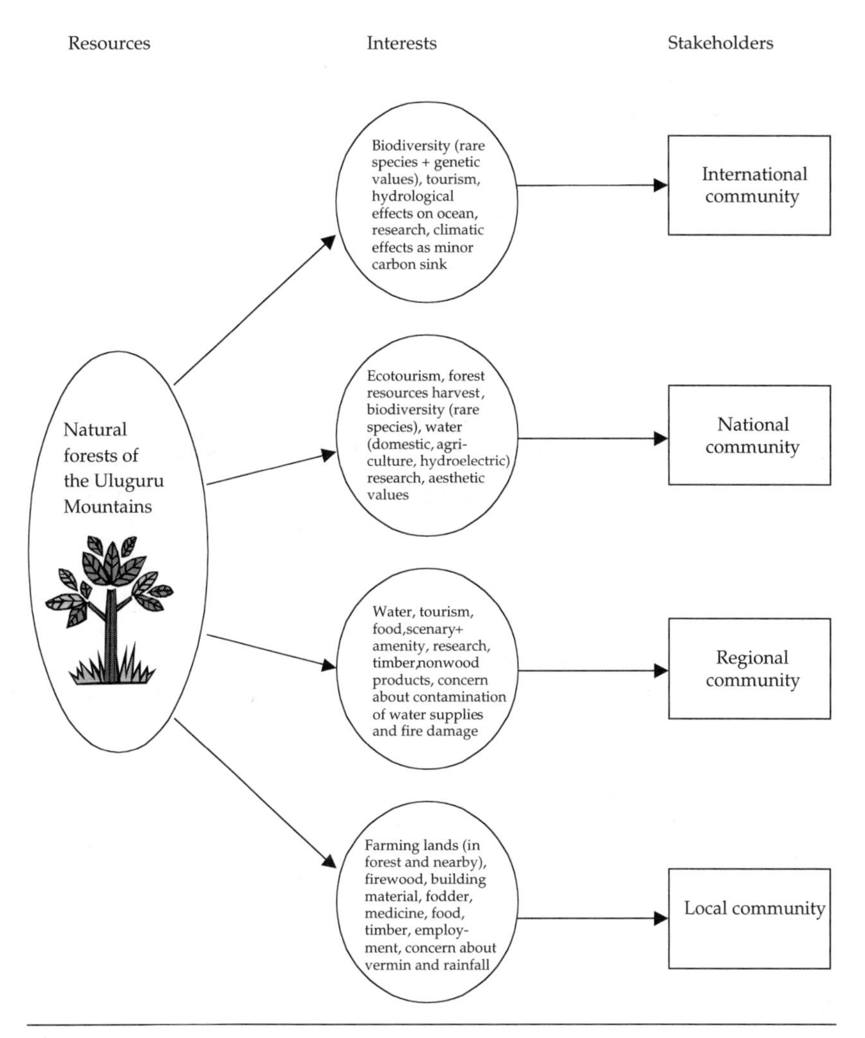

Figure 5.1.
Stakeholder interests in the forest of the Uluguru Mountains, Tanzania.

at a global level. In a recent exercise to design an ICD program for the Uluguru Mountains in Tanzania, a similar approach was employed to explore interests at various levels in relation to the biodiversity and associated ecosystem functions and services of the natural forests of the Ulugurus (Franks 2000). Figure 5.1 summarizes the output of the exercise.

This analysis raises the fundamental question, "conservation for whom?" The conservation interests of communities at local and regional levels relate primarily to ecosystem functions and services (notably water) and their relevance to livelihood security, while conservation interests at national and international levels relate primarily to biodiversity in the stricter scientific sense of the term.[1] Discussion of such issues is further complicated by different interpretations of the term *biodiversity.* Local stakeholders very often understand the term as referring to endangered species (i.e., largely an external agenda), while economists and planners tend to use the term in a broader sense to include the ecosystem functions and services that are of interest to local people.

Often the answer to the question "conservation for whom?" is "humanity." But there is no target condition for the forests of the Ulugurus that is optimal for all stakeholders. Thus, *as an issue of principle,* the conservation strategy for the Uluguru Mountains, and the ICDP that supports this strategy, should take account of both the biodiversity interests at global and national levels and the more development-oriented interests at regional and local levels.

By implication, as conservation strategies are developed, there will be a need for compromise. The ICD program in which CARE has been involved in Bwindi Impenetrable Forest in Uganda provides an example. In this case, programs for enabling access to forest products by local people, and to promote mountain gorilla–based tourism and revenue sharing with local communities,[2] were initiated because it was decided that there should be greater benefits for local people, as an end in its own right. At the time it was not clear that these programs would result in enhanced conservation of mountain gorillas, and indeed several well-known conservationists lobbied strongly against the two programs because of the increased risk of gorillas contracting potentially fatal human disease.

The following statement is a revised definition of ICD recently developed by CARE, the WWF, and the UNDP for an ICD training program in eastern Africa: "ICD is an approach to the management and conservation of natural resources in areas of significant biodiversity value that aims to reconcile the biodiversity conservation and socio-economic development interests of multiple stakeholders at local, regional, national and international levels." An obvious implication of this statement is that the overall goal of an ICD project should capture both the conservation and development impact resulting from the improved natural resource management practices supported by the project.

Nonetheless, like three of the CARE projects listed in table 5.1, the majority of ICDPs worldwide continue to frame their overall goal, and associated monitoring and evaluation indicators, in terms of biodiversity conservation. However biodiversity is defined, this is not the language that local stakeholders would use to describe the overall goal of natural resource conservation; the implication, at least from a local perspective, is that the agenda is being defined primarily by the interests of the international community. CARE's Bardia project illustrates a similar problem but in reverse, with an overall goal dominated by the interests of local communities. For a community forestry project with no significant external interests this would be fine, but not for an ICDP that by definition deals with a situation where there are substantial national and/or international interests in biodiversity.

CARE staff might respond that this is a CARE project, and CARE projects have overall goals framed in terms of livelihood security. Certainly this is CARE's primary interest in an ICD project, and should be reflected in CARE's own program goals. But CARE's interests should not dictate the goal of an ICD project, which must, for both pragmatic and ethical reasons, address a much broader range of interests (Blaikie and Jeanrenaud 1997).

This highlights the need to distinguish between the goals of the project and the goals of CARE as a partner in the project. Reflecting this thinking in CARE, several projects have, in the process of redesign, made substantial changes to their statements of overall goal to ensure that the full range of interests in the impact of the project are effectively addressed. Examples are the Bolivia and Uganda projects, where impact is expressed in terms of a set of social, economic, and environmental values, and the Tanzania project, which combines livelihood security and biodiversity in a single statement. Although some project design manuals warn against dual objectives, guidelines produced by the early proponents of the logical framework state that this is quite acceptable in cases where the goals are closely linked (Sartorius 1991).

Confusing Roles

Most ICDPs include a co-management process that seeks to promote some sharing of authority and control over the management and utilization of biodiversity resources between local stakeholders and central government authorities. Typically such a process involves conflict resolution, development and strengthening of institutions, and, frequently, the development of supportive policy at national and protected area levels (e.g., through the PA management plan). In many CARE-supported projects, CARE staff have adopted the role of "honest broker" in this process. Likewise, there are many ICDPs supported by conservation NGOs where the staff of the NGOs have adopted a similar role.

The question of CARE's interest in ICD programming has been much debated within CARE. The following statement, which should be read in conjunction with the generic ICD definition given earlier in this section, summarizes the conclusion of this discussion (CARE 2001): "CARE-supported ICD projects enhance the livelihoods of poor men, women and children by promoting social justice and equity in the use and sustainable management of natural resources. At the heart of this approach lies the reconciliation of national and international interests with the aspirations of local people."

How can CARE perform the role of "honest broker" in an ICD process and at the same time be faithful to its own mission as reflected in the statement just quoted? Surely there is a conflict of interest that potentially compromises both the integrity of the process and CARE's ability to pursue its specific objectives. At best, CARE staff can be members of a team with a cross section of interests that can collectively perform the "honest broker" function. The implication is that partnership between conservation and development agencies is an essential prerequisite for any ICDP.

This raises fundamental questions about the nature of a project. As projects become increasingly implemented through partnerships, CARE's role is changing from that of "lead implementing agency" to that of a "convener" and partner alongside other partners. Thus we are trying to move from the notion of "the CARE project" to the idea of a project with its own identity in which CARE is a partner. This change in role forces CARE to define more explicitly the "value added" by an international NGO in ICD programming.

The Significance of the Design Process

Typically an ICDP design process comprises several months of information gathering (secondary information supplemented by primary data collection where needed to fill gaps) followed by a design exercise, the output of which is summarized in a logical framework. Normally the design exercise comprises one or more workshops involving stakeholders at different levels. In total this design process may take anywhere from three to twelve months.

Broadly speaking, we can define three different levels of stakeholder participation in the design of CARE ICDPs:

Consultative: project designed by CARE (or consultants commissioned by CARE), with other stakeholders providing contextual information and commenting on the draft design.

Participatory for project partners, consultative for other stakeholders: project designed by representatives of the project partners, with other stakeholders providing contextual information and commenting on the draft design.

Participatory: representatives of all key stakeholders attend one or more workshops that generate the project design, typically employing a ZOPP (Zielorientierte Projektplanung or equivalent goal oriented Project Planning) methodology.

(Note: In this context, stakeholders are defined as individuals and organizations that have an interest in, or influence over, the project and the natural resources in question, while partners are defined very specifically as the organizations directly involved in project implementation that share project resources and control and accountability for delivering project outputs.)

Problem-Driven Planning

The project design exercise itself normally proceeds through a stage of problem analysis to a stage of developing the objectives that fit into the different levels of the logical framework. The problem analysis usually involves the development of a hierarchy of cause and effect (e.g., a "conceptual model"[3] or "problem tree"[4]). From this problem analysis the "core problem" is identified, which then forms the basis of the statement of project purpose. The causes of the core problem then provide the basis for developing the project outputs. With very structured methodologies such as ZOPP, the link between the problem analysis and the development of objectives is very direct. In other cases the link may be much looser. Whatever the approach, the development of the project design is essentially problem-driven.

Looking at CARE ICDPs, it is becoming clear that some significant weaknesses in design can be traced back to the problem-driven planning processes used in project design. Three constraints merit particular attention:

(a) Problem analysis tends to be framed in biophysical terms. Thus causes of the core problem, which form the basis of project strategy, are expressed in the form of direct physical threats, such as fire, poaching, and unsustainable harvesting of firewood. Having started with a biophysical perspective, project implementers can find it hard to switch to the very different perspective needed to explore underlying causes relating to policy, institutions, etc. In recent years

more sophisticated problem analysis tools have been developed to address this constraint (e.g., root cause analysis; cf. Wood, Stedman-Edwards, and Mang 2000), but these are not yet widely applied.

(*b*) Problem analysis tends to be inherently backward looking, focusing on historical patterns of legal and illegal resource use, and the deficiencies, or lack, of existing services and infrastructure. As a result, the process tends to place excessive emphasis on mitigation-compensation strategies and restoration of traditional resource management systems, while insufficient attention is given to opportunities for developing new management systems and institutional arrangements and innovative conservation incentives.

(*c*) The inherently reductionist approach of problem analysis, and in particular the requirement that the focus of the project be defined in terms of a single, clearly defined core problem, makes it hard to accommodate the perspectives and interests of the full range of stakeholders. Very often the discussion becomes a scientific debate, which excludes nonscientists (i.e., most local stakeholders), with the result that the conservation interests and perspectives of more powerful external stakeholders dominate the agenda.

The weaknesses of problem analysis are compounded by the fact that it is all too easy to import classic narratives and assumptions on the relationship between poverty and environmental degradation and force the linkage into a simplistic means-end relationship. Critical review shows many of these narratives and assumptions to be at best unreliable and at worst entirely false (Forsyth and Leach 1998; Ambler 2000; Blomley 2000; Risby and Blomley in prep).

Vision-Driven Planning

An alternative to problem-driven approaches are planning methodologies based on the development of a vision of a future desired state, and an analysis of the factors supporting and resisting the achievement of this vision. These "vision-driven" approaches, and related scenario planning approaches, are increasingly widely used at all levels, from community-based planning to strategic planning within major national and international organizations (Wollenberg, Edmunds, and Buck 2000; ARD 2001; Education Training Unit 2001).

Over the last couple of years CARE has been experimenting with a vision-driven approach for ICDP design/redesign, using a two-stage process (see box 5.1 for a summary of key steps). In general the pro-

Box 5.1. *A Vision-Driven Project Design Process*

Pre-Design Workshop (2 days)

- Identification of stakeholders in project design
- Development of common understanding on ICD
- Identification of information needs, and design studies for primary data collection (where necessary)

Design Workshop (4–5 days)

- Actor-oriented situation analysis
  - review of primary and secondary information and lessons learned
  - analysis of stakeholders in the natural resource
  - analysis of stakeholder interests in the natural resource
  - analysis of stakeholder influence over the natural resource
  - root-cause analysis for key negative influences (threats)
- Development of a long-term vision (e.g., 10 years)
- Development of statements of overall goal and project purpose
- Identification of supporting and resisting factors with respect to achievement of the project purpose, clustered under common themes
- Identification of interventions to address supporting and resisting factors
- Development of outputs and activities for each theme
- Identification and analysis of critical assumptions

cess has been well received. An example is the design of the Uluguru Mountains project mentioned earlier, in which a design team of project partners employed a process based on the strategic planning methodology used by many CARE country offices (Franks 2000).

Prior to the stage of vision development, this process involves a situation analysis, which includes an analysis of stakeholder interests in, and influences over, the natural resources in question. Negative influences are effectively equivalent to "threats," and in specific cases there is a need to further explore specific threats with a focused problem analysis (e.g., root cause analysis). Thus an analysis of threats and their causes is still a key element of the process, but in this case as part of a broader situation analysis rather than, as with some problem-driven approaches, as the basis of the whole design.

Although our experience in vision-driven approaches to ICDP design is still very limited, some key strengths are already emerging:

- Defining the project purpose in terms of a shared vision rather than a core problem allows for more effective discussion and negotiation of common interests, and less risk of domination by a specific interest group.
- The analysis of supporting factors captures opportunities that may be developed and exploited by the project that are often overlooked in a problem-driven approach.
- Delinking the situation analysis from the development of goals and outputs makes it easier to identify and focus on underlying causes and define the project design in these terms.

Problem-driven project planning methodologies may be very effective for relatively technical development projects where there is a fairly high degree of consensus among stakeholders on the objectives—for example, an irrigation project or a vaccination program. However, our experience in applying a vision-driven approach leads us to believe that this approach may be more appropriate for the more complex situation of an ICDP with a more diverse (and often conflicting) set of stakeholder interests. That said, in the hands of a facilitator with extensive ICD experience, the two approaches may well produce a similar result, but in the absence of such experience, a vision-driven approach may prove more robust.

The Need for a Longer Perspective

This discussion of the process of project design leads to the issue of time frame. In most of the situations where CARE supports ICDPs, the ICD approach is new, and at the heart of such a project lies a process for reconciling the interests of the major stakeholders in the natural resources in question. This process of reconciling interests begins, in effect, with design, and so in a very real sense the design of the project is actually the beginning of the process of the project itself (see also Brown this volume). For example, as with the CARE-supported ICDP in Vietnam, government authorities may be unwilling to address benefit sharing with local communities in a project design until they are convinced of the conservation value of such interventions. Thus it may take several years of dialogue with local communities, confidence-building measures, and exposure to experience elsewhere before partners will agree to the inclusion of increased benefit sharing in the project design. Likewise, it may take many months of working with local community institutions to reach a point where they are sufficiently knowledgeable, skilled, and empowered to effectively represent their interests in a project design process.

Again, the diverse and at times conflicting range of stakeholder interests associated with designing and implementing an ICDP makes a strong case for different treatment from that commonly used in a typical development project. Based on CARE's experience, we propose the following ICD process, which would take a minimum of six years for small-scale initiatives and up to twelve years for larger-scale initiatives. We would argue that no implementing agency should start such a process without a high level of confidence that such support will be forthcoming, although not necessarily from the same donor throughout.

Phase I: Initial enabling design, dialogue between stakeholders, pilot activities to test new approaches and build confidence by demonstrating impact on a small scale, sharing of experience with similar initiatives elsewhere, detailed design (2–3 years).

Phase II: Main implementation phase, including exit strategy (4–9 years depending on scale).

Phase III: (Optional) Further initiatives that support the interests of specific stakeholder groups if necessary—see box 5.2.

Box 5.2 *Reorienting the Development Through Conservation Project, Uganda*

The Development Through Conservation (DTC) project has been supporting the establishment of agreements between forest edge communities and park management for the use of limited nontimber forest resources in Bwindi Impenetrable National Park in Uganda. Since the inception of this process in 1993, CARE has viewed itself as a "neutral broker" helping to moderate between local community demands for greater benefit sharing and park management interests for biodiversity conservation. A recent review of the project highlighted the fact that while the agreements appear to be functioning well, the negotiation process (and therefore final outcome) for establishing the types of resources and the overall off-takes was inequitable and dominated by park management interests. Community

(continued)

> and resource user groups were largely passive participants in the negotiation process and were obliged to accept a final outcome well short of their expectations. The asymmetrical power relationship, caused by a concentration of knowledge about statutory and legal rights in the hands of the park management, calls into question both the equity of the process and the degree to which CARE can be said to have acted as a "neutral broker." The project is currently undergoing a redesign process, and in the next (and final) phase project activities will target community groups and legitimate civil society organizations representing poor park-dependent households, and help build their capacity to more effectively negotiate, advocate, and claim for their biodiversity entitlements. This is a delicate process, as it is not in the interest of the project to create a hostile situation where empowered communities degrade park resources. Strengthening of rights demands responsible actors, both within local communities and external service providers.

An Honest Broker for Project Design

In a previous section of this paper we argue that an external agency such as CARE may take on the role of "convener" of an ICD process, but no single agency on its own can perform the role of "honest broker." The necessary "honest broker" function should be provided by an appropriate combination of implementing agencies working in partnership.

The same principle applies to the design of the project. Whatever their own particular agenda, the implementing partners must facilitate a design process that genuinely addresses the full range of stakeholder interests. This is not to say that partners have to suppress their own interests. Within the project there may well be activities promoting the interests of particular stakeholder groups (e.g., education, advocacy, empowerment of local institutions), but it will be important to ensure a balance that does not destabilize the overall process.

That said, once an ICDP has succeeded in establishing and institutionalizing the necessary mechanisms for promoting effective stakeholder participation in management and decision making (or if such mechanisms already existed), external agencies such as CARE may become much more explicit in promoting their particular interests. After nine years of supporting a co-management process and the development of the underlying institutional framework, CARE's Development Through Conservation project in Uganda has now reached this stage (see box 5.2).

Landscape Approaches: The New Paradigm?

Linking the disciplines of political economy and ecology, a political ecology perspective, with its focus on actors at different levels and their differing interests in natural resources, appears to provide a sound basis for an ICD conceptual framework. In the previous sections we have argued that part of the problem with ICD programming is the distortion of goals and logic brought about in trying to apply rapid, problem-oriented planning methodologies that were originally developed for simpler, more technocentric, and less conflict-prone projects. Thus we argue for a longer design process using a vision-driven approach.

Key to the success of the vision-driven approach will be our success in developing a shared vision that is genuinely a "unifying theme" (as opposed to a set of disparate vision elements). Essential characteristics of this unifying theme are that it concerns the management of natural resources, that it involves all the key stakeholders in these resources, and that it provides a platform for the reconciliation of different stakeholder interests.

In conservation work in the European context, a new concept is emerging described as "multifunctionality" (Havnevik 2001), which is based on two key ideas:

(a) Moving away from a focus on production to managing rural landscapes in a manner that addresses the multiple objectives of a range of different stakeholders.
(b) Recognizing the crucial role of the residents of this landscape as stewards of these natural resources and primary stakeholders in conservation.

Although the context is very different from that of the tropical forests and impoverished communities of the developing world, there are still obvious parallels. A number of ICDPs are already moving in this direction, with the DTC project in Uganda being an example. The project has supported the development of an institutional framework that aims to strengthen community participation in protected area management and mainstream environment and biodiversity conservation within local government planning. In this way the project has encouraged a move toward addressing biodiversity objectives as an integral part of a natural resource management strategy within a larger rural landscape.

However, in contrast to the landscape approaches of some international conservation NGOs, the landscape is defined as much by institutional factors as by ecological criteria (Blomley and Franks 2001).

The concept of multifunctionality provides a holistic framework for natural resource management in a rural landscape. As such, the concept seems to have considerable relevance to our quest for a unifying theme for ICD programming that will strengthen the link between conservation and development objectives, and perhaps multifunctionality will encourage a somewhat different perspective on the challenges of conservation in low-income countries. Such an approach would seem likely to encourage current moves away from household-level interventions, toward the issues of planning, policy, power, and governance that increasingly appear to be key to the success of ICD programming.

Placing less emphasis on the household-level interventions typical of the first- and second-generation ICDPs, such as income-generating activities and improved farming practices, does not mean that we place less emphasis on the livelihood security of poor people. In fact quite the reverse—a major driving force behind the concept of multifunctionality in Europe has been the need to find more effective mechanisms for supporting the livelihoods of farmers in marginal areas. However, in such a landscape approach the role of household-level interventions will be rather different, being less about reducing threats and more about providing direct incentives to encourage specific land-management practices.

In effect, local communities are being asked to take on a stewardship role on behalf of external interests (e.g., national interests in water catchment and international interests in biodiversity). That being the case, there is a strong argument for national and international communities to fund direct payments to local communities for these services (e.g., support for community projects, tax breaks, direct payments). Although such fiscal incentives are difficult to administer, there is a growing consensus that they may prove more effective in terms of conservation impact than many of the traditional ICD livelihood interventions that are so often based on unreliable assumptions about livelihood strategies and their dependence on natural resources (Ferraro 2001; McShane and Newby this volume). That said, we believe there will still be a place for certain economic interventions at the household level, notably those that add value to forest and wildlife resources.

The ICD typology mentioned in the introduction to this paper presents "landscape approaches" as the third generation and new paradigm for ICD programming. Certainly these landscape approaches suggest a significant change in strategy, but does this really amount to a new paradigm? Central to this issue, from a CARE perspective, is the question of social justice. Do the new landscape approaches recognize poor households (at both local and regional levels) as primary stakeholders in conservation, whose interests and rights are to be considered on a par with those of stakeholders at national and international levels? Certainly they have the potential to do so, as we see with the emerging concept of multifunctionality, but perhaps not as currently formulated, defined largely by international interests (see also Maginnis, Jackson, and Dudley this volume).

Conclusion

Conservation of forests and wildlife must address the interest of multiple stakeholders. Interests at the international level (and to a lesser extent at the national level) tend to be expressed in terms of biodiversity conservation as an end in its own right. In contrast, the interests of local stakeholders, and poor households in particular, focus on the contribution natural resource conservation can make to enhancing their welfare. If the overall goal of an ICDP is stated purely in terms of biodiversity conservation, then in effect the international community has set the agenda and marginalized the interests of local stakeholders before the project has even started. Applying the same principle, there is a similar problem where the overall goal of an ICDP is framed purely in terms of sustainable livelihoods, thus marginalizing legitimate national and international interests in biodiversity.

If an ICDP is about reconciling the interests of multiple stakeholders in natural resource management and conservation, then the overall goal of such a project must surely reflect the multiple objectives of these stakeholders (while not predetermining the exact outcome of the process). The term *ICD* has become somewhat discredited by the poor performance of many ICDPs, but if we practice ICD in the manner described in this paper, then it remains, in our view, an appropriate term for these projects and programs, describing the generic goal.

In the past, ICDPs have tended to be dominated by the agenda of a lead implementing agency. This paper argues that such a position is incompatible not only with the principles of ICD programming but also with the principles of partnership, which require that all project implementation partners contribute to the definition of project objectives. Furthermore, in terms of process, we must recognize that the "honest broker" function that is essential in an ICDP cannot be performed by one agency alone. Thus we would argue that partnership between agencies representing the range of different interests in natural resource conservation is actually a prerequisite for establishing a successful ICDP.

Assuming consensus can be achieved on the conceptual framework, the challenge becomes operationalizing the ICD concept through a project mechanism. Here we come up against the constraints of problem-driven planning processes. Unless managed by a very experienced facilitator, this process tends to produce a backward-looking, simplistic analysis of the situation, and one that fails to capture emerging opportunities. Furthermore, the limitations of the approach are often compounded by the failure of the agencies leading the process to make a distinction between their own objectives and the objectives and requirements of a successful ICD program.

Vision-driven approaches to project planning appear to be more appropriate for ICDP design both in terms of situation analysis and the development of goals and strategies, and a number of agencies are now using such approaches both for project design and community-based planning. Key to the success of a vision-driven planning process is the mechanism for arriving at a shared vision. This shared vision needs to have ownership of the key stakeholders and provide a genuinely unifying theme that relates the multiple stakeholder objectives with the management and conservation of natural resources within the target area. Of course, this is an ideal situation. In reality the shared vision will continue to be developed throughout the project lifetime.

Landscape approaches appear to offer a suitable framework for developing such a shared vision and subsequent planning, so long as they provide a level playing field for the process of reconciling stakeholder interests, which is institutionalized at the relevant levels of government. Landscape approaches have been heralded as a new paradigm for ICD programming, and over the last five years a number of different landscape, or large-scale, approaches have been developed by conservation agencies (ecoregions, ecosystems, bioregional, heartlands, etc.).

Although clearly very useful in guiding the programming strategies of the parent agency and planning at a macro level, these particular landscape approaches, we find, do not yet provide the new approach to ICDP design and implementation that is required. However, in a European conservation context, we are seeing the development of rather different landscape approaches that may very well serve the purpose.

Where an ICD approach is being newly introduced, the process of reconciling differing stakeholder interests begins with the potential partners engaging in the situation analysis, visioning, and planning that constitute the project design. In a very real sense the process of the ICD project itself starts with the design, which is a very different situation from the typical development project where a relatively high level of consensus can often be achieved within a single design workshop. Donors and implementing agencies must recognize this difference and the implications: an extended design phase and a longer overall time frame.

With this paper, CARE is promoting an approach to ICD that takes more account of the interests of poor, natural resource–dependent households as primary stakeholders in the conservation of biodiversity. This is an issue of recognizing practical realities, but also an issue of principle. This approach challenges the classic ICD approach of most of the first- and second-generation ICDPs that implicitly (and sometimes explicitly) treat development benefits as a means to achieve a higher biodiversity conservation objective defined primarily by international interests. That said, we reemphasize the point that ICDPs are about benefits and values relating to the management and conservation of natural resources in areas of high biodiversity, and this principle defines the framework and boundaries for ICD project design.

A new paradigm for ICD programming is much needed. Further work is required to develop appropriate landscape approaches and vision-driven planning methodologies adapted to the ICD context that more effectively define shared vision and explore the underlying causes of environmental degradation and the linkage with processes of impoverishment. Not yet a new paradigm, but some of the key elements are coming together.

Endnotes

1. The term *biodiversity* refers to the variety and variability of living organisms at the three different levels of genes, species, and ecosystems.
2. Supported by the International Gorilla Conservation Program, a consortium of the WWF, the African Wildlife Foundation, and Flora and Fauna International.
3. See Margoluis and Salafsky 1998 and Salafsky and Wollenberg 2000.
4. Where there is some doubt over future funding, this is a critical assumption that should be included in the logframe at the level of project purpose (i.e., as a condition for achieving the long-term, overall goal).

References

Ambler, J. 2000. *Attacking Poverty While Improving the Environment: Towards Win-Win Policy Options*. Paper produced for the EC/UNDP Poverty and Environment Initiative. New York: UN Development Program.

ARD. 2001. Uganda: Conserve biodiversity for sustainable development. www.ardinc.com/htm/projects/p_cobs.htm (May 2002).

Blaikie, P. and S. Jeanrenaud. 1997. Biodiversity and human welfare. In K. Ghimire and M. Pimbert, eds., *Social Change and Conservation*. London: Earthscan.

Blomley, T. 2000. *Woodlots, Woodfuel, and Wildlife: Experiences from Queen Elizabeth National Park, Uganda*. Gatekeeper Series no. 90. London: International Institute for Environment and Development (IIED).

Blomley, T. and P. Franks. 2001. *Biodiversity Conservation Within the Context of Decentralised Governance: Towards Institutional Landscapes*. CARE. www.icd-net.care.dk (May 2002).

Brown, K. 1998. The political ecology of biodiversity, conservation, and development in Nepal's Terai: Confused meanings, means, and ends. *Ecological Economics* 24 (1):73–88.

CARE. 2001. *Integrated Conservation and Development Programming in CARE International*. www.icd-net.care.dk/lib/getImg.php?c_id=1606 (May 2002).

Education Training Unit for Democracy and Development. 2001. *Integrated Development Planning for Local Government*. www.etu.org.za/toolbox/docs/organise/webidp.htm (May 2002).

European Commission. 2001. *Manual: Project Cycle Management*. Brussels: European Union.

Ferraro, P. J. 2001. Global habitat protection: Limitations of development interventions and a role for conservation performance payments. *Conservation Biology* 15 (4):1–12.

Forsyth, T. and M. Leach, with I. Scoones. 1998. *Poverty and Environment: Priorities for Research and Policy*. Paper produced for the EC/UNDP Poverty and Environment Initiative. New York: UN Development Program.

Franks, P. 2000. Designing a forest conservation and management project for the Uluguru Mountains. Copenhagen, Denmark: CARE International. Unpublished manuscript.

Havnevik, K. 2001. Multi-functionality—A new trend with opportunities for everyone? *Forest, Trees, and People Newsletter* 44:74–75.

Hughes, R. and F. Flintan. 2001. *Integrating Conservation and Development Experience: A Review and Bibliography of the ICDP Literature*. London: International Institute for Environment and Development (IIED).

Kremen, C., A. M. Merenlender, and D. Murphy. 1994. Ecological monitoring: A vital need for integrated conservation and development projects in the tropics. *Conservation Biology* 8 (2):388–397.

Larson, P. S., M. Freudenberger, and B. Wyckoff-Baird. 1998. *WWF Integrated Conservation and Development Projects: Ten Lessons from the Field 1985–1996*. Washington, D.C.: World Wildlife Fund.

Margoluis, R. and N. Salafsky. 1998. *Measures of Success: Designing, Managing, and Monitoring Conservation and Development Projects*. Washington, D.C.: Island Press.

Newmark, W. D. and J. L. Hough. 2000. Conserving wildlife in Africa: Integrated conservation and development projects and beyond. *BioScience* 50 (7):585–592.

Pearce, D. and D. Moran. 1994. *The Economic Value of Biodiversity*. London: Earthscan.

Risby, L. and T. Blomley. In prep. *Environmental Narratives and False Assumptions in Protected Area Planning and Community Conservation Project Design.*

Salafsky, N. and E. Wollenberg. 2000. Linking livelihoods and conservation: A conceptual framework and scale for assessing the integration of human needs and biodiversity. *World Development* 28:1421–1438.

Sartorius, R. 1991. The logical framework approach to project design and management. *Evaluation Practice* 12 (2):139–147.

Wells, M. 1995. *Biodiversity Conservation and Local People's Development Aspirations: New Priorities for the 90s.* ODI Rural Development Forestry Network Paper 18a. London: Overseas Development Institute (ODI).

Wells, M. and K. Brandon. 1992. *People and Parks: Linking Protected Area Management with Local Communities*. Washington, D.C.: The World Bank.

Wells, M., S. Guggenheim, A. Khan, W. Wardojo, and P. Jepson. 1999. *Investing in Biodiversity: A Review of Indonesia's Integrated Conservation and Development Projects*. Washington, D.C.: The World Bank.

Wollenberg, E., D. Edmunds, and L. Buck. 2000. Anticipating change: Scenarios as a tool for adaptive forest management. Bogor, Indonesia: Center for International Forestry Research (CIFOR). www.cifor.cgiar.org (May 2002).

Wood, A., P. Stedman-Edwards, and J. Mang, eds. 2000. *The Root Causes of Biodiversity Loss*. London: Earthscan.

Wunder, S. 2001. Poverty alleviation and tropical forests—What scope for synergies? *World Development* 29 (11):1817–1833.

6

Making Biodiversity Conservation a Land-Use Priority

Agnes Kiss

Introduction

Since 1990, the World Bank has supported 226 conservation-related projects around the world, involving over U.S. $1 billion of IBRD/IDA[1] resources and U.S. $450 million of GEF[2] funds, as well as an additional U.S. $1.2 billion in cofunding from other national and multilateral donors, governments, NGOs, foundations, and private companies. More broadly, Conservation International estimates that the international community (governments, multilateral development banks, and conservation groups) spends at least half a billion dollars each year on conserving biodiversity in the tropics (Hardner and Rice 2002). Satchell (2000) reported that about U.S. $4 billion has been spent on conservation over the past decade, while a recent report from the Organisation for Economic Co-operation and Development (OECD 2000) estimated that in 1998, U.S. $778 million of bilateral official development assistance (ODA) in sectors such as agriculture, forestry, water, and general environmental protection sectors had a biodiversity focus.

Despite this high level of investment and effort, we (the conservation community collectively) can only point to some individual, localized successes. Taken as a whole, we have had little impact on stemming or even slowing the rising tide of biodiversity loss.

This raises the urgent question of whether our prevailing models are faulty, or whether the problem lies in the way they are being implemented. This paper argues that both are true, particularly when it comes to conserving biodiversity outside protected areas. Conventional conservation projects, including integrated conservation and development projects (ICDPs), which are the subject of this volume, have failed to address the true causes of biodiversity loss at the scale on which they operate. We have also focused too much on carrying out project activities and too little on creating incentives for conservation. Achieving better results in the future will require new approaches that encourage landholders to achieve conservation outcomes and reward them for doing so.

Who Controls Biodiversity?

The majority of the world's remaining biodiversity is found in tropical developing countries, while the great majority of biodiversity supporters are from industrialized countries. Governments and citizens of developing countries will determine whether the biodiversity found on the land that they own, occupy, or control will be preserved or lost. Donors and international conservation organizations that support biodiversity conservation around the world are largely external stakeholders, who try to influence the decisions and actions of the internal stakeholders, mainly by providing money and information.

Like most people, landholders and resource users in developing countries base their decisions mainly on their perceived self-interest, with a strong bias toward the short term. Unfortunately, the benefits of conserving biodiversity tend to be long term, indirect, and diffuse, while the benefits of activities that destroy or degrade biodiversity tend to be short term, direct, and easily captured by individuals. Most people do not routinely sacrifice short-term, personal gains in order to achieve long-term benefits for a wider community, regardless of whether they are living on the edge of survival or directing commercial enterprises. For example, Jensen, Torn, and Harte (1990) noted that California timber companies practicing sustainable management were ripe for takeover by companies intending to extract resources quickly. In the case of ICDPs, communities often become impatient waiting for benefits from enterprises such as ecotourism to materialize. To retain their interest and goodwill, projects typically provide short-term social benefits (schools, water supply, etc.) that are not directly linked to conservation actions or results. As a result, the original objectives of the project are deferred and often forgotten.

It has become popular to say that factors such as poverty and overpopulation are the root causes of biodiversity loss, which must be addressed if conservation is to be sustainable. But it is also overly simplistic: many species of domestic plants and animals, consumed by the same growing human population, continue to grow in numbers and expand their range with human assistance. Rather than poverty or human population growth per se, the fundamental cause of biodiversity loss worldwide is that those in a position to preserve it lack sufficient incentives to do so.

Targeting the Main Causes of Biodiversity Loss

Loss of natural habitats—as a result of conversion to agriculture or other uses—is the single greatest source of biodiversity decline and loss worldwide. Hardner and Rice (2002) determined that destruction of natural habitat over the past several decades has been driven mainly by logging for timber (followed by settlement) and by expanding production of beef, soybeans, palm oil, coffee, and cocoa. These pressures are increasing rather than decreasing around the world, and are not likely to diminish in the foreseeable future.

Therefore, the top priority for conservation investment must be to slow, halt, and even reverse this process of habitat loss in areas that are recognized as the most important sites for conservation (given that it is clearly not possible to do so everywhere that biodiversity is being lost). The key question therefore is: how can we encourage those who make the land-use decisions[3] in these areas to forego the benefits associated with destructive activities in favor of conserving biodiversity?[4]

In some areas, other sources of biodiversity loss, such as overharvesting of economically valuable plant and animal species, may be more important than habitat conversion. It is estimated that people in central Africa consume the equivalent of 4 million head of cattle in bushmeat each year (Bennet et al. 2002), and that in the Congo Basin alone, 25 million people consume over 1 million metric tons of bushmeat (Satchell 2000). Beyond the local level, there are huge international markets for tropical biological products such as bushmeat, timber, and fish. Invasion by alien species is another very significant cause of biodiversity loss in some areas, particularly on islands and in locations where the native vegetation is highly specialized and vulnerable to colonizers. For example, in South Africa it is estimated that one-quarter of the native plant species are directly threatened by invasive alien plants, many of which were intentionally introduced.[5]

Regardless of whether we focus on habitat loss, overexploitation, or alien invasions, the key question is how well our responses match the nature and scale of the problem. In particular, we must recognize that the forces behind biodiversity loss operate on a large geographic scale and involve the actions of huge numbers of people, and that local communities are sometimes the direct consumers, sometimes only acting as agents for distant buyers, and sometimes not involved at all.

Conservation Approaches: Experience to Date

Protection Versus Participation

Should biodiversity be protected *from* local communities, or *by* them? Protectionism (derogatorily named "fortress conservation" by some practitioners, particularly from southern Africa) includes exclusive protected areas, hunting bans, and regulations prohibiting trade in wildlife products. The participatory approach is based on encouraging people to accept or welcome the presence of biodiversity by utilizing it in some fashion. It may involve private landowners, but usually targets poor, rural communities who live in biodiversity-rich areas, often adjacent to protected areas. The two approaches may be seen as alternatives, but can also be complementary, as in the case of a protected area surrounded by a buffer area that is managed on community-based conservation principles.

A detailed discussion of the pros and cons, strengths and weaknesses, of these approaches is beyond the scope of this paper. Each has succeeded to some extent, but both must generally be considered to be failing overall. As currently practiced, neither is likely to result in preservation of a substantial part of the world's biodiversity over the long term. With regard to protected areas, the problem is that although they can be successful in a narrow sense (cf. Bruner et al. 2001), they are on the whole too few, too small, and too threatened to be relied on as the sole instrument for conserving biodiversity.

This paper focuses on conservation outside protected areas. Despite the popularity of ICDPs and other community-based conservation models, there is increasing evidence and growing acknowledgement that this approach is in most cases failing to achieve either conservation or lasting development benefits (e.g., see Brandon, Redford, and Sanderson 1998; Hackel 1999; Oates 1999; Wells et al. 1999; Newmark and Hough 2000; Kellert et al. 2000; and many others). Roe et al. (2000) summarized the conclusions from thirty years of experimentation as follows:

- Community-based conservation can work as a conservation approach, but only under a set of conditions rarely found in reality.
- In the absence of these circumstances, community-based conservation is likely to fail, sometimes dramatically.
- Community-based conservation can complement enforcement by improving communities' attitude and reducing pressure, but cannot replace enforcement.

PROJECT VERSUS NONPROJECT APPROACHES—IN THEORY

In a "project approach" to conservation funding, an organization or group of people is given financial and other assistance to help them carry out a specified set of activities expected to result in preservation of a particular segment or facet of biodiversity (often poorly defined). By contrast, in a "nonproject approach," the donor provides financial and other inputs to encourage and enable the beneficiaries to achieve specified goals, without necessarily specifying how they should do so. These goals may include maintaining a given area in a pristine natural condition, increasing or maintaining viable populations of particular species, reducing key pressures such as hunting levels or encroachment of protected area boundaries, or any other direct conservation objective.

The project approach is limited by the fact that biodiversity loss generally results from the actions of many people across wide areas over long periods of time, while projects by their nature target relatively small numbers of people in a relatively small area over a limited period of time. Kiss (1999) detailed this and other shortcomings of the project approach, such as a focus on activities rather than results, the tendency to establish a perverse "donor-recipient" mentality that promotes dependency rather than self-sufficiency, and the "magnet effect," which undermines conservation objectives by attracting more people to the project area. Most importantly, while a project might help people accomplish something they are already motivated to do, it is not likely to be an effective mechanism for creating the motivation to begin with.

Nonproject approaches, such as tax incentives, targeted subsidies, access to specialized niche markets, and so on, are better suited to creating motivation. They can also stimulate innovation and ingenuity as people seek solutions that yield the same reward for less effort and investment, and they can reach and influence the actions of large numbers of people across wide areas. The challenges of nonproject approaches include the need to design incentive systems correctly to ensure that they do not have perverse impacts and cannot easily be exploited by cheaters or free riders, and to set and monitor indicators and targets.

Project Versus Nonproject Approaches—In Practice

While the focus of this paper is on conservation outside protected areas, it is worth noting that most support for strengthening protected area management has been provided through the project model, with the associated short time frames and a focus on implementing park management plans and annual work programs. The result has often been a "boom and bust" cycle, in which staff increases and large capital inputs are made while the project is in place but cannot be maintained when the project closes. Nonproject alternatives include endowing trust funds to provide a modest but reliable funding over the long term and providing unprogrammed budgetary support based on the achievement of specific conservation results and impacts. Neither approach is common, but there are now a small number of trust funds supporting specific protected areas or protected area systems around the world.

In developing countries, conservation outside protected areas has been supported mainly through project vehicles, and has suffered from the constraints, limitations, and perverse effects discussed above. In addition, the real divergences and conflicts between the objectives of the donors and the beneficiaries are often glossed over in the rush to begin implementation. In ICDPs in particular, conservation and development objectives are assumed—often with little evidence—to be compatible or even synergistic (see Robinson and Redford this volume; Franks and Blomley this volume). Later, the project team becomes preoccupied with ensuring that project activities are implemented and funds accounted for, while recipients focus on getting benefits from the project. The conservation objectives get lost, as can be seen by how few projects actually monitor or evaluate their biodiversity impacts (see Salafsky and Margoluis this volume).

In industrialized countries, by contrast, conservation is mainly supported through nonproject approaches such as land purchase, easements, and subsidies (see also Glick and Freese this volume). These instruments focus on creating incentives and rewarding results, rather than on defining and implementing activities. This paper makes a case for greater use of similar nonproject approaches in developing countries and provides examples of how it is already being done.

Emphasizing nonproject vehicles does not mean abandoning project-based assistance altogether. Inadequate information, poorly defined property rights, a lack of organization or management capacity, and other factors can impede landholders from responding even when appropriate incentives are in place (see Tongson and Dino this volume). Project-based assistance can complement nonproject approaches by helping to overcome these obstacles. In the absence of appropriate incentives, however, this type of support will have little benefit and may even cause harm. For example, with the wrong incentives in place, devolving ownership or control over wildlife or forests from government to communities (or to private landowners or private companies) can trigger even more rapid decline by removing the last barriers to exploiting the resource for short-term, individual profit or survival.

Making Biodiversity a Land-Use Priority

As external stakeholders, we can help to make biodiversity a land-use priority in two ways: by helping to acquire or secure land specifically for conservation purposes, and by helping to make conservation an economically competitive land use from the perspective of existing landholders through economic and other incentives. The remainder of this paper examines these approaches, the types of incentives they create, and the conditions under which they are likely to be cost-effective and sustainable.

ACQUIRING LAND FOR CONSERVATION

An estimated 11.5 percent of the world's land area lies within official protected areas such as national parks and wildlife or forest reserves. Even when these protected areas have been established for other purposes such as watershed protection or recreation, they help protect biodiversity by restricting the conversion of natural habitat to other uses.

In many parts of the world, individuals, groups, and corporations have established various types of private protected areas (cf. Langholz 1999). For example, Colombia has a well-organized network of over 100 private reserves, and similar networks are found in Costa Rica, Brazil, Guatemala, and Australia, among others. South Africa's Natural Heritage Program, a cooperative venture between government, private landowners, and the business sector, has registered over 150 sites protecting more than 215,332 hectares, over 100 of them privately owned. In southern Africa, groups of ranchers have been combining their holdings into formally established conservancies to create large areas of wildlife habitat with the objective of generating income through tourism or sport hunting.

Conservation groups may buy land outright to establish or add to protected areas, or they may secure land for conservation without taking ownership by acquiring certain use and development rights through conservation leases or easements. Land purchase, leasing, and easements are the principal tools employed by some of the largest conservation organizations operating in developed countries, including The Nature Conservancy, the Conservation Fund, the U.K. National Trust, and the Royal Society for the Protection of Birds. In the United States alone, The Nature Conservancy has a system of more than 1,300 reserves protecting over half a million hectares, comprising the largest private natural reserve system in the world (Murray 1995). The Nature Conservancy also acquires land and transfers it to local or national government conservation agencies. The Revolving Fund for Nature, administered by the Trust for Nature in Victoria, Australia, purchases lands of conservation significance, places a binding covenant on them specifying permitted and prohibited activities to ensure that they are used for conservation in the future, and then resells them. In what amounts to an involuntary easement, under the U.S. Endangered Species Act the U.S. government can preempt a private landowner's rights to develop his or her land by declaring it to be critical habitat under a species recovery plan (cf. Innes, Polasky, and Tshirhart 1998).

Biodiversity conservation can also be a side benefit when land is secured for other purposes, such as maintaining watersheds. The government of New York City recently invested about $2 billion to buy land and easements to protect the Catskills watershed, thereby avoiding the need to pay about $9 billion to build and operate a water treatment plant (Chichilnisky and Heal 1998). Municipal authorities in Cuenca, Ecuador, used part of their water revenues to buy out land users in sensitive parts of the watershed and place the land under conservation.

Land acquisition and easements for biodiversity conservation are much less common in developing countries, probably for the reasons discussed in the final section of this paper. Nevertheless, conservation organizations are increasingly experimenting with this approach (see box 6.1). For example, Conservation International's Global Conservation Fund was set up to finance conservation land deals in developing countries.

Box 6.1. *Examples of Land Acquisition for Biodiversity Conservation in Developing Countries*

Costa Rica: Under the Forest Conservation and Management Through Local Institutions (BOSCOSA) Project, a forest conservation and management incentive fund (PROINFOR) gives landowners about U.S.$700 per hectare for putting land under conservation easement. The easement is initially for five years and may be extended for three more. Funds are paid into interest-generating accounts or bonds (Donovan 1994; Cabarle et al. 1992).

Costa Rica: The Monteverde Conservation League (MCL) was created in 1986 to purchase forest land surrounding the Monteverde Cloud Forest Reserve for conservation purposes. MCL now owns over 22,000 hectares, including the Children's Eternal Rain Forest, which was bought with contributions from European schoolchildren and is now the largest private preserve in Costa Rica (Rojas and Aylward 2001).

Philippines: An island off the coast of western Negros was purchased in 1997 by the Philippines Reef and Rainforest Conservation Foundation Inc., with the help of a loan from the World Land Trust.

Tanzania: An NGO, The Land Conservation Trust (TLCT), recently acquired title to the Manyara Ranch adjacent to Lake Manyara National Park. The area will be managed by TLCT in a way that provides for use and benefits to the community while maintaining critical wildlife corridors.

Kenya: The Wildlife Trust and the Friends of Nairobi National Park are using "wildlife conservation leases" negotiated with local Maasai landowners to maintain vital wildlife migration corridors south of the park. For the going rate of four dollars per acre (an average family has 100–200 acres), the landowners agree not to fence, cultivate, or sell the land for the period of the lease (currently on a year-to-year basis).

South Africa: The South Africa National Parks makes contractual arrangements with private, communal, or

(continued)

municipal landowners to incorporate their land into national parks. Examples include the Richtersveld National Park (a 100 percent contractual park), the Cape Peninsula National Park, and the Agulhas National Park.

Guyana: The Global Conservation Fund (Conservation International) has leased rights to 200,000 acres and established an endowment fund to cover royalties, management fees, and economic development activities for local communities (e.g., education, job training).

Making Biodiversity a Competitive Land Use

Economic Incentives

The essence of the strategy for conservation outside protected areas is to make biodiversity conservation a competitive form of land use in areas where we would most like to see biodiversity preserved. To do this, conservation must generate net income (and/or valuable products and services) equal to or greater than alternative land uses. Hulme and Murphree (1999) call the use of market-based approaches the "new conservation."

There is, however, a difference between establishing a *market for biodiversity* and *marketing biological products,* which may or may not be compatible with conservation. Markets (in land, timber, fish, bushmeat, ivory, etc.) have generally been the downfall, not the salvation, of biodiversity. Is it possible for a market-based approach to result in conservation rather than destruction? The answer lies in the types of incentives that various market-based models create for landholders and resource users. These models may be placed on a continuum with respect to how directly the commodity being marketed is linked to the objective of conserving biodiversity (table 6.1).

The pros and cons of these different approaches are discussed below. While most biodiversity projects in developing countries (including ICDPs) emphasize the "least direct" end of the range, approaches at the "most direct" end of the range are more likely to be successful and efficient in many cases.

1. Utilization through extraction and marketing of biological products: There is no doubt that biodiversity products can be very valuable. Huge industries have been built around harvesting of tropical hard-

Table 6.1 *Continuum of models linking biodiversity conservation objectives and the marketing of commodities*

(Least direct)
Utilization through extraction and marketing of biological products
↓
Utilization of biodiversity within relatively intact natural ecosystems
↓
Subsidies and other compensation for biodiversity-friendly land uses
↓
Direct payment for environmental services (biodiversity conservation as a side benefit)
↓
Direct payment for the service of maintaining natural habitat and/or conserving biodiversity
(Most direct)

woods, fishing, and hunting for food or sport. On a smaller scale, many animal and plant species are harvested for medicinal or other uses (e.g., rosy periwinkle as a source of leukemia medicine, *Prunus africanum* for prostate treatment, rhinoceros horn for many uses in oriental medicine, etc.). But commercial success usually leads to overharvesting of the source species, particularly when there is a regional or global market. The overriding incentive for those who have access to wild growing products is to exploit them as much and as quickly as possible, before someone else does the same.

Advocates of sustainable use as a conservation tool seek to create incentives for users to forego short-term "mining" of renewable resources in favor of managing them for long-term gains. Providing security and exclusivity of access is usually regarded as necessary but is certainly not sufficient. For example, Wunder (2001) found that in the highlands of Ecuador greater security of land tenure actually increased deforestation rates because farmers were more likely to be able to capture the long-term economic benefits associated with logging, followed by cultivation, and then conversion to pasture (see also Gartlan this volume). While many programs or producers claim to practice sustainable use, there are few clear demonstrations of truly sustainable commercial harvesting of noncultivated living organisms over any substantial period. In fact, many critics of the sustainable use concept hold that consumptive use of wild populations of plants or animals is not and cannot be sustainable, but inevitably leads to decline and loss of species and ecosystems, regardless of whether ownership or control is in the hands of government, communities, individuals, or corporations (e.g., see Redford 1992; Robinson and Redford 1994a,b; Barrett and Arcese 1995; Kramer, Van Schaik, and Johnson 1997; Alvard et al. 1997; Redford and Richter 1999; Bennett and Robinson 2000).

Extractive use of biological products may sometimes contribute to biodiversity conservation if it provides the only viable alternative to land transformation. On Hawaii, artifacts made of *Acacia koa* wood sell for thousands of dollars, leading some landowners to allow land previously converted to sugarcane or pasture to return to natural *A. koa* woodland to tap this lucrative market.[6] In this case relatively wealthy landowners are choosing between alternative sources of cash income. By contrast, in a comprehensive analysis of the CAMPFIRE program in Zimbabwe, Murombedzi (1999) found that the substantial income that community landholders derived from trophy hunting fees did not lead them to maintain wildlife habitat. Instead, they usually invested the income in expanding agriculture into the wildlife areas. In this case the incentive of income from wildlife is apparently not sufficient to overcome the bias toward farming as the primary source of livelihood.

2. Utilization of biodiversity within relatively intact natural ecosystems:

While still focusing on "products" of biodiversity rather than the biodiversity itself, economic utilization of biodiversity in situ is in principle less likely to result in overexploitation of individual species than extractive approaches. If successful, these issues can provide incentives for maintaining relatively intact natural ecosystems.

Ecotourism is a popular alternative-income approach because it is seen as environmentally benign and potentially very lucrative, and entry barriers are perceived to be low. The reality can be quite different. For low environmental impact and high socioeconomic reward, tourist numbers and infrastructure development must be kept low and per-tourist expenditures must be high. The pool of clients willing to pay more for less comfort and convenience is limited. Entry barriers are also higher than prospective ecotourism entrepreneurs may perceive because they fail to take into account costs such as provisioning, insurance, marketing, and training. Unfortunately, community-based ecotourism enterprises often fail, or earn only low and unreliable revenues from budget travelers passing through.

Aside from these considerations, ecotourism and biodiversity are not always synergistic. Many sites of high biodiversity richness are poor for tourism because they are remote, dangerous, very uncomfortable, or lack charismatic fauna and flora. Many so-called ecotourism operations also modify the natural habitat: for example, by constructing water holes, burning vegetation, diverting natural streams, etc. The biodiversity impacts of these modifications may be more significant than they seem. For example, constant observation can disrupt the feeding or breeding behavior of some species, and it has been found that the creation of artificial watering holes increases local densities of some plant and animal species at the expense of others.

Sport hunting is a form of tourism because the real value lies in the experience, not in the meat or skin. It can generate considerable profits from areas that offer poor prospects for other uses, including phototourism, and the population impact of well-managed sport hunting can be small because only a few animals (often males) are harvested. Unfortunately, good management has been very hard to achieve, largely because of the high potential for "rent seeking" on the part of regulating officials.

Bioprospecting is seen as biodiversity-friendly because it is the genetic information, rather than the organisms themselves, that is being exploited for profit. However, little money has been generated through bio-prospecting to date, and the level is likely to remain low because of the high level of risk and the limited pool of investors. Also, in the landmark 1991 bioprospecting deal between the Merck and Company and the Institute of Biodiversity in Costa Rica, only 10 percent of the $1 million up-front payment and 50 percent of any royalties from commercially successful products were earmarked for conservation purposes, mainly for biological inventories and training rather than forest protection (Heal 2000).

3. Subsidies and other compensation for biodiversity-friendly land uses:

Subsidies can help to close the gap between private and public interests by encouraging individuals to voluntarily take actions that benefit society but involve costs to themselves. Many governments and other organizations provide subsidies to encourage landowners to adopt environmentally friendlier land-use practices. Examples include the European Union's umbrella agri-environmental subsidies, the Tyr Cwmen system to subsidize conservation of heathlands in South Wales, and the U.S. Department of Agriculture's Wildlife Habitat Incentives Program (WHIP) and the Sustainable Agriculture Research and

Education (SARE) program, which subsidize habitat protection and soil restoration. The scale of such subsidies is enormous. Overall, between 1993 and 1997, fourteen European nations spent an estimated $11 billion for long-term set-asides and forestry contracts covering over 20 million hectares. In the United States, the Conservation Reserve Program spends about $1.5 billion annually to contract for 12–15 million hectares, and at about $8 billion per year, conservation-related subsidies are the third largest agricultural subsidy program (following wheat and corn). According to the Organisation for Economic Co-operation and Development (OECD) reports, conservation-contracting programs are among the fastest growing payments to farmers in high-income countries around the world (OECD 1999; Ferraro and Simpson 2000).

Conservation land-use subsidies are much less common in developing countries, but there are some emerging examples, usually funded by external donors and therefore usually serving both local and global interests. For example, the IDA/GEF-financed El Salvador Environment Project (in preparation) aims to subsidize land users to adopt practices that will generate local and national benefits such as flood protection, and also regional/global biodiversity benefits by creating corridors connecting protected areas to one another and to the Meso-American Biodiversity Corridor.

Because truly sustainable use of biodiversity is rarely a competitive land use on its own, many conservationists promote multiple use by subsidizing (e.g., through free or cheap inputs or technical assistance) commercial land uses that are relatively compatible with maintaining biodiversity assets. Common examples include mixed livestock and game ranching, agro-forestry, and various forms of "conservation farming." These uses inevitably result in some alteration of the natural ecosystem and biological communities in the area, but they can represent very viable options in those areas that are not of sufficient biodiversity priority to justify the costs of full protection.

Markets for "green certified" products such as organic produce, sustainably harvested timber, and shade coffee also represent a form of subsidy, in this case from consumers who are willing to pay a premium price for products produced in particular ways. Again, biodiversity can benefit even where conservation is not the main objective. An example is organic produce, which is mainly driven by health considerations but also benefits biodiversity by reducing the use of pesticides and fertilizers. The main constraint to this approach is the extent to which consumers are willing to pay the premiums (or, as in the case of the various types of "conscience coffees," which of several alternative environmental or social public goods they prefer to subsidize).

4. Direct payment for environmental services (biodiversity conservation as a side benefit):

In this model, "consumers" of environmental services are taxed or contribute voluntarily to generate funds to help maintain those services. Probably the best-known developing country example is the Environmental Services Payment Program in Costa Rica, which pays landowners in key watershed areas to maintain forest or reforest degraded slopes (see also Brandon and O'Herron this volume). The current rate is $40 per hectare based on delivery of four services: mitigation of greenhouse gas emissions, watershed protection, biodiversity protection, and natural scenic beauty. Part of the funding comes from hydroelectric plants and other water-using enterprises. Another case from Costa Rica involves a private contract between the La Esperanza Hydropower Project and the Monteverde Conservation League (MCL), which owns most of the 3,000-hectare watershed for the hydropower project. The hydropower project is paying $10 per hectare for the watershed service, as a means of assisting the MCL to protect the forest effectively (cf. Rojas and Aylward 2001).

An exciting recent development is the emerging global market for carbon emissions reduction units under the Kyoto Protocol of the UN Framework Convention on Climate Change (UNFCCC). This "carbon trading" follows the successful model of sulfur dioxide emissions trading launched in the United States in 1990. One global energy broker (Natsource) has estimated that 55 million tons of greenhouse gases have been traded since 1996, and that the market could expand to $200 billion in the next few years (Foroohar 2001). The linkage with biodiversity comes mainly through trading in "reduction units" for removing carbon from the atmosphere by sequestering it in the form of long-lived organic matter such as trees or soil reservoirs. When coupled with restoration or protection of natural habitats, payment for the service of carbon sequestration can yield biodiversity benefits. Ongoing projects involving developing countries include the following:

- The Costa Rica Environmental Services Program described above, in which a consortium of Norwegian power producers has paid $2 million for 200,000 tons of certified tradable carbon offsets.
- The Belize Rio Bravo Carbon Sequestration Project (implemented by the Program for Belize and The Nature Conservancy), in which corporate investors obtained carbon credits for purchasing approximately 6,014 hectares of endangered forest that were slated to be converted to mechanized agriculture.
- A deal brokered by The Nature Conservancy, in which General Motors will provide $10 million to restore a Brazilian rainfor-

est devastated by water buffalo ranching in exchange for credits for the carbon dioxide that the new forest will absorb over forty years.
- A deal in which the London-based Sustainable Forestry Management gave two Native American tribes in Montana $50,000 to reforest 250 acres devastated by fire in 1994, in exchange for the rights to the estimated 47,972 tons of CO_2 the trees are expected to absorb over the next eighty years. Based on current estimates, this investment could earn over $3 million (or nothing, if the project fails).

The World Bank is also a major player in this arena. In April 2000 the bank helped launch global carbon offset trading through the Prototype Carbon Fund (PCF), which mainly focused on carbon emissions by power producers and industry. It has since also launched a new Biocarbon Fund and a Community Development Carbon Fund, which aim to link carbon offsets with biodiversity conservation impacts and community-level economic development benefits.

5. Direct payment for the service of maintaining natural habitat and/ or conserving biodiversity:

In this final, most direct model, biodiversity is not a side benefit but the main objective. Like the watershed services and carbon offset markets, it is based on the premise that there are willing "buyers" of the biodiversity conservation service. The approach is far from common, but there are some examples, both from industrialized and developing countries. The former mainly target private landowners and endangered species (box 6.2), while the latter emphasize communities and habitats, demonstrating how the direct payment approach can be applied to community-based conservation (box 6.3).

A growing area of direct payment is *conservation concessions,* in which conservation organizations bid against logging companies to win logging concessions and take them off the market. Examples include several conservation deals (e.g., in Guyana and Peru) financed through Conservation International's (CI) Global Conservation Fund (GCF). A World Bank–financed forest conservation project under preparation in Papua New Guinea (PNG) will include establishment of a Conservation Trust Fund to pay communities (the legal owners of indigenous forests in PNG) to forego giving out logging concessions (see Seymour and Dubash 2000). The fact that logging concessions are relatively inexpensive in many parts of the world makes this a potentially viable instrument for large-scale application. For example, in the Guyana case, CI leased a 200,000-acre tract of forest for an application fee of $20,000

Box 6.2. *Direct Payment to Landowners for Conservation of Threatened and Endangered Species*

"Commodity Trading" of the Red Cockaded Woodpecker in the United States: The red cockaded woodpecker is listed under the U.S. Endangered Species Act (ESA), which mandates the development and implementation of a "species recovery plan," including obligatory preservation of critical habitat on private land. The species nests in forests owned by the International Paper Company (IPC), among a few other sites. Under an arrangement with the U.S. Fish and Wildlife Service, IPC is regarded as having met its obligation if it maintains a minimum number of breeding pairs in each area. Any pairs in excess of this number can be "banked," i.e., used by the company to offset ESA requirements on its other holdings, or even "sold" to other landowners to offset their requirements. Through this mechanism, breeding pairs of the woodpeckers have become a valuable commodity, giving the company an incentive to maintain more than the minimum target number. Recent press reports suggest that International Paper has been able to sell the banked "titles" for as much as $200,000 per breeding pair (see Heal 2000).

Increasing Populations of Grassland Birds in the Netherlands: Intensive farming practices in the Netherlands threaten many species of indigenous grassland birds. Farmers in certain crucial environmentally sensitive breeding areas receive financial compensation for using less intensive forms of agriculture that favor the birds but reduce crop production. This approach has been difficult to monitor and enforce and has bred conflict between conservationists and farmers. In recent experiments, some farmers have been paid instead for producing clutches of certain species (higher amounts for the rarest species). This results-oriented approach is proving successful (yielding significantly higher hatching success), is easier to monitor, and is less expensive (about $40 equivalent per clutch compared with $100–$400 per clutch with the approach of compensation for income losses). In this approach, biodiversity becomes one product that farmers choose to produce on the farm (see details in Musters et al. 2001).

Box 6.3. *Direct Payment to Communities for Biodiversity Conservation Services in Developing Countries*

In India, the National Forestry Policy calls for communities to participate in protecting and rehabilitating forests. Under the World Bank–financed Madhya Pradesh Forestry Project, village committees in MP state are signing "memoranda of understanding" (MU) with the Madhya Pradesh Forest Department (MPFD), which allocates a specific area to that community (up to 300 hectares). In addition to the conventional provisions authorizing specified forest uses, these MUs provide for payment to communities for carrying out specified forest protection activities such as digging cattleproof trenches and posting community guards. The amount paid is related to the area of forest land being protected and the per hectare cost to MPFD of protecting that area in the absence of community participation. Funds are released only after the district forest officer has ascertained that the designated area has been satisfactorily protected during the previous year.

In Burkina Faso, under a recently approved GEF-financed project (PRONAPE), local communities will become concessionaires on government-owned land in exchange for taking on stewardship responsibilities. As concessionaires, the communities will be able to contract with safari hunting and tourism operators to earn revenues. Recognizing that such revenues are likely to take years to materialize, however, the project will begin by providing direct payment for land stewardship into a community trust.

In South Africa, the Working for Water program pays local entrepreneurs for conservation services in the form of clearing of invasive alien vegetation. This program is highly successful in part because it serves simultaneously as an employment and entrepreneurship capacity-building program for the poor.

and an initial rate of fifteen cents an acre annually (see Wilson 2002). However, conservation bidders may need to cover not only the direct value of the timber concession, but also opportunity costs such as taxes and employment that the logging alternative would provide. There can also be legal obstacles, because national law may, for example, require that concessionaires actually carry out logging or risk losing the concession rights. This was the case in the GCF initiative in Peru, but was resolved when the government passed a new forestry law that provided for "conservation concessions."

The direct payment approach is often perceived to be very expensive. However, Simpson and Sedjo (1996) and Ferraro and Simpson (2000) have presented persuasive arguments that it is no more expensive, and in fact is likely to be more cost-effective, than conventional project-based approaches. Taking a forest conservation project in Madagascar as an empirical example, they estimated that the cost of protecting the forest through indirect approaches (such as subsidizing beekeeping) was more than twelve times the cost of a direct subsidy, and more than 350 percent higher than the opportunity costs (it would be far cheaper to buy the land outright). While probably the first attempt to model this comparison with respect to biodiversity, their conclusions echo those of Faeth et al. (1991), who demonstrated that direct subsidies would be much more efficient than the existing indirect crop subsidy programs used to help small farmers in the United States.

Ferraro and Simpson (2000) also analyzed another type of indirect intervention commonly found in conservation projects in developing countries: support for agricultural intensification. This approach is aimed at producing higher yields and incomes on existing cultivated land, thus reducing peoples' need and motivation to expand further into natural habitats. They concluded that this too would likely cost more than direct conservation payments. Aside from this, several studies have shown that support for agricultural intensification, like land security, can have the opposite effect: where there is good access to markets, increasing the profitability of agriculture by subsidizing agricultural inputs or infrastructure may stimulate people (residents and newcomers) to expand the area of cultivation even more to take advantage of the opportunity (cf. Angelsen and Kaimowitz 1999; Helmuth 2001).

NONECONOMIC INCENTIVES

Economic incentives are not always the only factor in landowners' decisions to conserve biodiversity. Langholz (1996, 1999) found that many private landholders in Costa Rica placed economic incentives below considerations such as enhanced protection of their assets and social prestige as a reason for choosing to put their land into a private reserve scheme. In the Ben Udyam sustainable forest management project in Nepal, obtaining longer-term, more secure use rights to forest areas was a stronger incentive for participation among community members than was income generation (Margoluis and Margoluis 2000).

"Homesteaders" laying claim to land by converting it from a natural state to another use is a familiar story. Even in Costa Rica, which has some of the most conservation-friendly policies in the world, the law recognizes occupancy with "improvements" as a step in gaining land title, and the government will provide compensation for clearings, buildings, and other investments if the land is later expropriated. Much less common is the reverse situation, in which people can lay claim to land by restoring natural habitat. This was, however, the principle behind the Ndirande Mountain Rehabilitation Project in Malawi, funded by the U.S. Agency for International Development, under its COMPASS program. In an effort to restore the badly degraded Ndirande Mountain Forest Reserve near Blantyre, the Forestry Department demarcated plots of about one hectare, to be allocated to individual community members by a local committee. To retain the plots, beneficiaries were responsible for rehabilitating them, including planting both indigenous and exotic tree species recommended by the committee. The project was initially successful, as the new plot owners purchased and planted 83,000 tree seedlings, with an estimated 80 percent survival rate. Unfortunately, the project later collapsed as a result of alleged corruption on the part of some members of the project committee who were accused of diverting funds and of failing to consult with other community members.

Issues in Applying Nonproject Approaches in Developing Countries

Land purchase, easements, subsidies, and direct payments are the predominant models for conservation in industrialized countries but are relatively rare in developing countries. Some of the likely reasons lie in the nature of developing country economies and legal and social systems.

For example, nonproject approaches are easier to use when land is privately owned, either by individuals or by corporations, than when it is held communally and/or without a legal title. In southern Africa, where a higher proportion of land is privately owned and land rights are clearer than elsewhere on the continent, easements and other contractual models are more common. In communal situations, there must first be a reasonably effective, legally recognized organizational structure to negotiate and implement contractual arrangements. Even so, a few individuals who refuse to comply with the agreement can undermine the whole initiative. Even in Namibia and South Africa, where wildlife conservancies are increasingly common on privately owned land, conservancies involving communally held land have often been difficult to put in place. There are, however, some successful examples, such as the Mdluli Tribal Trust, which has obtained title to traditional community lands inside the Kruger National Park in South Africa and is maintaining it as a wildlife area.

Another significant complication is that rural populations in developing countries mostly earn their living directly from subsistence agriculture, livestock, or extraction of natural resources. When land is dedicated to conservation through purchases, easements, or direct payments, people may become dispossessed if their land rights are not secured. This amounts to involuntary resettlement, which conservation groups and official donors strongly prefer to avoid (see Singh and Sharma this volume). Even if the local people become the direct beneficiaries of conservation-related payments, large numbers of people may become unemployed (direct conservation activities will rarely employ more than a small fraction of them). This is likely to lead to social disruption, and many of the people are likely to continue their previous activities covertly in any case. Therefore, a land acquisition or direct conservation payment initiative in a developing country may still need to be part of a broader package of development assistance, including measures to generate alternative employment (see box 6.3). In contrast to the typical ICDP, however, this employment would be focused away from the conservation area and would not normally involve use of its biodiversity.

In industrialized countries, the financial incentives provided to landowners for entering into easements or changing land-use practices are often partly or entirely in the form of property tax breaks. This is not meaningful in countries where property taxes do not exist or routinely go unpaid. Therefore, the financial incentives must be provided as direct cash transfers, which are usually more difficult to obtain than tax relief. There is also the issue of enforcement and timing of payments. The weak judicial systems typical of many developing countries can make it difficult to obtain and enforce long-term legal commitments, so the most likely method is to provide the payments over time. However, it may be necessary to front-load payments, particularly if the short-term opportunity costs to the landholders are high. A balance must be struck between effective short-term incentives and sustainable long-term incentives.

Finally, the donors who provide most conservation funding tend to feel a responsibility to ensure that the funds they provide are used in socially equitable and constructive ways. This is difficult to do when landholders receive direct cash payments and are free to use them as they wish. Even where direct payment is being tried, the tendency is to introduce project-related mechanisms to control how the beneficiaries use the funds. For example, in the Madhya Pradesh, India, example (box 6.3), only one-third of the payment to communities is untied; two-thirds must be used to finance approved community development plans. The wildlife conservation leasing program in Kenya provides an innovative alternative. The local community puts a high value on education, but children are often pulled out of school because there is no money to pay school fees. While recipients are not told what they must do with the lease payments, the payments are made in installments at the time that school fees are due, and are usually used for this purpose. The result is that school enrollment (particularly of girls) is higher in areas participating in the program than in neighboring areas.

The constraints described here need to be addressed, but they are not insurmountable, as is demonstrated by the numerous ongoing nonproject, incentive-based initiatives in developing countries. While these approaches will not work in all cases, the same is clearly true for the project-based approaches that currently dominate conservation practice and funding. In most cases both project- and nonproject-based approaches merit consideration.

Conclusion: Making Community-Based Conservation Work

The premise of community-based conservation is that, in order to succeed in the long term, biodiversity conservation must yield economic benefits for local communities. This paper fully supports that premise, particularly when it comes to biodiversity outside protected areas, where private or communal landholders make the daily land-use and resource-use decisions that determine whether biodiversity will survive or be lost. If they do not see biodiversity conservation as a priority, the land will be converted to other uses.

While this is the principle behind ICDPs, in practice it has proven difficult to make direct and concrete linkages between the conservation and development objectives. As a result, few ICDPs can demonstrate clear biodiversity impacts, and even the extent and sustainability of their development impacts are often debatable.

The problem lies not in the concept of integrating conservation and development, but in the reliance on the "project model," which fails to create real incentives for achieving conservation outcomes. The alternative explored in this paper is to use nonproject approaches to provide direct, positive incentives for conservation. The beneficiaries (actually, service providers) can then use this income to improve their living conditions and invest in their own development. Project-based assistance can be used to help communities overcome specific constraints and obstacles, but cannot substitute for direct incentives as a means to motivate people to maintain biodiversity on their lands.

Endnotes

1. International Bank for Reconstruction and Development/International Development Association.
2. Global Environment Facility (a grant fund administered by the World Bank and implemented by the World Bank, the UN Development Program, and the UN Environment Program).
3. For purposes of this discussion, the term *land* includes territory that is partially or entirely underwater, e.g., wetlands, lakes, coastal areas.
4. For discussion, see Steinberg 1998.
5. S. Noemdoe, Republic of South Africa Working for Water Program, personal communications.
6. Monoculture *A. koa* plantations have apparently not yet emerged.

References

Alvard, M. S., J. G. Robinson, K. H. Redford, and H. Kaplan. 1997. The sustainability of subsistence hunting in the neotropics. *Conservation Biology* 11:977–982.

Angelsen, A. and D. Kaimowitz. 1999. Rethinking the causes of deforestation: Lessons from economic models. *The World Bank Research Observer* 14 (1):73–98.

Barrett, C. B. and P. Arcese. 1995. Are integrated conservation-development projects (ICDPs) sustainable? On the conservation of large mammals in sub-Saharan Africa. *World Development* 23 (7):1073–1084.

Bennet, E., H. Eves, J. Robinson, and D. Wilkie. 2002. Why is eating bushmeat a biodiversity crisis? *Conservation in Practice* 3 (2):28–29.

Bennett, E. L. and J. G. Robinson. 2000. *Hunting of Wildlife in Tropical Forests: Implications for Biodiversity and Forest Peoples.* Environment Department Papers no. 76. Washington, D.C.: The World Bank.

Brandon, K., K. H. Redford, and S. E. Sanderson, eds. 1998. *Parks in Peril: People, Politics, and Protected Areas.* Washington, D.C.: Island Press.

Bruner, A. G., R. E. Gullison, R. E. Rice, and G. A. B. da Fonseca. 2001. Effectiveness of parks in protecting tropical biodiversity. *Science* 291 (5501):125–128.

Cabarle, B., J. Bauer, P. Palmer, and M. Symington. 1992. *BOSCOSA, The Program for Forest Management and Conservation on the Osa Peninsula, Costa Rica*. Project Evaluation Report. San José, Costa Rica: USAID.

Chichilnisky, G. and G. M. Heal. 1998. Economic returns from the biosphere. *Nature* 391:629–630.

Donovan, R. 1994. Forest conservation and management through local institutions (Costa Rica). In D. Western and R. M. Wright, eds., *Natural Connections: Perspectives in Community-Based Conservation,* 215–233. Washington, D.C.: Island Press.

Faeth, P., R. Repetto, K. Kroll, Q. Dai, and G. Helmers. 1991. *Paying the Farm Bill: U.S. Agricultural Policy and the Transition to Sustainable Agriculture.* Washington, D.C.: World Resources Institute.

Ferraro, P. J. and R. D. Simpson. 2000. *The Cost-Effectiveness of Conservation Payments.* Resources for the Future Discussion Paper 00-31. Washington, D.C.: Resources for the Future.

Foroohar, R. 2001. The green game. *Newsweek,* 27 August, 62.

Hackel, J. D. 1999. Community conservation and the future of Africa's wildlife. *Conservation Biology* 13 (4):720–734.

Hardner, J. and R. Rice. 2002. Rethinking green consumerism. *Scientific American,* May 2002, 89–95.

Heal, G. M. 2000. *Nature and the Marketplace: Capturing the Value of Ecosystem Services*. Washington, D.C.: Island Press.

Helmuth, L. 2001. Economic development: A shifting equation links modern farming and forests. *Science,* 12 November, 1283.

Hulme, D. and M. Murphree. 1999. Communities, wildlife, and the "new conservation" in Africa. *Journal of International Development (Policy Arena)* 11:277–285.

Innes, R., S. Polasky, and J. Tshirhart. 1998. Takings, compensation, and endangered species protection on private lands. *Journal of Economic Perspectives* 12 (3):35–52.

Jensen, D. B., M. Torn, and J. Harte. 1990. *In Our Own Hands: A Strategy for Conserving Biological Diversity in California.* Berkeley: California Policy Seminar, University of California.

Kellert, S. R., J. N. Mehta, S. A. Ebbin, and L. L. Lichtenfeld. 2000. Community natural resource management: Promise, rhetoric, and reality. *Society and Natural Resources* 13:705–715.

Kiss, A. 1999. Making community-based conservation work. Paper delivered at annual meeting of the Society for Conservation Biology, July, at Greenbelt, Md.

Kramer, R., C. van Schaik, and J. Johnson, eds. 1997. *Last Stand: Protected Areas and the Defense of Tropical Biodiversity*. Oxford: Oxford University Press.

Langholz, J. 1996. Economics, objectives, and success of private nature reserves in sub-Saharan Africa and Latin America. *Conservation Biology* 10:271–280.

Langholz, J. 1999. Conservation cowboys: Privately owned parks and the protection of biodiversity in Costa Rica. Ph.D. diss., Cornell University.

Margoluis, R. and C. Margoluis. 2000. *Lessons from the Field: Linking Theory and Practice in Biodiversity Conservation.* Washington, D.C.: Biodiversity Support Program.

Murombedzi, J. C. 1999. Devolution and stewardship in Zimbabwe's CAMPFIRE programme. *Journal of International Development (Policy Arena)* 11:287–293.

Murray, W. 1995. Lessons from 35 years of private reserve management in the USA: The preserve system of The Nature Conservancy. In J. McNeely, ed., *Expanding Partnerships in Conservation,* 197–205. Washington, D.C.: Island Press.

Musters, C. J. M., M. Kruk, H. J. De Graaf, and W. J. Ter Keurs. 2001. Breeding birds as a farm product. *Conservation Biology* 5 (2):363–369.

Newmark, W. D. and J. L. Hough. 2000. Conserving wildilfe in Africa: Integrated conservation and development projects and beyond. *BioScience* 50 (7):585–592.

Oates, J. 1999. *Myth and Reality in the Rain Forest: How Conservation Strategies Are Failing in West Africa.* Berkeley and Los Angeles: University of California Press.

OECD (Organisation for Economic Co-operation and Development). 1999. *Handbook of Incentive Measures for Biodiversity: Design and Implementation.* (Summary case studies accessible at http://www.oecd.org/env/eco/biod.htm) (August 2001).

OECD (Organisation for Economic Co-operation and Development). 2000. *Aid Targeting the Rio Conventions: First Results of a Pilot Study.* Paris: Organisation for Economic Co-operation and Development.

Redford, K. H. 1992. The empty forest. *BioScience* 42:412–422.

Redford, K. H. and B. D. Richter. 1999. Conservation of biodiversity in a world of use. *Conservation Biology* 13:1246–1256.

Robinson, J.G. and K.H. Redford. 1994a. Community-based approaches to wildlife conservation in neotropical forests. In D. Western and R. M. Wright, eds., *Natural Connections: Perspectives in Community-Based Conservation,* 300–322. Washington, D.C.: Island Press.

Robinson, J. G. and K. H. Redford. 1994b. Measuring the sustainability of hunting in tropical forests. *Oryx* 28:249–256.

Roe, D., J. Mayers, M. Grieg-Gran, A. Kothari, C. Fabricius, and R. Hughes. 2000. *Evaluating Eden: Exploring the Myths and Realities of Community-Based Wildlife Management.* Series no. 8: Series Overview. London: International Institute for Environment and Development (IIED).

Rojas, M. and B. Aylward. 2001. The case of La Esperanza: A small, private hydropower producer and a conservation NGO in Costa Rica. Land-Water Linkages in Rural Watersheds Case Study Series. Rome: Food and Agriculture Organization of the United Nations (FAO).

Satchell, M. 2000. Hunting to extinction. *U.S. News and World Report,* 9 October. http://www.well.com/user/dafidu/hunting.html (September 2001).

Seymour, F. and N. Dubash. 2000. *The Right Conditions: The World Bank, Structural Adjustment, and Forest Policy Reform.* Washington, D.C.: World Resources Institute.

Simpson, R. D. and R. A. Sedjo. 1996. Paying for the conservation of endangered ecosystems: A comparison of direct and indirect approaches. *Environment and Development Economics* 1:241–257.

Steinberg, P. F. 1998. Defining the global biodiversity mandate: Implications for international policy. *International Environmental Affairs* 10 (2):113–130.

Wells, M., S. Guggenheim, A. Khan, W. Wardojo, and P. Jepson. 1999. *Investing in Biodiversity: A Review of Indonesia's Integrated Conservation and Development Projects*. Washington, D.C.: The World Bank.

Wilson, E. O. 2002. *The Future of Life*. New York: Knopf.

Wunder, S. 2001. *The Economics of Deforestation: The Example of Ecuador*. Hampshire, U.K.: Palgrave Macmillan.

7

Yellowstone: A 130-Year Experiment in Integrated Conservation and Development

Dennis Glick and Curtis Freese

Introduction

Yellowstone National Park, created by an act of the U.S. Congress in 1872, is generally recognized as the world's first national park. Its incredible collection of geothermal features inspired early explorers, politicians, and officials of the Northern Pacific Railroad to seek the permanent protection of the Yellowstone Plateau, an act that has inspired conservationists throughout the world.

Yellowstone may be not only the world's first national park, but also the first integrated conservation and development project (ICDP). As with many ICDPs, nature tourism was promoted as the economic engine for park establishment and maintenance. Recognizing the economic potential the region held for tourism, an agent of the Northern Pacific Railroad Company sent a note to Washington, D.C., with the suggestion: "Let Congress pass a bill reserving the Great Geyser Basin as a public park forever" (Sax 1980:6). Based on urging by the railroad, which had a vested economic interest in establishing the park, and reports from expeditions extolling the wonders of the area, the Congress soon established Yellowstone National Park.

The ICDP concept in Yellowstone was embedded in the idea that tourists transported to the park by the railroad would support concessionaires and that these business owners, through their franchise fees, would finance resource protection and development. But the railroad line, concessionaires, and tourists failed to materialize as quickly as anticipated, and in 1878 the U.S. Congress appropriated the first federal funds for the park's development (Haines 1977). Thus ended the first ICDP in Yellowstone—part success, part failure. Since the establishment of Yellowstone, other adjoining lands have been set aside for public use and new land management agencies created. Many of these agencies also have elements of integrated conservation and development embedded in their founding legislation and policies.

The United States has been a leader in the field of conservation, as exemplified by the creation of Yellowstone Park. But has it been a leader in integrating conservation and development? In this paper we examine whether, after 130 years, progress has been made in applying the ICDP model to the cradle of wildland conservation, the Greater Yellowstone Ecosystem, and propose priority issues that must be addressed if the ICDP model is to be applied effectively in the future.

Our Approach

An ICDP attempts to create a mutually beneficial relationship between the socioeconomic well-being of people and the conservation of biodiversity for a particular biogeographic area or wildland. The words "integrated" and "project" imply a planned and purposeful linkage of socioeconomic benefits and biodiversity benefits, rather than the two occurring independently of each other.

If we use these criteria, Greater Yellowstone currently has few ICDPs, and all on a small scale. Interestingly, while the "ICDP" label is fairly common in international conservation circles, it is largely unknown to conservationists in the United States who have not worked internationally. We know of no project that carries that label in Greater Yellowstone, though the principles of ICDPs are certainly relevant to efforts to protect the ecological integrity of this region.

If, however, we broaden the concept of an ICDP, we begin to see interactions between socioeconomic interests and biodiversity in Greater Yellowstone that may be instructive for understanding the conditions under which development and conservation may be mutually beneficial. Thus, we examine largely unplanned as well as planned linkages of biodiversity conservation and socioeconomic development. To do this, we review how the monetary and nonmonetary values of biodiversity in Greater Yellowstone affect biodiversity and, conversely, how the conservation of biodiversity affects these values. We then evaluate policies and behavior in both the public and private sectors and their relative effects on development and conservation. Our primary interest is in actions that operate, or have the potential to operate, at a large scale within the Greater Yellowstone, rather than cases that have a minor or local impact. In the process, we try to identify where an ICDP for the Greater Yellowstone Ecosystem should focus its efforts.

Overview of the Greater Yellowstone Ecosystem

Within its original 1872 borders, Yellowstone National Park encompassed 800,000 hectares of roadless wildland, surrounded by millions of additional hectares of uncharted wilderness. As early as 1882, conservationists realized that, despite its size, Yellowstone was not large enough to preserve many of the wildlife species it was created to protect. General Philip Sheridan, when touring the park, was dismayed by the impact of market hunting on big game just beyond park borders. Shortly thereafter, President Benjamin Harrison created the Yellowstone National Park Timberland Reserve, the first national forest in the country (Haines 1977). Today, Yellowstone National Park is part of a complex of public lands that includes seven national forests (4,464,000 ha), two national parks (1,008,000 ha), state of Montana lands (144,000 ha), U.S. Bureau of Land Management lands (72,000 ha), and three national wildlife refuges (36,000 ha), for a total of 5,724,000 hectares (Glick, Carr, and Harting 1991), all within the Greater Yellowstone Ecosystem (figure 7.1).

The name "Greater Yellowstone Ecosystem" was coined as a way of defining the range of the grizzly bear *(Ursus arctos)* population in the region (Craighead 1991). The geographic definition of the Greater Yellowstone varies from an area of about 7.2 million hectares (Glick, Carr, and Harting 1991) to 10.4 million hectares (Noss et al. 2001). Defining ecosystem boundaries is not a precise science, but most people recognize Greater Yellowstone as a region centered on the Yellowstone Plateau, surrounded by mountain ranges such as the Gallatin, Absaroka, Gros Ventre, Wind, Beartooth, and Salt River. Three of the West's major river systems—the Green, Yellowstone, and Snake—flow from this area. We generally use the 7.2-million-hectare definition here, of which about 80 percent is public land and 20 percent private.

The landscape is crisscrossed by dozens of political and administrative boundaries. Portions of the Greater Yellowstone overlap three states—Montana, Wyoming, and Idaho—and twenty counties within these states. In addition, more than twenty-five federal, state, and local agencies have jurisdiction over various pieces of the Greater Yellowstone.

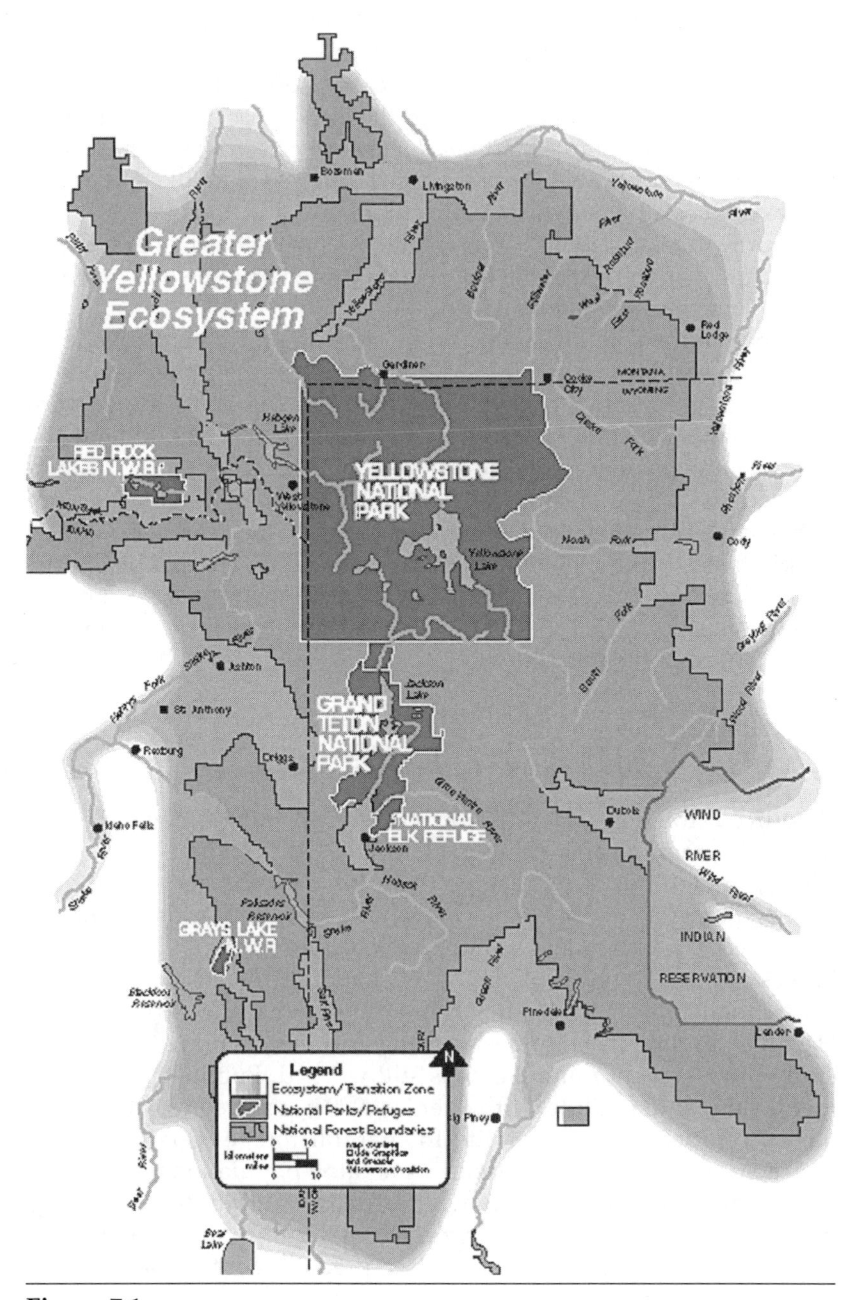

Figure 7.1.
The Greater Yellowstone Ecosystem, U.S.A.

Nearly every species of wildlife that existed when Yellowstone Park was created is still there today. Several species listed nationally as rare or endangered, like the grizzly bear and the bald eagle *(Haliaeetus leucocephalus)*, are relatively common in the park. Species that had been extirpated, like the gray wolf *(Canis lupus)*, have been successfully reintroduced. Many of the natural ecological processes that keep Greater Yellowstone healthy, such as wildfire, predation, and floods, are allowed to occur with relatively little human interference. This is in marked contrast to much of the rest of the contiguous United States, which, even on public lands, is intensely managed.

Although the core of Greater Yellowstone is sparsely inhabited, the privately owned river valleys and the fringes of the public lands are experiencing explosive population growth. In the 1990s the western United States was the fastest-growing region in the nation, and the Rocky Mountain region was the fastest-growing area in the West (growing twice as fast as the nation as a whole). Greater Yellowstone has not been spared from this influx of new residents. In the year 2000, 365,000 people lived in the region's twenty counties (U.S. Bureau of Census 2001). Although most residents reside in small towns and a couple of midsize cities (i.e., 25,000–50,000 residents), rural areas proportionally saw the most population growth.

Threats to Biodiversity in the Region

Resource extraction such as timber cutting, oil and gas development, livestock grazing, and mining all impact the region's ecological health, though not to the degree they did a decade ago. More than 12,000 kilometers of roads have been built on public lands (primarily for timber extraction) and billions of board feet of timber have been harvested. But logging in the seven national forests has declined substantially, particularly after 1990 (Hansen et al. 2002) (figure 7.2). Hardrock mining has also been greatly curtailed, with few of the literally thousands of mining claims still worked. Public-land grazing is still a threat to wildlife, with over 200,000 head of cows, sheep, and horses competing with wildlife for natural forage. Confrontations between livestock and expanding predator populations, including the recently reintroduced gray wolf, usually result in the removal or death of wildlife. Again, however, closer scrutiny of grazing practices and growing support for a halt to predator control has reduced livestock impacts.

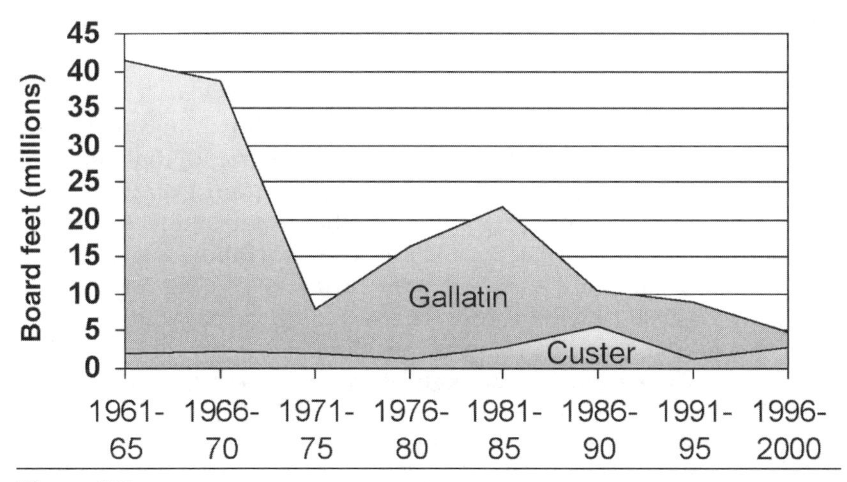

Figure 7.2.
Timber harvest levels in the Gallatin and Custer National Forests in the Greater Yellowstone Ecosystem, 1961–2000. *Source:* USDA Forest Service, 2001.

The threat of greatly increased oil and gas exploration and development is looming on the horizon. Previous attempts to drill for oil and natural gas on public lands have generally been kept in check by environmentalists, but high fuel prices and electricity bills could erode public opposition to drilling on public and private land.

Two emerging and important environmental issues can be described as "loving-it-to-death" problems. One, discussed later, is the conversion of rural agricultural and natural areas to low-density subdivisions (rural residential developments where houses are constructed at a low density, such as one per 1–5 ha), which can degrade terrestrial and aquatic habitats, increase human-wildlife confrontation, and reduce tolerance for natural processes such as wildfire. Also of concern is the exponential increase in outdoor recreation and tourism-related developments. Visitation to Yellowstone National Park now tops 3 million people annually, but the park's budget and infrastructure have not kept pace with the growth in tourist numbers. In the national forests, a proliferation of off-road vehicle use on nonwilderness lands shatters the stillness of even once remote regions. Mountain bikers, hikers, and backcountry skiers are increasing in numbers. Several ski resorts have expanded in recent years, gobbling up important wildlife habitat. While all of these site-specific problems are of concern, their cumulative impact is of even greater concern.

Socioeconomic Development Trends

One of the greatest challenges for a Greater Yellowstone ICDP is the task of defining socioeconomic goals for the region. The economy of the twenty-county region has been undergoing rapid change, from one based on extractive industries and agriculture to one based on more diversified sources of income. At the center of this change is rapid growth in the human population. The population of the Greater Yellowstone increased 55 percent from 1970 to 1997. The populations of the five fastest-growing counties in the region increased 107 percent over this period. The slowest-growing counties, largely on the eastern periphery of the region in the high plains, grew less than 15 percent over the twenty-seven-year period (Hansen et al. 2002).

From 1970 to 1995, 51 percent of all growth in personal income in the Greater Yellowstone was from retirement income, past investments, and rent; 41 percent was from service-related industries; 11 percent from government; 6 percent from construction; 3 percent from manufacturing not related to resource sectors; and 1 percent from resource extractive industries. Income from agriculture declined both in absolute and relative terms (Rasker and Alexander 1997). The proportion of total personal income from mining, oil, gas, farming, and ranching activities declined from 19 percent in 1970 to 6 percent in 1995 (Hansen et al. 2002).

A common belief is that the new service jobs being created in the Greater Yellowstone consist primarily of low-paying jobs in the tourist industry (e.g., workers in fast-food restaurants, motels, and ski resorts), compared to the relatively good pay of the extractive industries such as mining and forestry. Salaries in the service industry in this region, however, are 20 percent higher than the average for all industries in the Greater Yellowstone. The reason for this is that many of the so-called service sector jobs include professionals such as physicians, computer programmers, and architects (Rasker and Alexander 1997).

What do these changes mean in terms of basic economic indicators for the twenty counties of the Greater Yellowstone? Overall, the counties showed a modest increase in per capita income between 1980 and 1995 (table 7.1). The region, however, lags behind the United States as a whole in both absolute terms (i.e., income level) and the rate of income growth during this period. This disparity is even greater if one examines the average earnings per job. Average earnings in the Greater Yellowstone in 1995 were $18,259, a decline of 14.9 percent from 1980 (adjusted for inflation), whereas average earnings in the United States in 1995 were $28,910, an increase of 6.6 percent from 1980 (Rasker and Alexander 1997). Thus, despite the growth in relatively high-paying

Table 7.1 *Changes in personal economic indicators in the Greater Yellowstone, 1980–1995*

Per capita personal income (in constant 1995 $)				*Average annual*	*% below poverty*	
Place	*1980*	*1995*	*% change*	*unemployment rate, 1990s (%)*	*1980*	*1990*
Greater Yellowstone	16,743	17,635	5.3	4.9	13	13
Teton County, Wyoming	26,195	39,134	49.4	2.5	8	8
Teton County, Idaho	14,315	13,370	–6.6	3.9	18	18
United States	18,551	23,196	25.0	6.3	12	13

Source: Rasker and Alexander 1997.

service jobs, overall, wages have declined, as many people who migrate into the region are willing to take low-paying jobs in order to live in a high-quality environment.

It seems clear that population migration into the Greater Yellowstone is not a result of economic opportunity. Rather, as Rasker and Alexander (1997:25) state in commenting on growth in the western United States in general, "Today much of the population growth can be explained instead by 'migration first—then jobs,' where people first decide where they want to live, and then either look for a job, create jobs for themselves, or live off investment and retirement income." More specifically, several studies confirm that, in the western United States, counties with ready access to national parks, wilderness areas, and other wildlands and outdoor recreation opportunities, compared to those without these amenities, experience substantially higher rates of population growth, higher income and growth in employment, and lower levels of unemployment (Rudzitis and Johansen 1991; Lorah 1996; Rasker and Hackman 1996; Rudzitis 1999; Rasker and Hansen 2000).

Most of the new businesses in the region are "footloose" in that they are small and the owners are able and prepared to locate wherever they wish (Rasker and Glick 1994). A recent survey of 500 business owners and managers in the northern portion of the Greater Yellowstone found that cultural and natural amenities were more important than traditional "profit maximizing" reasons (e.g., tax incentives, labor costs) for both attracting new businesses and retaining existing ones. Of fifteen business location variables surveyed, "scenic beauty" was the highest ranking, followed by "quality environment" (Johnson and Rasker 1995). Analysis by Rasker and Hansen (2000) revealed that up to 60 percent of the variability in population growth among counties of the Greater Yellowstone could be explained by ecological variables. The role of ecological variables is, in part, captured by what is known as the "one-hour rule" among newcomers to the region: they want to work within an hour's drive of good fishing, hunting, skiing, and hiking (Hansen et al. 2002).

Some socioeconomic variables are also significant, particularly education of the workforce, percent employment in business services, and the presence of an airport with daily commercial service to larger markets. Combined, ecological and socioeconomic variables accounted for 79 percent of the variation in population growth. Rasker and Hansen (2000:9) sum up the effects of these variables as follows: "Most of the GYE counties that are still dependent upon resource extraction are nearly stagnant in population growth and economic expansion. In contrast, those counties featuring both high natural amenities and socioeconomic traits associated with the new economy are among the fastest growing in the country." Despite the economic importance of natural amenities, many regional economic development organizations such as local chambers of commerce as well as most elected officials seem to be either unaware or unwilling to accept this reality.

Although recreation and tourism are only part of a much more diverse service industry in Greater Yellowstone, in the national park gateway communities of Jackson Hole, Gardiner, Cooke City, and West Yellowstone, they are the major economic activities. There are a growing number of tour operators who embrace the concept of ecotourism. This is particularly true of the now multimillion-dollar wildlife-watching industry, such as the wolf observation tours that have sprung up since the introduction of wolves to Yellowstone Park in 1996. In some cases the proprietors of these ecotourism businesses have been active and outspoken proponents of environmental protection efforts.

The large-scale and commercially lucrative tourism industry of the Greater Yellowstone can create tensions between tourist businesses and conservationists. The current controversy over the use of snowmobiles in Yellowstone National Park is a case in point. Snowmobile rentals in some gateway communities, such as the town of West Yellowstone, are a significant economic activity during the winter months. The use of snowmobiles in Yellowstone has mushroomed since they were first allowed in the park in 1963. The park now averages nearly 75,000 snowmobiles a winter. Research has documented the significant impacts of these two-stroke machines on air quality, on wildlife, and on the aesthetic experience of Yellowstone visitors in the winter (U.S. National

Park Service 2000). This debate has pitted many in the gateway communities against the National Park Service and against environmental organizations advocating a ban on snowmobiles.

The sport fishing and hunting industry is large and diversified, ranging from fee hunting and fishing operations on private land to guided expeditions on public land. Added to this mix are tens of thousands of individuals and families venturing out on their own to hunt an elk *(Cervus elaphus)* or hook a trout. While healthy fish and wildlife populations depend on healthy habitat, many of the professional hunting and fishing guides, and many hunters and fishers, have not been active in conservation issues. Indeed, some have opposed the reintroduction of extirpated species such as wolves and cutthroat trout *(Salmo clarkii)*, which could affect game species like elk and rainbow trout *(Salmo gairdnerii)*.

The Major Players Affecting ICDPs in Yellowstone

Public Land Management Agencies

Greater Yellowstone is somewhat unusual in that the missions of the government agencies charged with managing large portions of the area (U.S. Forest Service, National Park Service, U.S. Fish and Wildlife Service, and others) include elements of integrated conservation and development principles. In theory, if not in practice, ICDPs could be facilitated not just by NGOs but also by large, relatively well-funded government agencies, which have the letter of the law behind their conservation and development activities. The potential exists for one of the planet's grandest ICDPs if these agencies could first reform their own land management practices in a manner that sustains ecological processes and if they successfully coordinated their actions with those of other public and private land managers. Thus far they have been less than successful at both of these endeavors.

The National Park Service, which manages Yellowstone and Grand Teton National Parks, is governed by an Organic Act, which states that "the purpose of parks is to conserve scenery and the natural and historic objects and the wildlife therein and to provide for the enjoyment of the same in such a manner and by such means as will leave them unimpaired for the enjoyment of future generations." The dual goals of this mission—resource protection and visitor use—have led to numerous conflicts in Yellowstone and Grand Teton as well as many other parks.

The U.S. Forest Service, which manages seven national forests in the Greater Yellowstone, operates under the 1960 Multiple Use–Sustained Yield Act, the Forest and Range Renewable Resources Act, the 1976 National Forest Management Act, and the Wilderness Act (Harting and Glick 1994). A review of these policies underscores the potential of implementing forest management activities that provide sustainable forest products while protecting and even enhancing the natural environment. The fact that the western boundary of Yellowstone Park is clearly discernible from satellite imagery because of the massive clearcuts on the adjoining Targhee National Forest underscores the failure of these policies. The U.S. Fish and Wildlife Service and its National Wildlife Refuge System are guided by various mandates, such as the 1956 Fish and Wildlife Act, the Refuge Administration Act, Executive Order 12996, and the 1973 Endangered Species Act. Although wildlife conservation is the dominant purpose of these mandates, the rights of citizens to benefit from hunting, fishing, and other recreational benefits are clearly indicated as important goals (U.S. Fish and Wildlife Service 2001). The Montana Department of Fish, Wildlife, and Parks and the Wyoming and Idaho Departments of Fish and Game all have historically managed wildlife with a strong emphasis on satisfying residents' demands for quality fishing and hunting and on generating revenues from out-of-state hunters and fishers. Again, all of these agencies have within their mandate elements of ICDPs.

PRIVATE LAND OWNERS

Most of the nearly 1.6 million hectares of private land in the Greater Yellowstone is in agriculture (farms and ranches). The traditional farming and ranching communities pride themselves on being good stewards of the land. Again, though their operations are not called ICDPs, many would describe them in terms that reflect a concern for the environment and long-term economic sustainability. In practice, however, many of these farms and ranches have had significant environmental impacts and have been less than economically viable. A study conducted in the late 1980s found that nearly 40 percent of the public rangelands in the Greater Yellowstone, which are leased to private landowners for livestock grazing, were in very poor, poor, or fair ecological condition (Greater Yellowstone Coordination Committee 1987).

In recent years an increasing number of these farms and ranches have sought to diversify their operations and add value to agricultural products. Fee hunting and fishing have been pursued by an increasing number of landowners. Although these ventures can generate considerable revenue, their ultimate impact on improving the ecological integrity of privately owned lands and adjoining public lands is largely unknown. A few livestock operators have also experimented with raising "predator-friendly" wool and beef, which guarantees the purchaser of these products that nonlethal predator control was used to protect herds from wild predators. These operations have shown some signs of success, but on a very limited scale.

Low commodity prices for agricultural products, skyrocketing land values for rural residential development, and a lack of regulatory controls over rural land use have set the stage for the dramatic conversion of farms and ranches to urban sprawl and subdivisions in many parts of Greater Yellowstone. Four hundred thousand hectares of private land have been subdivided for rural residential development in the region (Glick, Carr, and Harting 1991). Although most of these subdivisions have yet to be developed, the potential exists for a massive transformation of Greater Yellowstone's wide-open spaces.

Nongovernmental Organizations

The globally significant natural amenities of Greater Yellowstone, combined with the perceived inadequacies of resource protection on both public and private lands, have spawned the creation of dozens of nongovernmental conservation organizations. Some, such as the Greater Yellowstone Coalition (with over 100 member organizations), maintain a Greater Yellowstone–wide perspective on development issues affecting both public and private land. Most of these organizations, however, focus on a particular sector or issue, such as a geographic area (e.g., the Jackson Hole Conservation Alliance), particular agencies or public lands (e.g., the Montana Wilderness Association, Yellowstone Park Foundation), particular groups of organisms (e.g., Trout Unlimited, Predator Conservation Alliance), or the conservation of primarily private land (e.g., The Nature Conservancy) or public land (e.g., American Wildlands). Promoting or facilitating the integration of conservation and development is generally not within the mission or programmatic activities of these groups, though more conservationists in the region are incorporating economic arguments into their advocacy programs.

One conservation group, the Corporation for the Northern Rockies, was created specifically to link conservation and development. They have targeted the ranching community for programs that seek to improve the environmental compatibility of ranch operations while increasing ranch revenues. One of their efforts, Predator Friendly Inc., helps sheep ranching operations to develop and implement strategies for nonlethal control of predators such as coyotes *(Canis latrans)* and wolves. These animals prey on sheep but can be deterred with guard animals and better herd management. In return, the Corporation for the Northern Rockies assists these wool producers in marketing their product as "predator friendly wool," which can sell for a higher price than regular wool. Another attempt to link business interests and conservation in the Greater Yellowstone is the Greater Yellowstone Business Council, an NGO in the early stages of development but with already substantial financial backing.

Thus far, the few non-governmental efforts to promote ICDPs have not been on a scale large enough to have a measurable impact on the ecological integrity or socioeconomic development of the Greater Yellowstone. While non-governmental conservation organizations have been effective in protecting wildlife and wildlands, particularly in the core of the Greater Yellowstone (the national parks and national forests), their efforts have been modest in fostering environmentally benign development activities on the ecologically important private lands and in communities that influence public land management (such as the gateway communities).

The Regulatory and Policy Framework Affecting ICDPs

Natural resource development activities on the private and public lands of Greater Yellowstone are subject to a dizzying array of local, state, and federal regulations and policies. Resource extraction on public lands, for example, is subject to federal environmental laws and policies that include the National Environmental Policy Act, National Forest Management Act, Endangered Species Act, Clean Water Act, Clean Air Act, 1872 General Mining Law, Migratory Bird Act, Wilderness Act, as well as state policies and regulations such as the Montana Environmental Policy Act, Montana Streamside Management Act, Idaho Forest Practices Act, Montana Water Quality and Non-Degradation Law, Stream Segments of Special Concern designation in Idaho, and others. Unfortunately, this byzantine legal framework is ineffectual when it comes to assuring the melding of conservation and development.

Western water law is a good case in point. This law, originally drafted in the 1800s when the West was still a frontier and government officials were seeking to promote the settlement and development of the "wide open spaces," promotes the waste of one of the West's most precious resources—water. Landowners can lose their rights to surface waters if they do not use their allocated quotient of water, even if it is not needed or could be reduced through conservation practices. Consequently, the dewatering of streams for crop irrigation remains a significant environmental problem, with disastrous impacts on fish and other aquatic and terrestrial species.

Although some federal and state regulations can affect land-use practices, much of the oversight of private land development comes from local government. Of the twenty counties of Greater Yellowstone, however, few have effective land-use plans and ordinances (Greater Yellowstone Coalition 2000). Combined with an extremely entrenched ethic of private property rights, this has resulted in little control of private land development patterns and trends. Consequently, prime agricultural soils are being covered with pavement. Important wildlife habitat is being cleared for rural housing. And pristine water bodies are being polluted by septic systems and agricultural runoff.

Major Markets Affecting ICDPs

Five major overlapping market and nonmarket values, all dependent primarily on the private and public land base (and its waters), dominate the linkage between socioeconomic development and conservation in Greater Yellowstone: agricultural commodities, extractive commodities, private land development, environmental amenities, and biodiversity. The main agricultural commodities are livestock, wheat, potatoes, and hay. The principal extractive commodities are timber, gas, oil, and minerals. Real estate transactions, particularly the sale of open land for rural residential development, are the dominant development activity. The main environmental amenities that attract people to the Greater Yellowstone are the mountains, forests, fish and wildlife, scenery, and the recreational opportunities these provide. Although biodiversity could be included as an environmental amenity, we distinguish it as a separate value because (1) the full range of biodiversity—genes, species, habitats, ecological and evolutionary processes—is usually not a key component of the environmental amenities values and markets described above, and (2) this full range of biodiversity is the principal goal on the conservation side of the ICDP equation.

Use of the land for agricultural and extractive commodities is losing ground to markets for land development, environmental amenities, and biodiversity. Agricultural income in the twenty Greater Yellowstone counties declined from 13 percent of total income in 1970 to 3 percent of total income in 1995. Income from resource extraction declined from 6 percent to 3 percent over the same period (Rasker and Hansen 2000). Lands that traditionally have been agricultural (i.e., where agricultural production was the primary means of support and defined the land value) or for resource extraction are being converted to environmental amenity-based uses, or residential or commercial development. There is virtually no land, private or public, that in recent years has gone the reverse direction, from environmental amenity or biodiversity uses to agriculture or resource extraction.

The market for environmental amenities is diverse in terms of the uses and changes it brings to the land. On public lands in the Greater Yellowstone, the most notable change is probably the large decline in timber harvesting on national forest lands as public demand for these lands for recreational use and biodiversity conservation has greatly increased (figure 7.2).

On private lands the result is both a change in land use and ownership. At the one extreme are buyers who have the desire and financial means to keep a large ranch intact. The other extreme is where the seller or buyer subdivides the land into small building lots, or "ranchettes," often of two to eight hectares. This growth is evidenced by a 400 percent increase in rural residences in Greater Yellowstone since 1970 (Hansen and Rotella 2002). Whether big ranches or ranchettes, some agricultural use is often maintained on the land, but not as a major or necessary source of income.

Ranch sales in the mountainous region of western Montana are probably indicative of trends in Greater Yellowstone. From 1994 to 2000, the per-hectare value of land sold for agricultural purposes grew less than 6 percent annually, whereas the per-hectare value of land sold for environmental amenity purposes (often referred to as "recreational" purposes in local real estate markets) increased 11 percent annually. Moreover, as ranches in a region begin to shift from agricultural use to an environmental-amenity use, per-hectare values appreciate rapidly, often 40–80 percent in one or two years. An important feature of this trend is that buyers of western Montana ranches are now about 90 percent nonresidents (i.e., not from Montana), so we are dealing with a national and, to some degree, international market (Norman C. Wheeler & Associates 2001).

Biodiversity values and markets are evident on both public and private lands. Sometimes they supplant recreational values as well as agriculture and resource extraction. Yellowstone National Park's recreational resource management, whether in the form of restricted entry to areas used by wolves or attempts to curtail snowmobile use, is in large part driven by the goal of protecting biodiversity. Recent declines in timber harvesting in national forests and acquisition of private forest lands by the U.S. Forest Service via land exchanges and outright acquisition have been driven by both environmental amenity values and by the goal of conserving intact forest ecosystems and endangered species such as the grizzly bear.

Investments by the nonprofit sector, driven by the goal of maintaining environmental amenities and biodiversity, are becoming increasingly important in land markets. A recent example is a grant from the Doris Duke Charitable Foundation that will direct $6.5 million toward land acquisition and conservation easements in the Greater Yellowstone region (McMillion 2001). Seven land trusts or equivalent organizations now operate in the area. Some, such as local land trusts, generally target the maintenance of open space and other environmental amenities; others, such as the Rocky Mountain Elk Foundation, focus on certain species and their habitats; and others, such as The Nature Conservancy, are primarily investing in biodiversity conservation. The principal investment mechanism for land trusts in the Greater Yellowstone is conservation easements; fee-simple acquisition and leases are less commonly used. To date, over 120,000 hectares of private land in the Greater Yellowstone have been protected through donated or purchased easements.

A variation on the nonprofit market for biodiversity is aimed at particular imperiled components of biodiversity. Defenders of Wildlife, for example, established a trust fund in 1987 to compensate ranchers who lost livestock to wolves as well as to assist ranchers with special fences, guard dogs, and other means to keep wolves and livestock apart. Compensation to ranchers in the region totaled $303,410 from 1987 to 2003 (Defenders of Wildlife 2003).

Biodiversity and Development Patterns and the Importance of Private Lands

As we have learned more about what is required to conserve biodiversity in the Greater Yellowstone, we have both increased the geographic scale of conservation efforts and refined conservation targets within that larger scale (see also Robinson and Redford, Franks and Blomley, and Maginnis, Jackson, and Dudley, all this volume). One hundred years ago, park managers and conservationists were not concerned about questions of long-term genetic viability of large mammal populations or maintaining ecological corridors to other wildlands. Wolves were among those biodiversity features not considered worthy and were extirpated from the park in the first decades of the twentieth century.

Today, we recognize that to conserve species, genetic diversity, ecological processes, and other elements of biodiversity, we need to think and act at the scale of Greater Yellowstone, an area at least eight times larger than Yellowstone National Park, and of areas beyond the Greater Yellowstone, including linkages to and interaction with wildlands hundreds of kilometers away. As the geographic scale has expanded, so have the number and diversity of landowners and institutions responsible for managing the land and its biodiversity. The responsibility for conserving Yellowstone's biodiversity does not lie solely with Yellowstone National Park; it involves several public institutions and thousands of private landowners.

As a glance at the map of the Greater Yellowstone reveals (figure 7.1), one overarching effect of this growth in the scale of our geographic concern is that private lands have become an increasingly important component of the conservation landscape. In fact, it is becoming apparent that private lands are disproportionately important relative to the area they cover. This, we suspect, is a pattern common to many of the world's protected areas, which are often disproportionately located in areas of low biological productivity (Huston 1993; Scott et al. 2001).

Greater Yellowstone is centered on the Yellowstone Plateau and surrounding mountains, where the high elevation makes for a short growing season and the volcanically derived soils are poor in nutrients and water-holding capacity. In contrast, private lands are generally at lower elevations and include extensive bottomlands and alluvial soils. Because of their long growing season and relatively high nutrient levels, private lands in Greater Yellowstone harbor high levels of biodiversity (Hansen and Rotella 2002).

More than 60 percent of the wildlife in the western United States depends on riparian areas to meet their habitat needs. At the same time, riparian areas are highly desirable for new home construction. Just north of Yellowstone Park, over eighty kilometers of the Yellowstone River corridor have been subdivided (Greater Yellowstone Coalition 1998). The same phenomenon of new home development is occurring at the interface between private lands and adjacent national parks and forests. Ready access to a national forest or national park out your back door is highly desired. The effect of these development patterns is evident from an analysis within the Montana and Wyoming portions of the Greater Yellowstone that showed 57 percent of deciduous forest and 40 percent of conifer forest below 2000 meters (6,560 ft) were within two kilometers of a home in 1997 (Hansen et al. 2002). This pattern of housing development disproportionately magnifies, compared to a random development on private lands, impacts on biodiversity.

Several examples illustrate the strong influence of private lands on biodiversity conservation. Fifty percent of all grizzly bear mortality in recent years has occurred on private land (Johnson 2000). Yellowstone

National Park's population of pronghorn antelope *(Antilocapra americana)*, which historically migrated between summer habitats in the park and winter habitats at lower elevations on private land in the Yellowstone River valley, has declined to about 200 individuals and is now considered vulnerable to extinction. The causes are not yet clear, but higher antelope mortality may have resulted from farming and rural residential development in ungulate winter ranges (Goodman 1996; Hansen and Rotella 2002). Many other mammals, such as bison *(Bison bison)*, elk, and moose *(Alces alces)*, migrate from high-elevation public lands in the summer to low-elevation, largely private lands in the winter. Elk that summer in Yellowstone Park, for example, often migrate up to 100 kilometers to their winter range in lowland valleys (Hansen and Rotella 2002; Yellowstone National Park 2003). Rural sprawl and wildland recreation also provide avenues and vehicles for the spread of non-native plant species, one of the most serious and intractable threats to the region's terrestrial and aquatic ecosystems. Rural residential development also leads to the contamination of ground and surface water from septic systems and can alter important natural ecological processes such as wildfire, which is often suppressed in these newly populated areas.

A study of hot spots for bird species in Greater Yellowstone found that only 7 percent were on preserve lands (national parks, national forests, wildlife refuges); most were on or near private lands (Hansen and Rotella 2002). Moreover, land use on private lands, such as the placement of rural homes, was more intense near hot spots. Thus, for example, high-productivity sites on private lands are dominated by deciduous forest-cover types of aspen *(Populus tremuloides)*, cottonwood (*Populus* spp.), and willow (*Salix* spp.), and these habitats support high densities of nesting birds. However, people also like to nest in or near these forest types; some 50 percent of the aspen, cottonwood, and willow forests on private land are currently within 2.1 kilometers of a rural home. This can have multiple negative effects on bird populations and nesting success. Some are obvious, such as forest clearings to make room for home sites and access roads. A less obvious effect is that brood parasites, particularly the cowbird *(Molothrus ater)*, and avian predators, such as the black-billed magpie *(Pica pica)*, are especially common near human residences, which results in lower nesting success for many bird species that nest near rural residences (Hansen and Rotella 2002).

In short, our growing knowledge of conservation biology and of what is required to restore and conserve the biodiversity of Greater Yellowstone points to a strong emphasis on private lands and their ecological relationship with public lands. This also enables us to more intelligently target management interventions within this large area.

Private Lands and Public Amenities: The Core of the ICDP Challenge in the Greater Yellowstone

A universal concern among conservationists about ICDPs is that success in economic development will draw more people to the project area, which in turn may negatively affect the ability to meet conservation goals for the area (see also Gartlan this volume). The causes of demographic and economic change in Greater Yellowstone place a new twist on this concern. Economic development opportunities do not appear to be the main reason for the explosive population growth; this region lags behind the rest of the United States in per capita income and wages. Rather, the success in meeting conservation goals, beginning with the creation of Yellowstone National Park in 1872, is the main factor drawing people to the region. People come to Greater Yellowstone because of its diverse and abundant environmental amenities. Economic development follows.

These new residents largely recreate on and enjoy the region's public lands—its national forests, national parks, and national wildlife refuges—and its public rivers and streams. Although this increased usage can degrade the biodiversity and wilderness quality of public lands, it can also be of conservation benefit as recreational values begin to outcompete commodity production values of timber, minerals, and livestock grazing for the attention of policy makers and land managers. Many of the recreational values are monetary, a portion of which moneys goes directly to the public land agencies in the form of user fees. A much larger portion, however, goes to local businesses, as recreationists, both residents and tourists, pay for a broad array of services and products, from hunting outfitters, river rafting trips, and restaurants to backpacks, mountain bikes, and gasoline.

Other values, however, are nonmonetary in the sense that residents enjoy the region's environmental amenities, whether a scenic viewshed or a hike in a national forest, without directly paying for them. They indirectly pay, however, because they are willing to accept a lower wage for the benefit of living in a region that is rich in environmental and other quality-of-life amenities.

The recreational, conservation, and aesthetic values, both monetary and nonmonetary, of public lands for both a national and regional constituency should increasingly shift the management emphasis from commodity production to noncommodity values. This is particularly important on the region's 4.4 million hectares of national forest lands that are open to a broad array of potential uses. Increased recreational use, however, creates its own suite of conflicting interests and stresses on the environment. Thus, important challenges continue to lie ahead in managing Greater Yellowstone's public lands in a way that benefits biodiversity, communities, and the U.S. public at large.

We believe, however, that the region's nearly 1.6 million hectares of private lands are where the most significant changes are occurring in terms of the conservation of the region's remarkable biodiversity and wildlands and in the socioeconomic well-being of its communities. Private lands must therefore become the cornerstone of a Greater Yellowstone ICDP.

Whereas the Greater Yellowstone ICDP began with the private sector helping set aside more than 800,000 hectares of public land as Yellowstone National Park, 130 years later the tables have turned. The public amenities offered by the region, particularly its public lands, waterways, and wildlife, are a major driving force of change in the private sector. Amenity-based values are attracting people, and although these people largely recreate on the public lands, they buy and live on private lands. This in-migration has developed most rapidly over the last twenty to thirty years, spawned by rapid growth in financial wealth in the United States and in telecommunications technology, both of which allow people to increasingly live and work wherever they wish.

In Yellowstone, as in many other mountainous regions of the western United States, private lands are being converted from agriculture to residential development, with adverse and disproportionately large—relative to public lands—impacts on biodiversity (Ingram and Lewandrowski 1999; Rasker and Hansen 2000). Left unchecked, erosion of the region's biodiversity and other environmental amenities by rampant population growth and development will eventually undermine the socioeconomic well-being of the region's communities. The 130-year, largely spontaneous ICDP experiment in Greater Yellowstone will then have failed to meet either long-term conservation or development goals.

What, then, are the key factors that need to be addressed in formulating an ICDP approach for Greater Yellowstone's private lands?

Private Land Markets and Management for Public Benefit in Greater Yellowstone

A central challenge in conserving the earth's biodiversity, whether in Yellowstone or the Congo Basin, is how to get the private sector to manage land, water, and natural resources in ways that yield the broad and long-term societal benefits of biodiversity (Pearce and Moran 1994;

Freese 1998; see also Kiss this volume). The solution to this challenge in the Greater Yellowstone consists of four interrelated components, three largely market based and one nonmarket:

1. The purchase of large areas of private land of high biodiversity value by individuals/corporations for whom biodiversity conservation is an important management goal
2. The creation of programs and markets that foster philanthropic giving for acquiring private land, for purchasing conservation easements, and for other forms of private-land investment for conservation purposes
3. The development of government programs for acquiring private land, for purchasing conservation easements, and for other forms of private-land investment for conservation purposes
4. The development of governmental regulations and governmental incentive/disincentive programs that guide and support the three preceding activities

1. Acquisition of Large Areas of Private Land by Conservation-Minded Buyers

We need to encourage the acquisition of large parcels of land by conservation-minded buyers who are willing and able to place wildland and wildlife conservation before commodity production, subdivision development, or other activities that compromise biodiversity conservation. The conservation benefits of this are twofold: (1) preclusion of housing subdivisions and (2) tolerance and support of biodiversity restoration and management on the land. To ensure the durability of these benefits, conservation-minded owners need to donate strict conservation easements on their lands.

As noted previously, large properties in Greater Yellowstone are being acquired by wealthy owners from outside the region. Some of these have little need to make a living from the land and are attracted to the region for its wildlife and open-space values. They often, however, lack an understanding of sustainable land management techniques. This has resulted in a burgeoning business for ranch management consultants.

Moreover, some of these new landowners come with an appreciation for the open space and a desire to be "cowboys." They maintain cattle operations as a hobby, but are little, if at all, aware of the crucial role their lands play in biodiversity conservation. Thus, a concerted educational and awareness program is needed to help enlist these new landowners in managing their land for conservation.

The relatively high turnover rate among these new ranch owners points to the need for conservation easements. Although the current owner of a ranch may be strongly committed to conservation practices, this is of limited value if he or she sells the ranch to someone who is not similarly committed. Even if an individual purchases a ranch purely as an investment, in general the current market premium on ranches that are ecologically intact with abundant wildlife provides a positive incentive for these individuals to manage the land for biodiversity values. However, owners who view the land as a potential investment will be disinclined to donate a conservation easement because the restricted use options imposed by the easement decrease the land's market value.

For people who plan to retain ownership, this investment value is of less concern, and a conservation easement is an economic benefit because it will lower property taxes. The inheritance tax is similarly much lower for land that carries a conservation easement.

Compared to traditional cattle ranchers, new ranch owners in the Greater Yellowstone are clearly bringing a different mix of economic activities to the region—expensive new homes, more airline travel, and more disposable income in general. There are winners and losers in this transition, as the new ranch owners spend their money in different places and ways than the traditional cattle rancher. Whether this is "good" or "bad" for development depends on whether you are one of the winners or one of the losers (see also Brown this volume).

Another socioeconomic transition of concern among hunters, hikers, and other recreationists is that new ranch owners tend to restrict access to their lands more than traditional ranchers. Ranch owners, both old and new, are also beginning to charge access fees to hunters who want to hunt on their land, whereas in the past sportsmen often had free access. Hunting and fishing on private land therefore becomes increasingly a pursuit of the economically well-off—those who are able to pay. As a strong hunting and fishing culture exists in Greater Yellowstone and surrounding communities, this transition may be undesirable in the development component of the ICDP model.

Mechanisms for facilitating and guiding free-market mechanisms for the benefit of conservation can be promoted and developed. For example, conservation organizations can publicize the conservation importance of particular private lands and work with new landowners to encourage and assist them in adopting sound conservation practices. Another mechanism is in the marketplace itself; a few real estate firms now cater specifically to conservation-minded buyers. Conservationists should work with these realtors with the twin goal of helping identify potential buyers who are truly conservation minded and to identify lands for purchase.

Finally, conservationists can work with professional ranch management companies to ensure that they are well informed of biodiversity conservation needs and practices on the lands they are contracted to manage.

2. Philanthropic Giving Through Nonprofit Organizations for Land Investments

The second route for private sector investment in private-land conservation is through philanthropic giving by individuals, foundations, and corporations. As noted previously, the last two decades have brought significant growth in the Greater Yellowstone in both nonprofit conservation organizations that invest in private lands (i.e., land trusts) and in the donor dollars they receive for this purpose. The major factors that will determine the effectiveness of this approach for biodiversity conservation in Greater Yellowstone will be the amount of philanthropic giving and how land trusts invest these moneys.

The first factor will in large part depend on how much and in what way philanthropic dollars are spent in the United States in general. This is particularly so with regard to donors from the high technology sector, which has produced numerous millionaires and billionaires over the last decade. The result is the creation of new charitable giving foundations, many of which include conservation as one of their causes. The task of conservationists is to demonstrate the importance and effectiveness of investing philanthropic dollars in private-land conservation. As we noted earlier, donors are probably willing to dig deeper into their pockets for the acquisition of land than for supporting other conservation strategies because of the directness and permanence of land acquisition. We believe tens of millions of philanthropic dollars are potentially attainable annually for private-land investments in Greater Yellowstone.

How land trusts decide to spend this money is the second key question, and will affect how willing donors are to give more. Perhaps the first need is for the conservation organizations and land trusts of the region to develop a common vision for private lands in the Greater Yellowstone, a vision that will provide a compelling blueprint for how donor dollars will be wisely invested to yield the biggest conservation impact. This vision should answer two fundamental questions: (1) which lands are most important for protecting Greater Yellowstone's biodiversity? and (2) what type of investments in management of these lands is needed?

The importance of identifying priority areas for private-land conservation is obvious, and scientists and conservation organizations in the Greater Yellowstone have made considerable progress in this area. It is less clear, however, how we should invest in and manage these lands. Investments by land trusts in the Greater Yellowstone are generally of two types: (1) conservation easements, whether purchased from or donated by the landowner; and (2) the purchase of fee-simple title to the land, with subsequent donation to, or buyouts by, a government land management agency. Both of these mechanisms are extremely useful, but they also include constraints and compromises. Conservation easements, unless they carry extremely strict conservation provisions, may permit activities that are less than ideal for biodiversity conservation. For example, most conservation easements continue to permit cattle grazing, the use of roads, and the limited construction of residential and/or agricultural structures. In some cases this may compromise use of the area by native wildlife, particularly large predators. Transference of land to government agencies is limited by government authorization and funding to pay for or accept the land and to manage it effectively.

Because of these constraints, some private lands in Greater Yellowstone may be managed most effectively if they are owned by nonprofit conservation organizations. This form of land investment, however, is rare in the region. Apart from the higher cost of land acquisition compared to easements or acquisition followed by subsequent government buyout, land ownership imposes the additional financial burden of management. Nonprofit ownership therefore requires the creation of a sizable endowment for long-term management. The creation of incentives that encourage landowners to sell their property to conservation organizations could be an important component of this. A bill recently submitted before the U.S. Congress, for example, would provide a tax credit to individuals who sell their land to nonprofit conservation organizations (The Nature Conservancy 2000).

Greater investments through the nonprofit sector into private lands should benefit socioeconomic development in three primary ways. First, almost all philanthropic conservation funding for the Greater Yellowstone comes from outside the region, thereby providing a new source of revenue and jobs. Second, the purchase of conservation easements will enable many landowners to stay on the land, rather than being forced to sell and move elsewhere. Third, the conservation of wildlife and open space will help meet quality-of-life values that are important to the region's residents.

3. Government Purchase of Lands and Easements

Government agencies in Greater Yellowstone acquire private lands either by outright purchase or by exchanging government land in one area for private land in another. Land exchanges are one way of consolidating both public lands and private lands into large contiguous blocks of land, thereby replacing the checkerboard pattern of public and private lands that is common in the ecosystem. This has been important, for example, in the U.S. Forest Service's attempts to protect large parcels of important grizzly bear habitat. Land exchanges, however, often involve significant compromises, because the public lands that are given to the private sector—often timber companies with real estate development interests—in the exchange generally have conservation value.

How land exchanges affect development depends on which socioeconomic sector is being considered. In Greater Yellowstone, for the timber industry, private-public land exchanges involving forestlands mean that there will continue to be jobs in the woods for timber workers. However, people who live near, recreate in, or in other ways highly value the public forest lands being traded are often opposed to such exchanges. Some public lands traded to the private sector fall outside of the Greater Yellowstone, which complicates the assessment of both the conservation and development impacts of these exchanges. What is good for one region's ICDP may be bad for another's.

We believe that the outright purchase of private lands by government agencies for conservation purposes is generally preferable to land exchanges within the Greater Yellowstone. The value of most public land in this region is too great, both in terms of conservation and development goals, to be traded to the private sector for commodity production.

Thus, we should expand existing government programs, such as the Land and Water Conservation Funds (which are derived from a surcharge on offshore oil drilling revenues), that provide money for acquiring and placing easements on important private lands. An important factor to remember is that private funding often begets government funding. Once conservation organizations begin to acquire land or easements in an area, state and federal funding is easier to obtain. Often this comes in the form of matching grants.

Opportunities for funding from local governments should also be pursued. A $10 million bond for maintaining open space, recently approved by the citizens of Gallatin County on the northern edge of Greater Yellowstone, is a means for local citizens to support and direct public funding toward meeting conservation goals on private land. This may be a viable option for some of the other nineteen counties in the Greater Yellowstone, particularly those facing problems associated with rapid population growth and urban sprawl.

4. Regulations and Zoning

During a recent study tour of Yellowstone by national park professionals from other countries, one individual was astounded that the government had not imposed regulations limiting further human in-migration and settlement on private lands around the park. However, such an approach would be politically impossible considering the entrenched private property rights ethic in the American West. Governmental policies and regulations must take more circuitous routes in directing human settlement in the Greater Yellowstone.

Left alone, there is no reason to believe the three markets for private-land investment and conservation outlined above will optimize the dual ICDP goals of biodiversity conservation and socioeconomic development. In short, current market mechanisms are incapable of addressing the environmental externalities created by private sector developments in the region. Market-based incentives will not, for example, keep people from building new homes along rivers, on ridge tops, and next to prime bird-nesting habitat, although such home placement creates widespread negative environmental externalities for the public. Conservation-minded land buyers, philanthropic moneys, and government acquisitions can help protect such public amenities, but they will not be sufficient by themselves.

Local, state, and federal regulations must play a role in guiding development if the dual goals of a Yellowstone ICDP are to be met. On Greater Yellowstone's private lands, the role of federal and state agencies is primarily in the arena of fish and wildlife management. Since fish and wildlife belong to the public even if they inhabit private lands, government agencies have the responsibility and authority to manage them, and they do a reasonably good job. The area of greatest potential improvement is in managing more broadly for biodiversity on private lands, where innovative measures are needed. For example, federal regulations prohibit the shooting of migratory passerine birds, but a far greater impact on migratory passerines is home construction in bird-nesting habitat. State or federal regulations that prohibit the disturbance or destruction of these critical habitats would benefit these bird populations more than current hunting prohibitions. Federal regulations prohibiting the destruction of wetlands on private lands provide a precedent for such action.

Another opportunity for government action is government-initiated growth management. This generally takes the form of land-use planning (done at the county level) and associated land-use regulations that usually include zoning (which specifies which land uses are allowed in different areas), as well as other land-use-related regulations such as building setbacks along rivers and streams. Another important component of government-initiated growth management is the integration of transportation planning with the goals of county land-use plans. One effect of this lack of coordination between planning and transportation agencies is that private lands protected through easements and purchase are still being rendered unsuitable for wildlife because of surrounding residential- or transportation-related developments that fragment the rural landscape.

To rectify this situation, county planning must be elevated to the importance it deserves. This includes adequate funding and staffing and the political will needed to fully enforce these regulations. Strong and informed citizen support for planning is a prerequisite for achieving these goals. Conservation organizations can play an important role in empowering effective citizen involvement in planning issues. Although there are many environmental advocacy groups in Greater Yellowstone, most are focused on public land protection. Broadening the scope of their advocacy work to include private-land development issues could help to build a broad and strong constituency for government-sponsored growth management efforts.

Conclusion

Yellowstone National Park and the national forests that surround it largely began with the ICDP concept: the park would attract tourists and their money and the national forests would produce timber and revenues. Both would be managed to protect wildlife and natural resources for the American public. Although the balance between resource use and resource protection has sometimes been distorted and resource conservation challenges remain, one can cite several reasons for calling this grand ICDP a success. The very existence of Yellowstone National Park, of nearby Grand Teton National Park, and of the seven national forests in the Greater Yellowstone is perhaps the most compelling evidence. Further proof is that all species and habitats native to the Greater Yellowstone are still found there. And the fact that the high quality of life in the Greater Yellowstone—based largely on the environmental amenities of the public lands—has become a magnet for both residents and newcomers suggests that conservation and development have been well integrated.

But problems remain, and there is cause for concern that both the biodiversity conservation and the development goals of the ICDP model will be severely compromised if current trends in population growth and development continue. The conversion of rural open lands to subdivisions, combined with our increased awareness of the critical role that these lands play in maintaining Greater Yellowstone's biodiversity, lead us to conclude that private lands, though they constitute only 20 percent of the region's area, occupy center stage in the future of the Yellowstone ICDP.

A major challenge in applying the ICDP model in Yellowstone and, we suspect, elsewhere, is to define biodiversity and development goals (see also Robinson and Redford this volume; Franks and Blomley this volume). Whereas we believe conservationists can generally agree to the biodiversity conservation goals, defining success in terms of development in Greater Yellowstone is much more difficult. There are generally both winners and losers in any given socioeconomic development scenario. To ensure that biodiversity goals are upheld, we need to swiftly and strictly protect Yellowstone's private lands from environmentally destructive human development. Private sector investment by conservationists, both through conservation-minded individual buyers and philanthropic moneys, combined with much more stringent government-sanctioned growth management on private land, holds the most promise for such protection. Ultimately, our goal should be well planned and managed growth on the private lands where appropriate, and the prohibition of development on lands of critical ecological importance.

National—international—treasures such as the Greater Yellowstone have two major stakeholder groups: those who live locally and directly benefit socially and economically from the region's amenities, and the public at large who view Yellowstone as a natural heritage that belongs to everyone. We are reminded of this duality when Yellowstone's gateway communities demand the park be managed in a way that benefits their economy, but park managers and conservationists respond that the park belongs to the U.S. public and its major purpose is not to service local communities. In setting ICDP goals for Greater Yellowstone, our decisions should be tempered by concern for the well-being of local people and communities and their importance in managing the region's resources. First and foremost, however, our development goals should be molded to meet the interests that society at large has in preserving an area of such national and global importance.

Mustering the will, cooperation, and capital needed to ensure the sound management of both private and public lands in the Greater Yellowstone will require vision. If private and public land managers, businesses, and citizens can develop a common and sound vision for the region's future, both biodiversity and development needs should be achievable.

References

Craighead, J. J. 1991. Yellowstone in transition. In R. B. Kieter and M. S. Boyce, eds., *The Greater Yellowstone Ecosystem: Redefining America's Wilderness Heritage*, 27–40. New Haven, Conn.: Yale University Press.

Defenders of Wildlife. 2003. Compensation statistics. http://www. defenders .org/wildlife/wolf/wcstats.pdf (September 2003).

Freese, C. 1998. *Wild Species as Commodities: Managing Markets and Ecosystems for Sustainability.* Washington, D.C.: Island Press.

Glick, D., M. Carr, and B. Harting. 1991. *An Environmental Profile of the Greater Yellowstone Ecosystem*. Bozeman, Mont.: Greater Yellowstone Coalition.

Goodman, D. 1996. *Viability Analysis of the Antelope Population Wintering Near Gardner, Montana.* Final Report to the National Park Service. 7 pp.

Greater Yellowstone Coalition. 1998. *Subdivided Land in Park County, Montana* (map). Bozeman, Mont.: Greater Yellowstone Coalition.

Greater Yellowstone Coalition. 2000. *Greater Yellowstone Report,* vol. 17, no. 4. Bozeman, Mont.: Greater Yellowstone Coalition.

Greater Yellowstone Coordination Committee. 1987. *The Greater Yellowstone Area: An Aggregation of National Park and National Forest Management Plans.* Billings, Mont.: U.S. Forest Service.

Haines, A. 1977. *The Yellowstone Story.* Yellowstone National Park, Wyo.: Yellowstone Library and Museum Association.

Hansen, A. J., R. Rasker, B. Maxwell, J. L. Rotella, J. D. Johnson, A. Wright Parmenter, U. Langer, W. B. Cohen, R. L. Lawrence, and M. P. V. Kraska. 2002. Ecological causes and consequences of demographic change in the New West. *BioScience* 52 (2):151–162.

Hansen, A. J. and J. J. Rotella. 2002a. Rural development and biodiversity: A case study from Greater Yellowstone. In J. Levitt, ed., *Conservation in the Internet Age,* 123–140. Washington, D.C.: Island Press.

Hansen, A. J. and J. J. Rotella. 2002b. Regional source-sink dynamics and the vulnerability of species to extinction in nature reserves. *Conservation Biology* 16:1112–1122.

Harting, B. and D. Glick. 1994. *Conserving Greater Yellowstone: A Blueprint for the Future*. Bozeman, Mont.: Greater Yellowstone Coalition.

Huston, M. A. 1993. Biological diversity, soils, and economics. *Science* 262:1676–1680.

Ingram, K. and J. Lewandrowski. 1999. Wildlife conservation and economic development in the West. *Rural Development Perspectives* 14 (2):44–51.

Johnson, J. D. and R. Rasker. 1995. The role of economic and quality of life values in rural business location. *Journal of Rural Studies* 11:405–416.

Johnson, V. 2000. *Rural Residential Development Trends in the Greater Yellowstone Ecosystem since the Listing of the Grizzly Bear 1975–1998.* Bozeman, Mont.: Sierra Club.

Lorah, P. 1996. Wilderness, uneven development, and demographic change in the Rocky Mountain West, 1969–1993. Ph.D. diss., Department of Geography, Indiana University, Bloomington.

McMillion, S. 2001. Preservation for posterity. *Bozeman Daily Chronicle,* 18 Aug., A1, A10.

The Nature Conservancy. 2000. Sightings: Capital gains. *The Nature Conservancy,* November/December, 7.

Norman C. Wheeler & Associates. 2001. *For Land's Sake,* no. 24. Bozeman, Mont.: Norman C. Wheeler & Associates.

Noss, R., G. Wuerthner, K. Vance-Borland, and C. Carrol. 2001. *A Biological Assessment of the Greater Yellowstone Ecosystem*. Corvallis, Oreg.: Conservation Sciences.

Pearce, D. and D. Moran. 1994. *The Economic Value of Biodiversity*. London: Earthscan.

Rasker, R. and B. Alexander. 1997. *The New Challenge: People, Commerce, and the Environment in the Yellowstone to Yukon Region*. Bozeman, Mont.: The Wilderness Society.

Rasker, R. and D. Glick. 1994. The footloose entrepreneurs: Pioneers of the New West? *Illahee* 10 (1):34–43.

Rasker, R. and A. Hackman. 1996. Economic development and the conservation of large carnivores. *Conservation Biology* 10:991–1002.

Rasker, R. and A. Hansen. 2000. Natural amenities and population growth in the Greater Yellowstone Region. *Human Ecology Review* 7 (2):30–40.

Rudzitis, G. 1999. Amenities increasingly draw people to the rural West. *Rural Development Perspectives* 14 (2):9–13.

Rudzitis, G. and H. E. Johansen. 1991. How important is wilderness? Results from a United States survey. *Environmental Management* 15:227–233.

Sax, J. 1980. *Mountains Without Handrails*. New Haven, Conn.: Yale University Press.

Scott, J. M., F. W. Davis, R. G. McGhie, R. G. Wright, C. Groves, and J. Estes. 2001. Nature reserves: Do they capture the full range of America's biological diversity? *Ecological Applications* 11 (4):999–1007.

U.S. Bureau of the Census. 2001. *Census of Population and Housing*. Washington, D.C.: U.S. Bureau of the Census.

USDA Forest Service. 2001. Region one timber sales program statistics. 30 January. Data sheets: Forest 8–Custer; Forest 11–Gallatin. Washington, D.C.: USDA Forest Service.

U.S. Fish and Wildlife Service. 2001. Legislative mandates and authorities. http://refuges.fws.gov/policyMakers/mandates/ (June 2001).

U.S. National Park Service. 2000. *Record of Decision for Yellowstone and Grand Teton National Parks*. Washington, D.C.: U.S. National Park Service.

Yellowstone National Park. 2003. Elk population issues. *Yellowstone's Northern Range: The Official Website of Yellowstone National Park*. http://www.nps.gov/yell/nature/northernrange/ch62.htm (September 2003).

8

Parks, Projects, and Policies: A Review of Three Costa Rican ICDPs

Katrina Brandon and Michelle O'Herron

Introduction

Costa Rica is frequently cited as embracing policies that promote "sustainable" or "compatible" development. This paper traces activities that fall under the broad heading of integrated conservation and development projects (ICDPs) in three regions of Costa Rica to see the extent to which the national policy context has shaped project outcomes. Each region has had activities under way for at least fifteen years, with improved conservation in one or more adjacent protected areas as an important and stated project goal. There are a number of differences among the three sites in terms of biological, social, and organizational criteria. The three sites are Corcovado National Park and the Osa Conservation Area; Santa Rosa National Park and the Guanacaste Conservation Area; and the Gandoca-Manzanillo Wildlife Refuge and the La Amistad-Caribe Conservation Area[1] (figure 8.1).

This paper reviews the policies that support conservation and rural development within Costa Rica and the implementation of these policies. It reviews the activities at each site and summarizes the successes and challenges faced in achieving positive ICDP outcomes within each area. It also reviews the overall importance of policy conditions for each of the sites and activities supported by the ICDPs. It concludes with

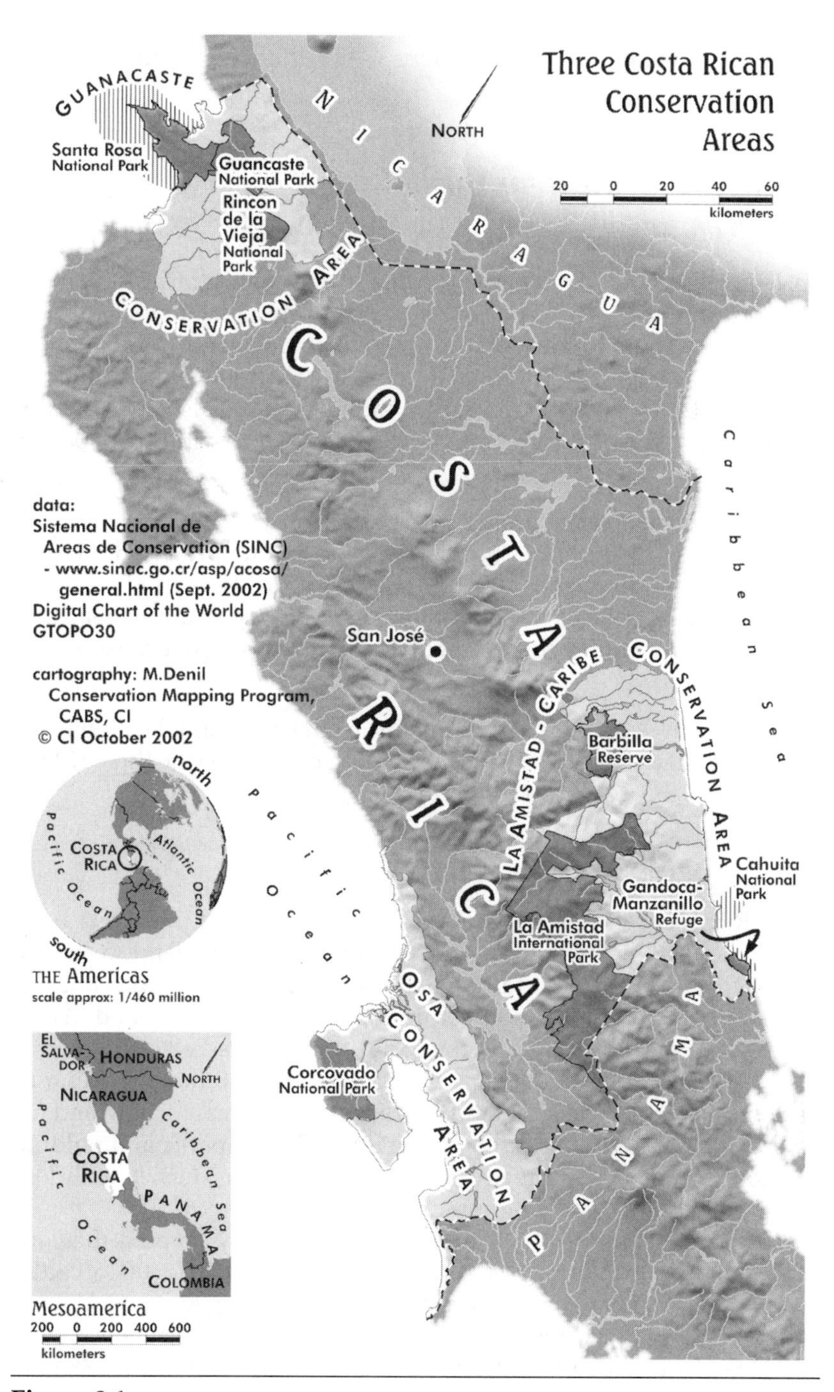

Figure 8.1.
Costa Rica and the three ICDPs.

lessons on linking policy-level actions with regional-level conservation activities, especially ICDPs.

The Policy Context for Costa Rican Conservation

Costa Rica is known as a country with a strong commitment to sustainable development—including strong social programs and safety nets and a commitment to conservation. High literacy rates, low infant mortality, and life expectancies equal to those in many developed countries all demonstrate that public sector investments in people have paid off. The strength of Costa Rican conservation policy lies in three specific elements:

- Creating an extensive protected area network
- Developing institutions and mechanisms linking protected areas with surrounding lands
- Introducing policies that promote market–based mechanisms, such as payment for environmental services, to support conservation

Along with Costa Rica's investments in education and development, these three elements have been the defining context for the three ICDPs described in this paper (Brandon in press).

CREATION OF A PROTECTED AREA NETWORK

Costa Rica's current park system did not really begin in earnest until 1970, when a new law allocated over 10 percent of the land to strictly protected areas and an additional 17 percent to buffer zones and/or forest reserves. Park creation continued during the 1970s and 1980s, but a severe economic crisis in the early 1980s affected all public sector activities, including the nascent park service. This limited funding to compensate landowners for land expropriated for park creation and for paying park staff. As a result, coordination among parks and wildlife and forestry agencies, already weak, declined even further with less money, fewer staff, and deteriorating morale.

Government policies both directly and indirectly favored the conversion of forests to export-oriented commodities such as coffee, cattle, cacao, sugar, and bananas. Deforestation levels during that period were the highest in Latin America, reaching 847,403 hectares between 1966 and 1989. Between 1979 and 1992, primary forests declined by 38 percent and secondary forests by 8 percent. By 1997, deforestation rates dropped to about 16,000 hectares per year. By 1999, 24.8 percent of the land in Costa Rica was under some category of protection (World Bank 2000b:12). Strictly protected areas (IUCN categories I–IV) amount to 14.2 percent of the national territory in eighty-five sites, protecting approximately 723,000 hectares. Costa Rica also had one of the most rapid population growth rates, reaching 3.8 percent per year by the 1960s. Far from being an easy context for park establishment, Costa Rica's park system began amid levels of deforestation and population growth that were among the highest in the world.

Institutional Framework to Link Protected Areas with Surrounding Lands

A National System of Conservation Areas, known as SINAC, was created in 1987 to integrate protected area management and decentralize decision-making power to regional levels. SINAC is currently composed of eleven administrative units, called ARCS (regional conservation areas) or "megaparks." Each ARC includes three general zones: core areas such as national parks, buffer areas (forest or indigenous reserves), and areas for intensive use (e.g., agriculture). A director, who is responsible for park management, community outreach, and intergovernmental coordination, manages ARCs. Staff from the forest department, wildlife department, and parks service were merged into one unit. The philosophy underlying this reorganization was that sectors dealing with natural resources would coordinate their actions internally, with one another, and externally—i.e., with other government agencies—if consolidated. ARCs must link protected areas with the lands around them, encourage local participation in decisions, and coordinate and organize actions by government agencies to support conservation.

Supportive Environmental Policies

Consolidating government management, innovative financing, and private sector involvement were key policies that Costa Rica enacted. Disparate environmental functions scattered among ministries were consolidated to form a Ministry of Environment and Energy[2] in 1986.[3] Innovative financing for conservation, using mechanisms such as debt-for-nature programs (debt-swap) and intensive international fund-raising, was promoted early on. Financing for each ARC to facilitate fund-raising efforts and improve the regional disbursement of funds for conservation was decentralized. Hopes, money, and awareness were raised through the creation of the National Biodiversity Institute (INBio) in 1989, dedicated to bioprospecting—using biodiversity as the basis for new pharmaceutical products. Recent policies introduced environmental services payments (ESP), administered through the National Fund for Forestry Financing (FONAFIFO). Landowners who sign agreements with FONAFIFO receive cash payments for reforesting, managing natural forests under approved harvesting regimes, or protecting natural forests.[4] FONAFIFO pays owners for these services through a tax on fossil fuels, selling carbon bonds to utility companies, and with support from international donors.[5] The ESP transfers urban sector resources that impose high costs on the economy from foreign imports and pollution to the rural sector, providing environmental benefits at local, national, and international levels. The Private Wildlife Refuge Program, passed in 1992, provides a tax break to property owners, technical assistance for management, and help removing squatters for a minimum of a ten-year agreement. Most private reserves are not a part of this program, however (Langholz, Lassoie, and Schelhas 2000). These reforms at the policy level are impressive, and they form the context for the ICDPs described in the next section.

Scaling Up: Moving from Parks and ICDPs to Corridors and Compatibility

This section briefly reviews three parks and actions undertaken by ICDPs in or adjacent to them over the past twenty years:[6] Corcovado National Park and the Osa Conservation Area; Santa Rosa National Park and the Guanacaste Conservation Area; and the Gandoca-Manzanillo Wildlife Refuge and the La Amistad-Caribe Conservation Area (figure 8.1). Focusing on these three sites allows for a better understanding of the policy context and external issues, factors that are key in understanding the challenges faced by ICDPs (Wells and Brandon 1992).

Corcovado National Park and the Osa Conservation Area

Corcovado National Park (CNP),[7] located on the Osa Peninsula in southwestern Costa Rica, was created amid dramatic social change and conflict, conditions lasting to the present. Migrants came to the lowland areas of the Osa region in the early 1930s to work on large banana plantations, in 1937 to mine gold, and between 1947 and 1960 with the construction of the Pan American Highway. Conflict over land began in the 1950s with the presence of Osa Forest Products (OPF), a U.S. company that became a symbol of corruption, scams, and land-hording Americans. The OPF dealt with local disputes over land ownership by burning houses and evicting and shooting at squatters to keep the land available for agriculture, ranching, and logging. The international scientific community pressured the government to protect the Osa region from development, while the national legislature drafted legislation to kick OPF out of Costa Rica. To prevent the precedent that this legislation would set against foreign investment, the president hastily decreed CNP in 1975. Park creation required land trades with OPF and the relocation of several hundred inhabitants from the park. The adjacent Reserva Forestal Golfo Dulce (RFGD) was established in 1979. No funds were initially allocated for promised compensation at either site.

CNP expanded again in 1980, and thirty additional settlers were evicted without compensation, generating even more local hostility. Another wave of migration took place in 1984, when a banana company in Golfito closed operations and unemployed workers invaded CNP to mine gold. When the number of mining families soared to 800, it was evident that the government could not control the area. In an unprecedented action, the Costa Rican Congress approved compensation for the miners—in-kind (mostly food) and cash payments—in exchange for their exiting the park. This compensation cost the government nearly $3 million—three times the annual budget for all the parks—and was a drain on other government programs. Ironically, the compensation had more to do with public sentiment on behalf of unemployed banana workers (who became the gold miners) than with a belief that those who squatted within parks were entitled to formal compensation (Umaña and Brandon 1992).

The creation of other reserves on the peninsula was also conflictive. The Guaymí Indigenous Reserve was created in 1985 for Indians who migrated to the Osa in the 1960s. Untitled settlers were removed to establish the reserve and did not receive any compensation until 1992. By 1995, about 120 Guaymí lived in the reserve. They have kept 80 percent of their reserve forested (Alonso 1999). In 1994, Piedras Blancas National Park (PBNP) was established, linking the peninsula and the mainland. As of the late 1990s, PBNP was still just a "paper park." Compensation to titled landowners led to complaints from locals that payments were a way to channel money to elites. Deforestation rates climbed in the park, as people quickly logged their lands before they officially became parkland.

The Osa Peninsula Conservation Area (ACOSA) was established in 1989, incorporating Corcovado National Park, Golfo Dulce Forest Reserve, Sierpe-Térraba Mangroves Forest Reserve, Golfito Wildlife Refuge, Piedras Blancas National Park, the Isla del Caño Biological Reserve, and, until 1994, the Guaymí Indigenous Reserve (Cuello, Brandon, and Margoluis 1998). ACOSA protects 410,402 hectares, or over 80 percent of the Osa Peninsula. CNP, called the "jewel in the crown" of Costa Rican parks, includes mountains, swamps, cloud forest, marsh, mangrove, and beach and an estimated 1,000 tree species, 140 mammal species, including jaguars and Baird's tapirs, 40 fish species, 367 bird species, including the largest population of scarlet macaws in the country, 177 reptile and amphibian species, including the American crocodile and caiman, and about 6,000 insect species (Franke 1993; Minca and Linda 2000). ACOSA also includes many important marine and coastal ecosystems, including the largest mangrove system and one of the last and most intact pacific coastal rainforests in Central America. Management of the adjacent RFGD and other protected areas is vital to biodiversity conservation. CNP is not ecologically viable if it becomes an island or if wide-ranging species cross the park boundary only to be hunted.

Threats to ACOSA fall into two broad interrelated categories: localized threats owing to poverty and rapid social change, and threats resulting from poor, weak, or inefficient government policies. Despite waves of migration to the Osa region, until recently, long-term settlement has been limited and turnover among settlers high, because of poor soils, very wet weather, and poor or absent services. Illegal logging, clearing, mining, and hunting are the most serious threats to ACOSA. Deforestation for timber, settlements, road clearing, and farm establishment cause habitat fragmentation and disrupt connectivity. Improvements in infrastructure such as roads and electricity, without planning, have led to a new wave of migrants, rich and poor, foreign and local, and more tension over ownership. Land conversion for oil palm, rice, and cattle ranching is common in the lowlands, while vacation homes for foreigners and planned resort development lead to land conversion and speculation. Artisanal mining is less of a current threat within CNP, but mining concessions in the buffer of CNP cause pollution and bring in hunting (Alonso 1999). Local miners feel hostility toward the government since they believe that their operations are less damaging than operations by larger international companies that conducted commercial mining with heavy equipment there. Illegal commercial and subsistence hunting and pet trade collection occur throughout ACOSA, including harvesting of turtle eggs from the beach and unregulated fishing (Stem 2001). Most coral reefs (90 percent) in the Golfo Dulce have been destroyed by mining sediment (Donovan 1994).

ACOSA has four offices in the Osa Peninsula's larger towns and eight management programs: coastal zone management, ecotourism, research, land surveying, environmental education, agroforestry, protection, and mining oversight. ACOSA is heavily dependent on international funding and has received money from the Global Environmental

Facility (GEF) to support sustainable development, ecotourism, and research. There is little evidence of integration or coherent planning in the different units that make up ACOSA or among non-governmental organizations (NGOs) working there.

NGOs and various development projects have been active in or near the Osa Conservation Area for nearly twenty years. INBio has four field offices in the Osa Peninsula that employ locals. The Osa-Golfito Project, funded by the European Union, has promoted agricultural production to reduce pressure on regional forest resources ($22 million for 2.5 years) (Minca and Linda 2000). Other groups include the TUVA Foundation and Fundación Neotrópica, which led the BOSCOSA project along with the World Wildlife Fund (WWF). Beginning in 1987, BOSCOSA provided environmentally sustainable economic alternatives to deforestation to locals in the buffer zones of CNP. It was a highly participatory grassroots effort to provide technical assistance and create linkages between and among local groups and between local programs and national programs. It was considered very successful in its first years, and many of the participatory approaches used were exemplary.[8] The focus has shifted to forestry, agriculture, environmental education, and training, with varying degrees of success in 13,000 hectares of the 77,000 hectares of the RFGD (Cuello, Brandon, and Margoluis 1998; Minca and Linda 2000).

Despite the existence of numerous NGOs and lots of money for activities around the Osa region, local people believe that they have not benefited from the various NGO projects. A formal evaluation of BOSCOSA in 1995 noted that the project's impact was greatly reduced because it attempted to deal with the whole buffer zone surrounding CNP (Hitz, Alpizar, and Montoya 1995). While the project received $7 million between 1987 and 1995, there is little evidence of this on the ground, leading to hostility toward conservation NGOs and a perception that the shift from development to conservation has left the people on the Osa Peninsula poorer (Minca and Linda 2000; see also Child and Dalal-Clayton this volume). Efforts that they believe should have paid off, such as tourism, have had few benefits; while ecotourism in Costa Rica boomed, CNP only had 103,089 visitors in the seven years between 1992 and 1999 (PROCIG 2001). Lack of coordination among projects in an area is often a problem with ICDPs, but this has been a serious problem for ACOSA. Even worse, there is not "any correspondence between key threats to CNP—hunting, deforestation, and mining—and buffer zone activities" (Cuello, Brandon, and Margoluis 1998:177). In 1997, Alvaro Ugalde, one of the founders of Costa Rica's park system, said in an open letter to the president in 1997, "Corcovado is a park that is rapidly dying, a park in extreme peril" (qtd. in Cuello, Brandon, and Margoluis 1998:186).

Guanacaste Conservation Area

Santa Rosa National Monument was created in 1966 to preserve La Casona, an old hacienda where Costa Ricans, in 1856, defeated invading U.S. mercenaries. Land for the monument was expropriated from cattle ranches belonging to Nicaragua's dictator, Somoza. In 1971, the monument became Santa Rosa National Park (SRNP). Rincón de la Vieja to the southeast of SRNP was established as a park in 1974 (Boza and Mendoza 1981), and Hacienda Murciélago was expropriated from Somoza in 1978 to expand the SRNP. In 1989, the Costa Rican government created Guanacaste National Park across the Pan American Highway from SRNP. This 10,400-hectare park protects 0.1 percent of the original expanse of tropical dry forest remaining in Costa Rica and destroyed elsewhere in Meso-America (Franke 1993; Sánchez-Azofeifa 1997). Together, these units make up the Área de Conservación Guanacaste (ACG), protecting approximately 120,000 terrestrial hectares, including more than 50,000 hectares of tropical dry forest and 70,000 marine hectares. The ACG has grown over the years, currently covering about 90 square kilometers, from 17 kilometers out into the Pacific Ocean all the way to the foothills of the Caribbean coastal plain. It contains more terrestrial biodiversity than can be found in the entire continental United States in its dry forest, cloud forest, and rainforests (Janzen 2001a). It is estimated that the ACG protects some 253 species of birds, 115 species of mammals, 110 species of reptiles and amphibians, and 10,000 species of insects, or about 60 percent of the species that occur in Costa Rica. The ACG was declared a UNESCO World Heritage Site in 1999 because of its natural and cultural value. A campaign to purchase the Rincon Rainforest is under way. This corridor would link lowland dry forest to adjacent rain and cloud forests, improve connectivity with other ecosystems, and increase the area for many species, from small seasonal migrants such as butterflies to wide-ranging carnivores such as jaguars (Janzen 2002).

Historical threats to the ACG included deforestation for agriculture and ranching, loss of habitat, unsustainable harvest of natural resources, anthropogenic fires, and uncontrolled tourism and development. Management, selective hiring, education, and fire control have helped mitigate these threats. Guanacaste has long been subject to human use; forests were converted to pasture during the late 1500s (Janzen 1986). Opening the Pan American Highway in the 1950s allowed easier access to markets and the expansion of a cattle industry based on direct government subsidies, leading to the conversion of nearly half of the region to pasture by the mid-1950s (Maldonado et al. 1995). Between 1979 and 1992, 119,712 hectares of primary and secondary forests were transformed into pasture—an increase of 151 percent in pastureland and enough to cause detectable changes in precipitation patterns (Canet et al. 1996). Covert U.S. activities along the border area against the Sandinista government in Nicaragua exacerbated lawlessness in the region.

Public relations campaigns in the 1980s by international conservation advocacy groups such as Rainforest Alliance Network (RAN) decried the "hamburger connection" and urged U.S. consumers to boycott Costa Rican beef. One observer noted: "The cattle industry never did recover from this campaign. There are lots of charales (fallowed pastures) that testify to the effectiveness of such a media blitz by RAN" (Frankie personal communication). An economic crisis and decline in international meat prices coincided with the campaign, causing Guanacasteco landowners to search for new government supports and subsidies or put their property up for sale. Pasture declined, but an invasive grass *(Hyperania rufa)* for grazing and the use of fire to control the grass remain as threats from the cattle legacy.[9] Ranchers set dry season fires, unnatural in the Guanacaste region, to eliminate the two-meter tall grasses and maintain open pasture. Yet the grass is adapted to fire and returns, hindering the growth of other plants. Tourism and development remain potential threats; the local airport in Liberia was upgraded, and the government has approved Cancún-style beach development along the Gulf of Papagallo. Environmentalists fought this development for years, fearing serious environmental consequences on marine populations, especially turtles (Honey 1999:131–181).

Dan Janzen, a world-renowned biologist, has worked tirelessly to promote his vision of decentralized management for the ACG and his belief that protected areas must provide direct benefits that peasants and politicians alike view as essential for development. The ACG is managed as a multiuse area, and limited and controlled resource extraction is allowed in certain areas. A high level of local employment has been a central goal, and the ACG operates as a single unit with "one budget, one director, one staff, one local board of directors, and one goal," composed of a resident staff of 110 Costa Ricans with programs in education, tourism, research and restoration, fire control, and policing (Escofet 2000; Janzen 2000, 2001a:42). Other activities include reforestation with gmelina, a fast-growing exotic that outcompetes the jaragua grass (Janzen 1999, note added Dec. 17, 2001). Begun in 1999 with Wege Foundation support, the forest understory is regenerating with native trees and vegetation beneath the gmelina. Visitation to the park has increased, providing some funds; for example, between 1992 and 1999, 330,549 people visited Santa Rosa National Park (PROCIG 2001).

Controversy involving the ACG and Dan Janzen is common. One example is the comptroller general's decision to revoke a 1999 deal between the ACG and the Del Oro juice company. The deal called for Del Oro to pay the park for the environmental services received, such as a source of pollinators and a sink for orange peel disposal. In addition to monetary payment, the park benefited by using orange peels to "smother" fields of jaragua grass and improve soil conditions (Escofet 2000). This deal was based on the principle of payment for environmental services, but turned into a politicized power struggle between competing juice companies and MINAE and debates over decentralization versus national control. The ACG has been a flashpoint for such debates, with one side viewing the ACG as a model of how a large protected area can be managed apart from the central government and the other side claiming that the ACG is *too* independent and that its strength detracts from the other ARCs. Janzen views these conflicts as rooted in the difference between the ACG's independent, horizontal, and site-specific management and the (slowly changing) traditional command and control, top-down, vertical management style of the government (see also Tongson and Dino, Gartlan, and Child and Dalal-Clayton, all this volume). He notes, "major Costa Rican forces do not want to see true decentralization emerge" (Janzen as qtd. in Escofet 2000:7).

Since 1995, it has cost about $44 million to establish the ACG (Janzen 2001b). It operates on a management endowment that is augmented by payments received for environmental goods. The ACG suffered a 30 percent cut in endowment funds in 1998, increasing its dependence on the government (Janzen 2002). This, and the firing of the ACG's director, has created management problems, and the staff feel the ACG has changed from a "wild area pioneering in changes that grows and evolves with an important level of independence, autonomy, and pride, to gradually becoming a traditional conservation area focused more on following procedures and rules tied to the restrictions, limitations and vices of the thinking of a traditional public institution" (personal communication). Janzen views the loss of financial independence from the central government and the return to being a guarded park as irreconcilable with long-term conservation (Janzen 2000, 2001a). Yet the line between being an unguarded and guarded park is not always clear. Last year, despite local support for the ACG, poachers who were apprehended and released got drunk and returned with gasoline and burned La Casona to the ground. Hopes run high that the new government administration that took office in 2002 will help turn things back around and again support decentralization (Gutierrez, Blanco, and Janzen, all personal communications).

Gandoca-Manzanillo Wildlife Refuge and La Amistad-Caribe Conservation Area

The Gandoca-Manzanillo Wildlife Refuge (GMWR)[10] was created in 1985 with support from ANAI, a conservation NGO, to block road construction through an area of great biological richness. GMWR protects 5,013 terrestrial hectares and 4,436 marine hectares, the last orey swamp, and last intact mangrove swamp on the Caribbean coast in Costa Rica, and coral reefs. The region has an exceptional number of endemic plant and animal species. It is home to 358 bird species and endangered species such as American crocodiles, tapirs, manatees, and sea turtles (Franke 1993; Lipton 1999). The Talamanca region has the most extensive montane ecosystems in the country and includes four major watersheds. La Amistad-Caribe Conservation Area (ACLA-C) is made up of fifteen separate conservation units, including La Amistad Biosphere Reserve, which was created in 1982 as a binational park sharing a border with Panama along the Talamanca Mountain Range and is a UNESCO World Heritage Site. Gandoca-Manzanillo as well as parts of La Amistad, Cahuita, and Barbilla National Parks, several indigenous reserves, the Hitoy Cerere Biological Reserve, and three wildlife refuges are among the other protected areas. Many of these are part of the Talamanca Biological Corridor (TBC). ACLA-C was one of the two conservation areas in the country with the most remaining forest cover in 1992 (PROCIG 2001).

Talamanca is the poorest region of Costa Rica, with an unusually heterogeneous mixture of indigenous, Afro-Caribbean, Chinese, and Ladino peoples. About 25,000 live in or around the TBC; most near urban areas earn money through some combination of agriculture, logging, ranching, and tourism. The area is rich in biodiversity and indigenous culture but has very poor soils and steep slopes, making it bad for farming. Lowland areas were cleared and settled in the early 1900s for banana plantations, rail construction to take bananas out, and worker housing. Widespread deforestation contributed to flooding in the 1930s, which in turn led to a fungal outbreak that caused the banana company to leave the area. Unemployed workers switched from wage labor to small-scale cacao production. In 1980, a new fungal outbreak devastated the cacao production, causing a 95 percent reduction in production in the region and triggering a local recession (Lipton 1999). Large multinational corporations like Del Monte, Dole, and Chiquita employ 71 percent of the indigenous population in this region (Lipton 1999; Parrish, Reitsma, and Greenberg 2001).

Deforestation and logging are problematic in the highlands, even though indigenous populations living there generally do not give permission for logging, nor receive benefits from it (Heffington and Mimbs 1999; Lipton 1999). Illegal clear-cutting without subsequent reforestation has resulted in erosion and downstream sedimentation that has damaged coral reefs and hurt tourism. Lowland deforestation results from logging, clearing for small-scale agriculture, and commercial banana production, and has led to flooding over the years (Parrish, Reitsma, and Greenberg 2001). Residents have expressed concerns that they are getting sick from chemicals used on the banana and plantain crops (Lipton 1999).

Road construction from the mid-1980s onward led to a tourism boom. The coastal areas have become popular weekend and vacation spots, leading to land speculation and clearing and a dramatic change in ownership of the land along this stretch of the coast; foreigners now own an estimated 90 percent (Venegas personal communication). The protected areas in the region with the greatest levels of visitation are the Gandoca-Manzanillo Wildlife Refuge and Cahuita National Park (Parrish, Reitsma, and Greenberg 2001).

Petroleum exploration was until recently the greatest threat to Gandoca-Manzanillo, since several companies have concessions for oil exploration (Venegas personal communication). In 2001, a U.S.-based company with close ties to the Bush administration began drilling off the Talamancan coast without the appropriate impact assessments. Citizens and NGOs began fighting this project on technical, legal, and environmental grounds. In February 2002, Costa Rica's highest court ruled on a technical point that effectively blocked exploration, and in March the National Technical Secretariat declared that oil exploration was not an environmentally viable development option for the region. The petroleum companies filed appeals of these decisions, but their appeals were rejected in May 2002 by the new government of President Pacheco (Loaiza 2002a, 2002b; National Resources Defense Council 2002).

ANAI is a grassroots organization founded in 1978 that works with the communities in Talamanca on ecotourism, cultural programs, biomonitoring, bird and turtle monitoring, organic farming assistance, education, and advocacy. ANAI was instrumental in the establishment and protection of the Gandoca-Manzanillo Wildlife Refuge (cf. Wells and Brandon 1992; Franke 1993). ANAI tried to halt deforestation by working with small farmers on diversification and cacao production, using small plots and labor-intensive methods to control fungal outbreaks. This intensification reduced the footprint of cultivated area and absorbed excess labor. There have been dual benefits for farmers and biodiversity because much of the cacao is grown as a shade crop by small farmers, and such methods maintain a high diversity of birds and other canopy dwellers (Parrish, Reitsma, and Greenberg 2001). This project has led ANAI into helping create the following:

- An organic cacao producer's cooperative called APPTA,[11] which has 1,500 members and is the primary producer of organic cacao in Central America, selling to Paul Newman's Own Organics for above-average market prices (Barquero 2001)
- The Talamancan Association of Ecotourism and Conservation in 1990, which promotes responsible tourism and cultural/local pride, offers tours and birding trips, hosts student groups, and runs a small-scale credit system[12]
- ASACODE, a peasant-run group that works with sustainable agriculture and forestry, and offers tours and lectures on their property in the TBC[13]

ANAI's current emphasis is on leatherback sea turtle conservation, where they have turned a 95 percent poaching rate of turtle eggs around to a greater than 90 percent survival rate (Barquero 2001; Asociación ANAI 2002).

La Amistad Conservation and Development Initiative (AMISCONDE), started in 1992, is designed to improve the economic context of the local communities in the buffer zones of La Amistad Biosphere Reserve. AMISCONDE provides loans to locals through a revolving credit fund that was eventually turned over to a local community organization. By 2000, the majority of the members had benefited from the loans, and they had a 99.5 percent recovery rate. A small percentage of the recovered loans are put into an environmental education fund, and the remainder is reinvested as new loans (Conservation International 2001; Duffy, Corson, and Grant 2001). The project cost nearly $3 million, or about $290 per beneficiary.

Local organizations banded together to form the Association of Organizations of the TBC, working on corridor planning and local participation and supported by the government and The Nature Conservancy. The TBC project has supported organic cacao farming, a carbon sequestration program (see also Kiss this volume), a reforestation project, education programs for children, and endangered green iguana farming to augment the wild population. The Nature Conservancy, the World Wildlife Fund, and the University of Rhode Island sponsored the PROARCA/Costas project from 1995 to 2001, providing technical and financial assistance and training local organizations to set priorities and link with national-level groups. This project included site planning for Gandoca-Manzanillo and materials for refuge guides (Coastal Zone Management Component 2000).

Summary of ICDP Impact at the Three Sites

The broader conditions for ICDP success related to their scope and scale include the need for long-term funding and time horizons, strong social stability, and political support for policy and project-level components. How well did the social context at each site support these elements? What are the key lessons from looking at twenty years of ICDP implementation at each site?

The rapid levels of change on the Osa region make it a poor place for ICDP approaches (Wells and Brandon 1992; Brandon 2002). Yet this has not stopped virtually all funds spent there from falling under the ICDP label; little money has gone for direct activities that support conservation, a long-term commitment to enforcement through direct employment and education, or direct and prompt land purchases (see Robinson and Redford this volume). While funding levels at times have been high, they have not gone "through" the conservation area—they have often competed with it. There has been little support or link between local-level initiatives and the broader policy context. A list of positive steps is impossible to generate; instead, the Osa region provides examples of classic mistakes:

- Resettlement without prompt compensation or payment of inholdings has been prevalent, creating high levels of resentment toward conservation.
- NGO coordination has been factionalized or absent, leading to numerous projects that lack any coherent strategy and few links between conservation and development.
- There is no adaptive management or learning process in place; so no one knows what has been tried or its outcome, and mistakes are repeated (see Salafsky and Margoluis this volume).
- Money has been spent in ways that are "flashy"—vehicles, consultants, studies, housing—but little local employment has been generated, nor have funds been invested in visible outcomes.
- Boundaries have not been clearly demarcated or patrolled, leading to biodiversity loss in ACOSA; for example, 20 percent (or 10,200 ha) of the GDFR had been logged by 1998 (Barrantes et al. 1999).

In hindsight, Corcovado National Park could easily have been expanded to include most of the forest reserve at a fraction of the money spent on ICDPs. Land could have been bought, and endowment for management and local employment created, which could have provided long-term financial stability for the park and directly linked benefits for local communities. Experimentation and innovation were possible—perhaps land claimed by gold miners could have been excised from the park and traded for other land. Yet money was spent in uncoordinated and misdirected ways unlinked to clear conservation outcomes. The absence of an institutional memory has meant that old ideas are introduced with new enthusiasm, but with little chance of success. The bottom line for biodiversity conservation is that things are perhaps worse on the Osa Peninsula than they have ever been, and there is no evidence as yet of an institutional structure that is likely to lead to improvements.

Activities in the Guanacaste region have been the most focused of the three sites, largely because of the long-term presence of Dan Janzen, who has steadfastly held to several principles, even when they were controversial. The following have been key elements of Guanacaste consolidation:

- Direct land purchase at competitive market rates, thereby limiting local hostility and making conservation "competitive" with other land uses (see also Kiss this volume)
- Decentralization, from local management committees to an independent financial base through the creation of an endowment
- High levels of long-term local employment as a priority for the ACG
- Strong local environmental education programs coupled with clear explanations of the environmental services generated by the ACG
- Publicity and a visible demonstration of the biodiversity significance of the area through Dan Janzen's use of personal prize money to support land purchases
- Decline in pressure from external factors: falling cattle prices and the end of the war with Nicaragua
- An inconsistent policy context even as the management of the conservation area has tried to be consistant, and tension created by changes in what is or is not permissible and what decisions can and cannot be decentralized

The social context in Guanacaste has been one of change, but of reduced pressure as people migrated out. While this has affected social stability, it has made conservation an attractive employer in the area. The creation of the endowment is significant. Janzen's views have acted like a rudder, setting clear strategic directions and management objectives.

Talamancan communities are more stable than those on the Osa Peninsula, so the local organizations in the Gandoca-Manzanillo and ACLA-C region have benefited from social stability amid significant changes in infrastructure, population density, and landownership patterns. After more than fifteen years, ANAI is well known, as are its staff, many of whom are local. In 1992, it seemed that ANAI was simply scattering diffuse investments over a large area, hoping something would pay off; activities covered an extensive geographic area, had many different project components, and had a low degree of intensity (Wells and Brandon 1992). Funding has been sporadic and not well orchestrated in the ACLA-C. Yet hard work and lots of luck led to a payoff when the community-run nurseries supported fungal-resistant strains of cacao, and farmers' inability to purchase agricultural inputs and the creation of APPTA (the producer's coop) meant that lots of pesticide-free cacao coincided with a market demand for this product. ANAI has scaled activities to funding opportunities, conservation priorities, and local desires. In a sense, ANAI has remained the "most true" to its objectives—even though the relatively small scale of activities relative to the region led outside groups to come in with their own ICDPs. The following are the key elements of ICDPs in Talamanca:

- ICDPs run by conservation organizations generally revolved around stabilizing and supporting (e.g., credit) existing activities, not introducing new ones.
- ACLA-C conservation areas are somewhat protected by steep topography that limits logging access, slowing deforestation.
- Communities have been stable enough for groups to demand that project funding be channeled through them, rather than directly to parks, capital city NGOs, or consultants.
- Direct links between conservation and development activities have been weak, but have been supported by long-lasting programs and outreach, environmental education, nascent ecotourism, and long-standing and indigenous communities.

Other lessons that have more to do with implementation issues were found in this study and have been noted in other papers in this volume. These include the following:

- *External impacts,* beyond what ICDPs could foresee or control, led to numerous changes in all three projects. Falling cattle prices in Guanacaste and the emergence of markets for organic cocoa in Talamanca were unforeseen impacts that changed local conditions to improve the chances for conservation. The Osa region had no such luck! These external impacts can give rise to opportunities that would not otherwise have been available (cf. Wells and Brandon 1992; Brandon, Redford, and Sanderson 1998).
- *Direct links work best* in gaining support for conservation. Direct employment generated through conservation activities provides the clearest link to conservation objectives.

- *Alliances are politically necessary, but they cannot be equated with successful implementation* (Brandon, Redford, and Sanderson 1998; Margoluis et al. 2000). The sites with the greatest number of "players"—the Osa and ACLA—have diffuse activities, lots of organizations, and unclear links between development actions and conservation objectives. The different groups involved all have their own agendas—leading to lots of different activities but no clear aim. In contrast, the activities at Guanacaste focus on increasing the size of the protected area, and in generating direct, local benefits linked to conservation. Activities at Guanacaste go through ACOSA, whereas activities at the other sites "happen" nearby.

Lessons on Linking Parks, Projects, and Policies

Beyond reviewing ICDP performance at the site level, what can we say about the policy context and how it has or has not affected ICDP performance at each of the three sites? What lessons are there for other countries interested in "scaling up" from park-based efforts to corridors and landscapes amid rapid change and development and high pressure for conversion? At the outset of this paper, three major elements of policy reform in Costa Rica—creation of a protected area network, an institutional framework to link protected areas with surrounding lands, and supportive environmental policies—were introduced as the three areas where Costa Rica has been innovative. The lessons that emerge on how these policies affected implementation of ICDPs are discussed below.

Creating a Protected Area System

Costa Rica is exemplary for all the land that has been protected—25 percent. Yet this number reveals only part of the story. Strictly protected areas form about 14 percent; yet many of these national parks are too small to be ecologically viable unless the protected areas that surround them, which fall into the less restricted category, remain intact. Protected areas next to national parks *must* have biodiversity conservation as a key management objective. Yet management of these areas, especially the forest reserves, has been weak to nonexistent; there is corruption and inefficiency, from permitting timber harvesting in inappropriate areas to outright uncontrolled illegal logging (Campos 2002). While national parks are highly respected by most Costa Ricans, other protected areas are not (Sanchez-Azofeifa et al. 1999; Campos 2002). This highlights the important role of areas where the management intent is strict protection. A recent study by Conservation International of ninety-three parks of over 5,000 hectares each, over five years old, and in twenty-two countries found that despite high levels of pressure, over 80 percent of the parks have as much natural vegetative cover today as they did when they were first established, and a substantial percentage have more (Bruner et al. 2001). Parks also suffered much less deforestation than surrounding areas. Similarly, in Costa Rica, land within "other" categories may be protected in name only. These findings are significant given recent evidence that the existing parks do not provide a good representation of biodiversity within Costa Rica (Powell, Barborak, and Rodriguez 1999). There is therefore a need to establish more protected areas and to ensure that the protected areas that do exist are functional.

Lessons from the Guanacaste experience are useful here: we must establish parks in the "right" way from the beginning, by making conservation competitive with other land uses (Brandon 2002; Terborgh et al. 2002; Kiss this volume). This means buying land, eliminating inholdings, and supporting enforcement through local employment. Although this may seem expensive, the costs are relatively small compared to the amount spent on ICDPs or "sustainable development" activities that fund international consultants but leave little on the ground. While there is an assumption that strict protection may be socially unacceptable, stable employment generated through conservation, coupled with environmental education and negotiation, may produce better results than well-intentioned ICDPs. Such ICD approaches in an inappropriate environment have had little success in helping the poor or contributing to conservation. In some places, such as the Osa Peninsula, they have led to outright hostility. A second lesson is that enforcement is an inevitable component of management; it will not always be effective, but it does not have to be draconian to work. Hiring local people—as many as funding allows and training them to be effective park advocates—is one of the best direct links. In the case of Guanacaste, the hundreds of people trained as parataxonomists, guards, environmental educators, and ecotourism guides did not stop disgruntled hunters from burning down one of the country's most historic sites. But it has provided a direct link between conservation and development and helped demonstrate the economic importance of the ACG to the area. The same has been true for the turtle protection initiatives and local tourism activities supported by ANAI—these have created an active, vocal, peasant constituency whose livelihoods are tied to biodiversity outcomes.

Institutional Framework to Link Protected Areas with Surrounding Lands

SINAC is an attempt to create a decentralized, intermediate-level institutional structure that can simultaneously coordinate across and effectively deal with both national-level ministries and local needs and municipalities. Dividing the country into eleven ARCs, each semiautonomous and somewhat responsible for its own fund-raising, has not brought about the intended results. The creation of SINAC has led to inequality across the country's ARCs; those with prominent advocates or "better" biodiversity have been more successful at capturing funds. There is little relationship between the importance of the area in terms of biodiversity and the funding received through the ARC. SINAC as a whole has not consolidated the necessary political power to thwart bad decisions by other governmental agencies or capture benefits from supportive environmental policies. The structure to do this is in place, however. Even though the way that the ARC system has been implemented has been somewhat "symbolic," getting a good framework in place is a significant first step (Barrett et al. 2001).

The three conservation areas mentioned here have had the highest financial contributions within the ARC system. Guanacaste, because of the attention and international connections created by Dan Janzen, has been the most adept at capturing international funds and using them to help the ACG stand on its own—financial resources help bring power. ACOSA, in contrast, has had no charismatic leader to help rally round the ARC. While the areas within it are famous, relatively limited support has come to ACOSA. The lack of a strong leader or a clear vision for the ARC has meant that donors have programmed most money as they have seen fit. This funding has not been coordinated in any way that addresses biodiversity conservation—it rests on a naïve assumption that if things are better off for some people in the region, the effect will ripple over to conservation. ACOSA has lacked the funding or ability to effectively coordinate internally or manage or coordinate the actors, events, and projects that are ostensibly linked to conservation. While millions of dollars were available for rural development and agricultural projects designed to reduce pressure on the Osa region's protected areas, in 1997 all of the protected areas within ACOSA, and the ACOSA administration itself, depended on fewer personnel than the total number that Corcovado used to have, leaving ACOSA as a collection of unprotected parks. In this case, biodiversity has been lost, the odds for conservation diminished, and investments in buffer zone activities have produced few tangible benefits for local people. Ironically, "progressive" funding purporting to support "rural development" in zones that are inappropriate for agriculture have brought meager social benefits and may have created antagonism that is misdirected toward conservation organizations. The ACLA-C conservation area has had relatively little formal support even though projects around it have been well funded. However, the ICDPs in the ACLA-C have all been channeled through community groups promoting activities that stabilized existing livelihoods, buffered against profound changes, and had linkages to biodiversity conservation.

SUPPORTIVE ENVIRONMENTAL POLICIES

The virtues of SINAC's decentralized approach have been offset by the lack of political power accorded to conservation at the national level. Costa Rica has taken many actions that, together, could solidly and broadly support conservation. It is a model in policy formulation—but not in policy implementation. Consolidation of different agencies into an environmental ministry should have consolidated national political power, but at the national level this ministry has been weak. Costa Rica has devised extremely creative financing mechanisms to support conservation, such as the ESPs, but the money is often misdirected owing to conflicting priorities and politics. For example, a third of the revenues from taxes on fossil fuels should go to conservation, but for one fifteen-month period, only $7 million of the $25 million that was supposed to go to conservation was paid out to forest owner associations (World Bank 2000b). How does this affect sites such as Guanacaste, Osa, and Gandoca-Manzanillo? It means that legitimate programs set up to encourage the private sector to support conservation are undermined (see also Kiss this volume). Incentives for local landowners to engage, in groups, in ecotourism, reforestation, or natural forest management are weakened. It means that the ARCs must continue to look to donors from outside Costa Rica for support. It also means that although the public believes that their taxes support conservation, the reality is different, and little local-level employment or few direct cash benefits are directly generated by the conservation sector. This negates key components of the intended resource transfer and ARC system.

The government's inconsistency on policies has also affected the tourism sector. While official rhetoric supports ecotourism as a basis for supporting local-level development, government policies actually support large-scale projects and put numerous obstacles in the way of small-scale tourism development (Honey 1999:131–181). Although tourism has surpassed all agricultural exports, providing $950 million, or 9 percent of GDP in 1991—between 40 percent and 70 percent of foreign tourists visited protected areas every year from 1992 through August 2000—at the national level nature is a marketing tool for tourism, not a national priority (World Bank 2000b:23; Zamora and Obando 2001). The few concrete benefits from tourism only provide limited incentives for conservation in the ARCs. At the local level, the lack of linkages among tourism money, jobs, and conservation often prevents local people from embracing conservation (cf. Stem 2001). The failure to support resource transfers through the ARCs diminishes the standing of the conservation sector.

The legal and judicial sector, and governance issues of corruption, inefficiency, and transparency, also undermine the ARC system. Nationally, there have been huge difficulties controlling environmental crimes; there are not enough prosecutors, communication in the judiciary is weak, and there is inadequate follow-up on cases. There are few convictions for environmental crimes. According to the environmental comptroller, enforcement of environmental laws has gotten worse, and there is a lack of resources. A lack of administrative ability is also true of SINAC—only one in ten logging plans is verified on the ground (Rivera 2000).

Conclusion

The policy actions taken in Costa Rica demonstrate that a great deal is possible amid situations of high resource-use pressure and habitat fragmentation. Despite a strong commitment to protected areas and conservation in the official rhetoric of the country, the challenges at the park level differ little from those of many other developing countries. Parks are invaded, illegal hunting and poaching take place, and funds for park management are lacking. ICDPs in Costa Rica have led to some successes, but it is likely that, on balance, the gains have largely been from luck rather than planning, and many ICDPs have caused more harm than good. Support was channeled through the ARC in Guanacaste, and for the most part the linkages have been fairly clear and direct, leading to hope that over time the benefits from the conservation area will be seen as outweighing the costs. In the Osa region, funding and projects purportedly designed to support conservation have undermined it thus far. In Talamanca, it has only been through a combination of luck and the long-term existence of local organizations that a link has been forged between conservation and development activities.

Costa Rican policies do provide some models for other countries. For example, the ESP is a model of implementing a "polluter pays" principle in the urban sector to support conservation in the rural sector. Thus far ICDPs have not benefited from the array of policy reforms that Costa Rica has enacted. On the contrary, in some cases they have diminished the intent of key policy reforms, such as the power of the ARC, and mechanisms for lasting, innovative financing. ICDPs have been expensive and have left few tangible results on the ground. Creation of protected areas with endowments for management, that pay market prices for land, that provide direct employment of local people in activities linked to conservation, and that have clear management objectives for sites all worked well when they were applied. What has been achieved in Costa Rica is remarkable, in that it has been done under such difficult conditions and within an overall policy context where "sustainable development is still not an official national objective" (World Bank 2000b:61). Perhaps if it were, and if the policies and laws were implemented as written, conservation could proceed as a competitive sector providing a range of services and benefits. If genuine political support for conservation existed, it would be even further reason to abandon the delinked parts of ICDPs and make sure that support was channeled directly through protected areas. This would undoubtedly improve the odds for conservation, and would likely lead to better outcomes for poor rural communities.

Endnotes

1. SINAC maintains Web-based information on each ARC.
2. This was called MIRENEM when created; it is now MINAE.
3. Terms of this are now changing with the new administration (Rodriguez personal communication).
4. For example, a joint loan/grant of $49.2 million from IRBD and the GEF targeting 50,000 hectares in ACLA-C, ACOSA, and Tortuguero by 2006 (World Bank 2000a).
5. See review of ICDP activities of ANAI in Talamanca and BOSCOSA (Wells and Brandon 1992).
6. Drawn from the authors' experience and Cuello, Brandon, and Margoluis 1998, unless otherwise cited.
7. In the late 1980s, FN produced the Osa 2000 plan, a vision exercise coordinated across stakeholders. High mobility meant many local leaders left Osa, and by 1995 no one remembered this exercise and there was little basis for planning or action (Wells and Brandon 1992; Donovan 1994; Cuello, Brandon, and Margoluis 1998).
8. Some benefited from a $38 million irrigation project supported by the Inter-American Development Bank and the Venezuelan government to grow nontraditional agricultural exports.
9. Ironically, the grass is not even a good source of cattlefeed (Quesada and Stoner in press).
10. See Wells and Brandon 1992 for a detailed account of ANAI and the area.
11. Asociación de Pequeños Productores de Talamanca.
12. ATEC 2002.
13. ASACODE 2002.

References

Alonso, M. 1999. *Rainforest Conservation and Development in the Osa Peninsula.* http://www.tuva.org/pdf/report.pdf (14 April 2002).

ASACODE (Asociación San Migueleña de Conservación y Desarollo). 2002. http://www.costaricanoni.com/asacode/index.html (14 April 2002).
Asociación ANAI. 2002. http://www.anaicr.org (16 April 2002).
ATEC (Talamanca Association for Ecotourism and Conservation). 2002. Talamanca discovery. http://www.greencoast.com (16 April 2002).
Barquero, Julio. 2001. Interview by C. August Brown. Hone Creek, Costa Rica, August.
Barrantes, G., Q. Jimenez, J. Lobo, T. Maldonado, M. Quesada, and R. Quesada. 1999. *Evaluación de los planes de manejo forestal autorizados en el período 1997–1998 en la Peninsula de Osa. Cumplimiento de normas técnicas, ambientales e impacto sobre el bosque natural.* San José, Costa Rica: Fundación Cecropia
Barrett, C. B., K. Brandon, C. Gibson, and H. Gjertsen. 2001. Conserving tropical biodiversity amid weak institutions. *BioScience* 51 (3):497–502.
Boza, M. A. and R. Mendoza. 1981. *The National Parks of Costa Rica.* San José, Costa Rica: INCAFO, S.A.
Brandon, K. 2002. Getting the basics right: Key actions in designing effective parks. In J. Terborgh, C. van Schaik, L. C. Davenport, and M. Rao, eds., *Making Parks Work: Strategies for Preserving Tropical Nature,* 443–467. Washington, D.C.: Island Press.
Brandon, K. in press. The policy context for conservation in Costa Rica: Model or muddle? In G. Frankie, A. Mata, and S. B. Vinson, eds., *Biodiversity Conservation in Costa Rica: Learning the lessons in a seasonal dry forest.* Berkeley and Los Angeles: University of California Press.
Brandon, K., K. H. Redford, and S. E. Sanderson, eds. 1998. *Parks in Peril: People, Politics, and Protected Areas.* Washington, D.C.: Island Press.
Bruner, A. G., R. E. Gullison, R. E. Rice, and G. A. da Fonseca. 2002. Effectiveness of parks in protecting tropical biodiversity. *Science* 291 (5501):125–128.
Campos, J. J. 2002. *Illegal Logging in Costa Rica, An Analysis for Discussion.* Turrialba, Costa Rica: Tropical Agricultural Research and Higher Education Centre (CATIE).
Canet, G., M. Chavarria, O. Gamboa, D. Garita, M. Jimenez, S. Lobo, P. Marín, L. Sevilla, Z. Trejos, and M. Valerio. 1996. *Informacíon estadística relevante sobre el sector forestal 1972–1995.* San José, Costa Rica: Ministerio del Ambiente y Energía, Sistema Nacional de Áreas de Conservación, Área de Fomento.
Coastal Zone Management Component of the Central American Environment Project. 2000, August. http://www.eco-index.org/search/results.cfm?ProjectID=71 (16 April 2002).
Conservation International FACTS. Updated June 2001. *Credit for Conservation: AMISCONDE Trust.* http://www.conservation.org/ImageCache/CIWEB/content/publications/amisconde_2epdf/v1/amisconde.pdf (14 April 2002).
Cuello, C., K. Brandon, and R. Margoluis. 1998. Costa Rica: Corcovado National Park. In K. Brandon, K. H. Redford, and S. E. Sanderson, eds., *Parks in Peril: People, Politics, and Protected Areas,* 143–191. Washington, D.C.: Island Press.
Donovan, R. 1994. BOSCOSA: Forest conservation and management through local institutions (Costa Rica). In D. Western and R. M. Wright, eds., *Natural Connections: Perspectives in Community-Based Conservation,* 215–233. Washington, D.C.: Island Press.

Duffy, S. B., M. S. Corson, and W. E. Grant. 2001. Simulating land-use decisions in the La Amistad Biosphere Reserve buffer zone in Costa Rica and Panama. *Ecological Modeling* 140:9–29.

Escofet, G. 2000. Costa Rican orange-peel project turns sour. *EcoAmericas* 2 (8):6–8.

Franke, J. 1993. *Costa Rica's National Parks and Preserves*. Seattle, Wash.: The Mountaineers.

Heffington, D. and J. Mimbs. 1999. Sustainable development in Costa Rica: An approach to the geography curriculum. *Social Education* 63 (2):80–85.

Hitz, W., E. Alpizar, and F. Montoya. 1995. *Evaluación Proyecto BOSCOSA, Fundación Neotrópica y Recomendaciones para la Transición*. Draft final report. San José, Costa Rica: USAID.

Honey, M. 1999. *Ecotourism and Sustainable Development: Who Owns Paradise?* Washington, D.C.: Island Press.

Janzen, D. H. 1986. *Guanacaste National Park: Tropical Ecological Cultural Restoration*. San José, Costa Rica: Editorial Universidad Estatal a Distancia.

Janzen, D. H. 1999. Use of gmelina plantations as a tool to restore ancient rainforest pastures to natural rainforest and simultaneously build a living endowment for a conserved rainforest. Proposal to the Wege Foundation from the Área de Conservación Guanacaste (ACG). Submitted by: Área de Conservación Guanacaste (ACG) in northwestern Costa Rica, 2 March.

Janzen, D. H. 2000. Costa Rica's Área de Conservación Guanacaste: A long march to survival through non-damaging biodiversity and ecosystem development. In *Norway/UN Conference on the Ecosystem Approach for Sustainable Use of Biological Diversity*, 122–132. Trondheim, Norway: Norwegian Directorate for Nature Research and Norwegian Institute for Nature Research.

Janzen, D. H. 2001a. Good fences make good neighbors: Área de Conservación Guanacaste, Costa Rica. *Parks* 11 (2):41–48.

Janzen, D. H. 2001b. Saving fractured oases of biodiversity. In R. A. Mittermeier, N. Myers, P. Robles Gil, and C. Goettsch Mittermeier, eds., *A Review of Hotspots: Earth's Biologically Richest and Most Endangered Terrestrial Ecoregions*, 327–330. Chicago: University of Chicago Press.

Janzen, D. H. 2002, 13 February. Rincon Rainforest. http://janzen.sas.upenn.edu/caterpillars/RR/rincon_rainforest.htm (13 February 2002).

Langholz, J. A., J. P. Lassoie, and J. Schelhas. 2000. Incentives for biological conservation: Costa Rica's private wildlife refuge program. *Conservation Biology* 14 (6):1735–1743.

Lipton, J. K. 1999. Social dimensions of biological corridors: The Talamanca Caribbean Biological Corridor, Costa Rica. Master's thesis, University of Texas at Austin.

Loaiza, N. V. 2002a. Setena rechazó búsqueda petrolera. *La Nación*, 2 March. www.nacion.com (16 April 2002).

Loaiza, N. V. 2002b. Petroleras defienden exploración. *La Nación*, 9 March. www.nacion.com (16 April 2002).

Maldonado, T., J. Bravo, G. Castro, Q. Jiménez, O. Saborio, and L. Paniagua. 1995. *Evaluación ecológica rápida región del Tempisque Guanacaste, Costa Rica.*

San José, Costa Rica: Fundación Neotrópica, Centro de Estudios Ambientales y Políticas.

Margoluis, R., C. Margoluis, K. Brandon, and N. Salafsky. 2000. *In Good Company: Effective Alliances for Conservation*. Washington, D.C.: Biodiversity Support Program.

Margoluis, R., V. Russell, M. Gonzalez, O. Rojas, J. Magdaleno, G. Madrid, and D. Kaimowitz. 2001. *Maximum Yield? Sustainable Agriculture as a Tool for Conservation*. Washington, D.C.: Biodiversity Support Program.

Minca, C. and M. Linda. 2000. Ecotourism on the edge: The case of Corcovado National Park, Costa Rica. In X. Font and J. Tribe, eds., *Forest Tourism and Recreation: Case Studies in Environmental Management*, 103–126. New York: CABI.

National Resources Defense Council (NRDC). 2002. http://www.nrdc.org/media/pressreleases/020513.asp (21 May 2002).

Parrish, J., R. Reitsma, and R. Greenberg. 2001. *Cacao as Crop and Conservation Tool: Lessons from the Talamanca Region of Costa Rica*. Washington, D.C.: Smithsonian Migratory Bird Center, Smithsonian Institution.

Powell, G. V. N., J. Barborak, and M. Rodriguez. 1999. Assessing representativeness of protected natural areas in Costa Rica for conserving biodiversity: A preliminary gap analysis. *Biological Conservation* 93 (1):35–41.

PROCIG (Central American Geographic Information Project). 2001. Análisis de población y agricultura asociado a Áreas Silvestres Protegidas y Áreas de Conservación. May. http://www.procig.org/reunion-ctg/documentos/costa-rica/procig-final/procig-final.pdf (16 April 2002).

Quesada, M. and K. E. Stoner. In press. Threats to the conservation of dry tropical forest in Guanacaste, Costa Rica: The case of the Tempisque Conservation Area. In G. Frankie, A. Matz, and S. B. Vinson, eds., *Biodiversity Conservation in Costa Rica: Learning lessons in a seasonal dry forest*. Berkeley and Los Angeles: University of California Press.

Rivera, E. 2000. Débil control en delitos ambientales. *La Nación*, 31 December. www.nacion.com (16 April 2002).

Rodriguez, C. M. 2002. Presentation. La Selva Biological Station, Costa Rica, 27 May.

Sánchez-Azofeifa, G. A. 1997. *Assessing Land Use–Cover Change in Costa Rica*. San José, Costa Rica: Centro de Investigaciones en Desarrollo Sostenible, Universidad de Costa Rica.

Sánchez-Azofeifa, G. A., C. Quesada-Mateo, P. Gonzalez-Quesada, S. Dayanandan, and K. S. Bawa. 1999. Protected areas and conservation of biodiversity in the tropics. *Conservation Biology* 13 (2):407–411.

Stem, C. J. 2001. *The Role of Local Development in Protected Area Management: A Comparative Case Study of Ecotourism in Costa Rica*. Draft summary of dissertation presented to the faculty of the Graduate School of Cornell University.

Terborgh, J., C. van Schaik, L. C. Davenport, and M. Rao, eds. 2002. *Making Parks Work: Strategies for Preserving Tropical Nature*. Washington, D.C.: Island Press.

Umaña, A. and K. Brandon. 1992. Inventing institutions for conservation: Lessons from Costa Rica. In S. Annis, ed., *Poverty, Natural Resources, and Public*

Policy in Central America, 85–107. Washington, D.C.: Overseas Development Council.

Wells, M. and K. Brandon. 1992. *People and Parks: Linking Protected Area Management with Local Communities*. Washington, D.C.: The World Bank.

World Bank. 2000a. *Project Appraisal Document on a Proposed IBRD Loan of US$32.6 Million and a Grant from the Global Environmental Facility Trust Fund of SDR 6.1 Million (US$8 Million Equivalent) to the Government of Costa Rica for the Ecomarkets Project*. Report No: 20434-CR. Washington, D.C.: The World Bank.

World Bank Operations Evaluation Department. 2000b. *Costa Rica: Forest Strategy and the Evolution of Land Use*. Washington, D.C.: The World Bank.

Zamora, N. and V. Obando. 2001. *Biodiversity and Tourism in Costa Rica*. San José: Instituto Costarricense de Turismo, ICT Sistema Nacional de Áreas de Conservación, SINAC Ministerio del Ambiente y Energía, MINAE Instituto Nacional de Biodiversidad, INBio.

9

Indigenous Peoples and Protected Areas: The Case of the Sibuyan Mangyan Tagabukid, Philippines

Edgardo Tongson and Marisel Dino

Introduction

This paper describes the experience of Kabang-Kalikasan ng Pilipinas (KKP; WWF–Philippines) in assisting the island's indigenous peoples in obtaining tenure and seeking recognition of their ancestral domain. A summary of the history and past interventions with the Sibuyan indigenous people is presented. The legal framework governing indigenous people's rights and protected areas is briefly discussed. This is followed by a description of project objectives, assumptions, and the methodology undertaken. Intervention results, issues, and obstacles encountered, as well as opportunities and challenges for the future, are presented. Finally, the lessons learned and recommendations are discussed.

The Setting

Located at 12°21′ north latitude and 122°39′ east longitude in the center of the Philippine archipelago, about 350 kilometers south of Manila, Sibuyan is the second largest island of Romblon Province (figure 9.1). It has a land area of approximately 45,600 hectares, more than half of which is covered with forest. As of May 2000, its population stood at

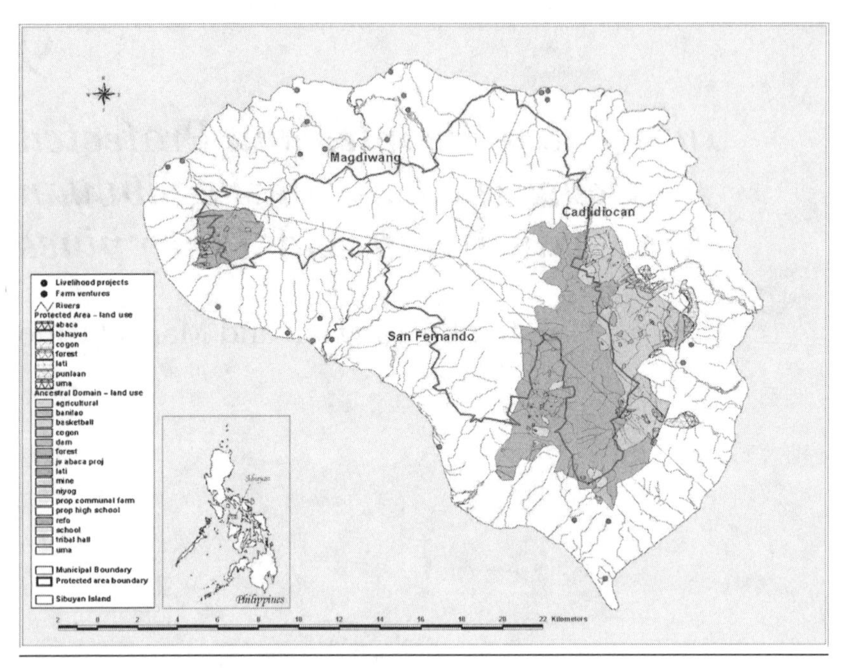

Figure 9.1.
Map of Sibuyan Island.

52,615, distributed mainly along the coastal plains and on the lower slopes of the island's mountainous forests. The majority of this population depend on subsistence-level agriculture and fishing.

At the heart of Sibuyan Island is the Mount Guiting-Guiting Natural Park (MGGNP). Conservation efforts for this strikingly beautiful mountain are of particular importance because it is the only remaining mountain in the Philippines with relatively intact habitats along its entire elevational gradient. These habitats include summit heathland and grasslands, mossy forests, montane forests, lowland evergreen forests, and forests over ultrabasic rocks (NIPAP 2000). Mount Guiting-Guiting's biodiversity is also one of the richest in the Western Visayas Biogeographic Region. Fifty-four species of plants and five species of mammals are endemic to the island. Additionally, 10 species of fruit bat and 131 bird species, 24 of which are migrants, have been recorded on the island (Goodman and Ingle 1992).

Despite its steep slopes and relative inaccessibility, Mount Guiting-Guiting's abundant resources have not been spared from the destructive means by which the local people meet their economic needs. Logging has been carried out on the island dating back to Spanish colonial times. Timber extraction from the 1970s to the early 1990s was accelerated to meet the huge demand of the mining pits in Masbate. At the same time, furniture making was fast becoming a backyard industry throughout much of the island. Blast fishing and cyanide poisoning brought in a sizable catch, but destroyed large tracts of coral reef. Upland dwellers engaged in swidden farming and poisoning of rivers to catch eels and shrimp. Substantial tracts of mangroves were cleared for use as household fuel and converted into fishponds.

Concern about the environmental threats posed by the unregulated and unsustainable use of the island's forest resources prodded the three municipalities of Sibuyan—Magdiwang, Cajidiocan, and San Fernando—to promote the conservation of Mount Guiting-Guiting. In February 1996, Mount Guiting-Guiting was proclaimed a protected area under the National Integrated Protected Areas System (NIPAS) through a presidential proclamation. With a total land area of 15,475 hectares and an additional 10,000 hectares as a buffer zone, the park straddles the three municipalities of the island. As a result, project interventions closely followed an ICDP approach of protected area security through co-management by indigenous peoples living in and around the protected area.

In 1997, KKP, in collaboration with other NGO advocates of indigenous people's rights such as Anthropology Watch (Anthrowatch), Legal Assistance Center for Indigenous Filipinos (PANLIPI), and the Philippine Association for Intercultural Development (PAFID), assisted the Sibuyan Mangyan Tagabukid in obtaining a community title and in planning for the sustainable management of natural resources within their ancestral domain. This assistance was made possible through a DGIS (Netherlands development assistance)-funded WWF (World Wildlife Fund) project. The activities focused on the research and the gathering of proofs for the purpose of applying for a communal title, delineation of the ancestral claim, capacity building, paralegal training, and livelihood support to farm activities.

The Sociopolitical Environment

The Sibuyan Mangyan Tagabukid

Residing in and around Mount Guiting-Guiting Natural Park are the Sibuyan Mangyan Tagabukid (SMT). This indigenous group has managed to retain a culture and tradition distinct from the lowland Sibuyan culture. Rituals such as *paminhi* (preplanting ritual) and *tugna* (preharvest ritual) denote respect for the spirits that play an important role in Sibuyan Mangyan culture. Several generations of kin identified to have previously inhabited the area and improvements introduced by their ancestors attest to the longevity of the indigenous peoples in the area. The SMT are primarily engaged in subsistence agriculture—making their living through swidden farming, charcoal making, gathering of minor forest products such as rattans, resins, vines, and honey, and fishing for freshwater fish and shrimp in the numerous water channels and tributaries on the mountain.

The declaration of MGGNP in 1996 led to prohibitions that threatened the traditional forest-dependent livelihoods of the SMT. The gathering of vines, rattan, honey, medicinal plants, and other nontimber forest products inside the protected area was curtailed. The SMT were threatened with eviction from their ancestral domain. Logging was virtually halted. Because of eviction threats and uncertain tenure, many SMT expressed fears of permanent displacement once the NIPAS law was fully implemented.

A 1994 survey conducted by the Protected Areas and Wildlife Bureau of the Department of Environment and Natural Resources estimated the indigenous population at about 1,557 individuals, roughly 3 percent of the island's total.[1] A KKP survey in 2000 updated the population count to 1,687 individuals residing in 315 households. The ancestral domain has an area of 7,905 hectares located in two noncontiguous areas within the municipalities of San Fernando and Cajidiocan (figure 9.1). The domain consists of settlements, farm lots, burial grounds, and forest. The rivers and ridges serve as natural boundaries for each settlement.

International Agreements and Legal Framework

The rights of indigenous peoples to their ancestral domain are embedded in several international agreements and dispositions:

- Agenda 21 in reference to indigenous peoples mentions their "participation in development decisions affecting them, including, where appropriate, their participation in the establishment or management of protected areas" (Agenda 21 1992, Ch 26).
- International Labor Organization (ILO) convention 169 spells out the right of indigenous peoples to manage natural resources on their territories, to exercise their customary laws, and to represent themselves through their own institutions (ILO 1989).
- The decennial World Congress on Protected Areas has emphasized the importance of local and indigenous communities, culminating in the 1992 Caracas recommendations (IUCN 1993).
- The Convention on Biological Diversity (CBD) on in situ conservation includes both protected areas (8[a]), its connection to

indigenous knowledge and practices (8[j]), and customary use of biological resources (10[c]).

In the Philippines, the constitution and pertinent national laws guarantee the rights of indigenous peoples:

- The Philippine Constitution of 1987 recognizes the rights of indigenous peoples to their ancestral lands and domains (Art II, Sec 5; Art XII, Sec 17; Art XIV, Sec 6; Art XIII).
- The National Integrated Protected Areas System Act of 1992 recognizes the claims and rights of indigenous communities over ancestral areas overlapping with the protected areas and promotes partnership in formulating and implementing plans and policies (Ch VII).
- The Indigenous Peoples Rights Act (IPRA) of 1997 is a comprehensive law that includes not only the rights of indigenous peoples over their ancestral domain but their rights to social justice and human rights, self-governance and empowerment, and cultural integrity. This landmark law meant usufruct and stewardship rights would be converted to rights of land ownership through Certificate of Ancestral Domain Titles (CADT).

Protected Areas and Indigenous Communities

That there are conflicts between protected areas and indigenous people is beyond dispute, primarily because the areas rich in biodiversity coincide with areas occupied or depended on by indigenous peoples. In many cases, public interests override local interests, and environmental values for humankind often interfere with the realization of local values.

All forestland in the Philippines is nominally state owned, and to most people this has been equated with "open access." Most official land maps depict these public lands as underutilized or marginal lands. This masks the true situation in these areas. Many public lands have indigenous inhabitants whose claim over native territories is not reflected in official maps. Hence, there is an existing underestimation of the social and political implications of establishing protected areas in these territories. Seen in this light, the opposition to the establishment of parks by indigenous communities is not surprising.

Since the passage of NIPAS in 1992, all newly designated protected areas in the country have been established on forestlands that overlap with ancestral claims by indigenous groups. Interestingly, all eight protected areas that make up the European Union–funded five-year

Table 9.1 *NIPAP protected areas and ancestral domain overlaps*

Indigenous group	*Tenure instrument*	*Protected area*
Calamian Tagabanwa	CFSA, CADC, CADT	Coron Island, Coron, Palawan
Tagbanwa	Pending CALC	El Nido, Palawan
Tagbanwa	CADC, CALC	Malampaya Sound, Palawan
Sibuyan Mangyan Tagabukid	CADT	Mt. Guiting-Guiting, Sibuyan Island, Romblon
Buhid, Tau Buid, Tadyawan Mangyan	CFSA, CADC	Mt. Iglit-Baco, Oriental Mindoro
Agta Tabangnon, Agta Cimarron	CADC	Mt. Isarog, Camarines Sur
Subanen	CADC	Mt. Malindang, Misamis Occidental
Kalanguya, Ibaloi, Kankana-ey, Karao, Ifugao	CADC	Mt. Pulag, Benguet

Source: NIPAP (2002).
CADC (Certificate of Ancestral Domain Claim); CADT (Certificate of Ancestral Domain Title); CFSA (Community Forestry Stewardship Agreement); CALC (Certificate of Ancestral Lot Claim)

National Integrated Protected Areas Program (NIPAP) were implemented inside or in areas overlapping with ancestral domains, as illustrated in table 9.1.

There are 181 pending claims over ancestral lands, with a total area of 2.5 million hectares distributed throughout the country. These lands have been identified, delineated, and officially recognized under the Department of the Environment and Natural Resources (DENR) Ancestral Domains Program and are undergoing mapping and land-titling procedures. However, these claims show substantial overlaps between indigenous territories and remaining forest cover in the country. This has serious implications for the process, design, and implementation of institutions and mechanisms that will work to conserve and save what is left of these forests—home as they are to rich biological and cultural diversity. In these areas, jurisdictional conflicts are common. There is a national law governing protected areas and another for ancestral lands. In these overlap areas the relevant laws that should apply are impossible to differentiate. The institutional responsibilities that govern these overlaps are unclear.

Fortunately, the premises behind the recognition of ancestral lands under both the NIPAS and IPRA laws are similar if not identical. These convergences provide an opportunity for the indigenous peoples and parks to develop an operational framework where collaboration instead of conflict can prevail.

Both the IPRA and the NIPAS Acts refer to areas within the ancestral domain whose use, because of their environmental significance, is determined by appropriate agencies. Under the IPRA, however, it is unclear to which agency this refers. Under the NIPAS, the DENR is designated as the lead agency. However, both laws recognize that the management of said areas is the exclusive responsibility of the indigenous peoples.

Recognizing the livelihood needs and aspirations of indigenous peoples, the NIPAS law provides for a development framework to govern the indigenous territory within protected areas. A development framework that is nested in a conservation area has inherent conflicts, as the history of ICDPs[2] show. Harmonizing seemingly conflicting objectives between conservation and development has been an elusive goal. Co-management regimes, a paradigm gaining increasing acceptance, attempts to reconcile the objectives of indigenous communities and those of protected areas.

Co-Management

Many good examples of co-management are found in Latin America, the Arctic, North America, and Australia (East 1991; Weaver 1991; Beltran 2000; Larsen 2000). There are exceptional cases where indigenous populations and lands are legally recognized and granted local autonomy and where decisions were made that benefited their long-term interests (cf. Herlihy 1997). Many Asian countries, the Philippines excepted, have no formal legal recognition of customary settlement, tenure, and use rights, although there is de facto acceptance of some use. For the region, the emphasis has been on buffer zones and alternative-income generation, combined with limited recognition of traditional use and presence.

Philippine policies on protected areas and indigenous peoples are among the most progressive in the region. Co-management examples are found in Bataan National Park, where two indigenous Aeta representatives on the Protected Area Management Board (PAMB) have raised awareness of Aeta concerns as well as mobilized communities to engage in volunteer patrolling, firefighting, and conservation (Griffith and Colchester 2000). However, conflicts between protected areas and indigenous peoples are more common in the newly designated protected areas. Because of these conflicts, there is mutual distrust between protected area offices and indigenous communities. This results in non-cooperation, turfing, and weak management responses.

Methodology

The overall goal of the Sibuyan project is to protect the biodiversity of Mount Guiting-Guiting Natural Park through the development of sustainable livelihoods. Within this goal, a major objective is to improve tenurial security of the indigenous Sibuyan Mangyan Tagabukid. This includes strengthening their social organization, culture, and customary laws, as well as assisting them to become responsible stakeholders in the management of the environmentally sensitive areas in which they live.

The project's key assumption is that land tenure security, coupled with development and natural resource management interventions that are identified, designed, and implemented by indigenous community-based organizations, will ensure sustainability and responsible management of resources. One expected result is improved resource security in MGGNP—a key aspect of the ICDP approach. The project is considered successful when: (1) the SMT have consolidated their social organization, culture, and customary laws; (2) they have improved their capability to manage natural resources in their territories; (3) they are protecting their biological resources; and (4) they have developed a management plan and monitoring and evaluation system for the ancestral territory.

RESEARCH, MAPPING, AND TENURIAL APPLICATION

Field interventions consisted of anthropological research and documentation, participatory mapping and planning, capacity building, legal assistance, farm support, and joint ventures. The procedures and steps in identifying and delineating the ancestral domain and applying for a community title are outlined in thirteen steps under the IPRA law (Ch VIII). To summarize, the steps include: (1) filing for petition for delineation, (2) delineation proper, (3) submission of proofs, (4) inspection by an NCIP (National Commission on Indigenous Peoples) representative, (5) evaluation and appreciation of proofs, (6) survey and preparation of survey plans, (7) identification of boundary conflicts, (8) submission of an NCIP investigation report, (9) map validation, (10) public notification, (11) endorsement of claim to the NCIP Ancestral Domains Office, (12) review and endorsement by the Ancestral Domains Office to the

NCIP board, and (13) approval by the NCIP board of the Certificate of Ancestral Domain Title application.

In 1998, KKP supported implementation of the activities prescribed under the IPRA. To hasten the process of delineation of the ancestral domain of the SMTs, KKP, AnthroWatch, and PANLIPI entered into a "memorandum of agreement" (MoA) with the NCIP. The MoA created a pilot project for the NCIP for the processing of CADT, using procedures provided by the "implementing rules and regulations" of the IPRA.

In September 1998, the SMT council of elders filed a petition with the NCIP for the delineation of their ancestral domain. Since then, KKP and PANLIPI have carried out community-wide information dissemination on the IPRA law as well as consultations about the delineation process. The council of elders convened to identify the landmarks indicating the boundaries of their ancestral domains on a topographic map. Sacred sites, worship areas, hunting, gathering, collecting and fishing grounds, swidden farms, and residential areas were presented on an indicative map. A census was conducted that put the number of applicant beneficiaries of the CADT at 315 households, or 1,687 individuals. In addition, various testimonials, written/historical accounts, anthropological data, genealogical surveys, pictures, and artifacts were gathered, documented, and submitted to the NCIP provincial office for verification and validation.

The NCIP provincial office conducted an inspection with the SMT community, adjoining communities, and other affected entities to verify the landmarks of the ancestral domain and the physical proofs supporting the claim. KKP and its partner NGOs assisted the SMT in preparing the survey plans, conducting the perimeter walk, and preparing flat maps with the necessary technical descriptions and a 3-D map. The resulting maps were consequently validated by the indigenous communities. Boundaries, markings, and the names of places were rechecked and appropriate corrections made. After validation, the 3-D map was assembled and displayed in the tribal hall for use by the members. A community resolution attesting to the veracity of delineation and the content of the map of the ancestral domain was likewise drafted. The ancestral domain maps were published in the provincial newspaper, *Romblon Today*. These maps were posted in prominent places within the locality such as municipal halls, barangay halls, and indigenous community centers. During the time of publication and posting, no petition of protest was submitted to the local NCIP office.

After publication, the NCIP provincial office endorsed the ancestral domain claim to the NCIP regional office for verification. After further review of the proofs and evidence, the claim was finally forwarded to the Ancestral Domain Office (ADO) of the NCIP by the regional office. After establishing and acknowledging the veracity of the claim, it endorsed the application and forwarded it to the NCIP board for its favorable action.

MANAGEMENT PLAN PREPARATION

The results of the delineation and research activities were fed into village workshops that led to the formulation of a comprehensive management plan, also known as the Ancestral Domain Sustainable Development and Protection Plan (ADSDPP). A KKP coordinator was hired to carry out organizational and institution-building activities to revive nonfunctional tribal councils and federate them into a CADT-wide organization that would be the implementing structure for the ADSDPP. An annual plan was then developed.

Interweaving the research, perimeter survey, mapping, and planning activities, KKP and PANLIPI organized paralegal training activities and orientation seminars on existing laws (i.e., IPRA, NIPAS, fishery code, forestry laws). The project sponsored study tours and cross visits and made it possible for SMT leaders to participate in meetings, conferences, and dialogues on indigenous issues. SMT cultural practices were documented and customary laws codified. The project initiated small-scale plantations (i.e., abaca, coffee, tree seedlings) through joint venture arrangements. The SMT presented their plans and concerns during consultation meetings with local government officials.

IMPACT EVALUATION

To gather community perceptions on the impact of project interventions, the project carried out focus group discussions in four cluster areas—Hagimit, Kabuylanan, Layag, and Gintak-an. The qualitative survey was used because it was deemed more appropriate for the indigenous community given their rich oral tradition. There were 77 respondents: 33 women (44 percent) and 44 men (56 percent). Two groups with 8–10 members per group were organized in each of the four clusters. One group consisted of community officers and paralegals, while the second group, which served as control group, consisted of new officers and nonmembers. This process facilitated validation of information by triangulating community responses. The census list of indigenous people was used to randomly select participants. Each group had a facilitator and a documenter. The questions focused on the effect of KKP-led development intervention among the Sibuyan Mangyan Tagabukid.

Results

Issuance of CADT

In January 2001, the NCIP board unanimously approved and signed the CADT application of the SMT covering 7,905 hectares.[3] The claim consists of two noncontiguous areas within the municipalities of Cajidiocan and San Fernando. Of the 7,905 hectares covered by the ancestral domain, approximately 5,000 hectares of forest overlap with MGGNP (figure 9.1). This overlap constitutes one-third of the protected area and 60 percent of the ancestral domain.

The speedy processing of the CADT was abetted by the smooth coordination existing between the assisting NGOs and the NCIP at the provincial, regional, and national levels. For the NCIP, the Sibuyan experience proved to be a model of NGO collaboration in the implementation of the procedures outlined under the IPRA, serving as an example for processing and expediting future applications.

Development of a Management Plan

The Ancestral Domain Sustainable Development and Protection Plan was completed in 2001. The plan embodies the framework for the long-term management of the ancestral domain. It is a comprehensive management plan that encompasses: (1) livelihood and land/forest resource use and protection programs; (2) infrastructure projects; (3) community strengthening, leadership, and self-governance; (4) enrichment and preservation of indigenous cultures and traditions; and (5) health and sanitation plans.

Apart from expressing indigenous communal rights and aspirations, the ADSDPP is a manifestation of the collective responsibility of managing the domain and all natural resources that lie therein. This management plan illustrates how the indigenous community views its own development in its own terms and how it will manage resources using its customary laws and traditions. In addition, the plan serves as a reference point for collaborative engagement and negotiations with other agencies that the SMT recognize as valuable partners in achieving their own objectives.

Impacts to Community Life

The focus group discussion results provided an initial indication of the impact of project interventions as perceived by the indigenous community in the upland areas (table 9.2). The results stem from community

Table 9.2 *Focus group results on the impact of land tenure interventions*

Problems	*Desired results*	*Interventions*	*Impact*
Illegal activities by lowlanders and uplanders	Stop or minimize illegal activities	Orientation on laws and rights (IPRA, NIPAS, forestry, fishery)	Illegal logging reduced because of enforcement by the IPs
Lack of food security and livelihoods	To protect the forest	Organized paralegal teams	Sense of identity
Lack of law enforcement	To possess a CADT	Research on indigenous culture and practices	Increased sense of ownership through CADT
Vices (drinking, gambling, and smoking)	To improve incomes to support expenses	CADT application	Greater appreciation of the importance of protecting own territory
	Peace and order in the community	LGU orientations on IP issues	Gambling and drinking reduced
		MOA with NCIP	Many more occupied in planting gabi, kamote, and bondo
		Survey of ancestral domain and mapping	Respect for and agreement with one another
		Preparation of the ADSDPP	Disputes and problems now resolved within the community
		Cross-visits to Bakun and Coron	IPs hold regular and special meetings
		Organized tribal councils	
		Register the ATSMATA with the SEC	
		Developed the annual plan based on ADSDPP	
		Codification of customary laws	
		Joint venture projects	

ADSDPP (Ancestral Domaine Sustainable Development and Protection Plan); ATSMATA (Association of Sibuyan Mangyan Tagabukid); CADT (Certificate of Ancestral Domain Title); IPRA (Indigenous Peoples' Rights Act); IPs (Indigenous Peoples); LGU (Local Government Units); MOA (Memorandum of Understanding); NCIP (National Commission on Indigenous Peoples); NIPAS (National Integrated Protected Areas System); SEC (Securities and Exchange Commission) perceptions of changes in its social, economic, political, and environmental conditions. The respondents observed a reduction in illegal activities (i.e., logging and river poisoning), citing increased vigilance and protection efforts by their members. Responses also indicated an increased sense of ownership in the domain area and greater appreciation of protecting their own territory. Gambling and drinking were also seen as having been reduced, and more indigenous people were preoccupied with planting root crops (i.e., gabi, kamote, and bondo). Disputes were resolved internally, and regular community meetings were held. However, there was still concern about occasional timber poaching by lowlanders and the difficulty of obtaining the original communal title. Community members expressed need for further support through livelihood projects, nonhybrid seeds, potable water systems, and infrastructure (i.e., schools, tribal hall, daycare center), as well as radios, antennas, and paralegal support to help them continue protecting their domain.

Discussion

The ICDP experience has identified many instances where efforts to support local people's aspirations have clashed with protected area development (Colchester 1994; Wells 1995; Wells et al. 1999; Hughes and Flintan 2001). A number of issues and challenges have been identified from the Sibuyan Island experience.

The Protected Area/Ancestral Domain Overlap

Of the 7,905 hectares covered by the ancestral domain, approximately 5,000 hectares of the forest overlap with the eastern portion of the protected area. How did this overlap come about? During the delineation of park boundaries, the Protected Area Office failed to appreciate the definition of the ancestral domain that goes beyond the mere existence of settlements and farm lots (see Ch II, Sec 3[a] of the IPRA). The delineation activity erroneously trespassed the burial grounds, sacred sites, and hunting and harvest areas of the indigenous community and included these within the park boundaries.

And while the general management plan (GMP) reportedly conducted "extensive consultation" with the SMT in the delineation of the park boundaries, the indigenous peoples claim that their participation in the delineation process of the protected area was limited to the hauling and planting of boundary monuments. Furthermore, they claim that no meaningful coordination with the indigenous communities was initiated to properly delineate the boundary of the protected area from the ancestral domain and that no meaningful consultation with the SMT was initiated to determine what their participation would be in the management of MGGNP. The conclusion drawn by the SMT was that the park was simply trying to keep them out and restrict access to forest resources. Frustrated and distrustful of the park's intentions, the SMT petitioned for the segregation of their ancestral domain from the protected area.

From the park authority's perspective, the area of overlap between the CADT and the protected area contains the most important areas of pristine forest left in Sibuyan. From the point of view of the DENR, and as stated in the park management plan, this area reflects the park's biological importance and is therefore zoned for strict protection (NIPAP 2000).

Interestingly, the park designated the overlap area as a sustainable use zone, but this is viewed by the park as an interim arrangement. According to the GMP, the overlap area will be upgraded to a rehabilitation zone after five years if no CADT is issued to the claimants. However, the GMP further states that if a CADT is issued within this period, "the management of the specific area will be jointly undertaken by the PAMB and the indigenous groups" and that "joint planning will be undertaken to identify the respective roles and responsibilities of both parties" (NIPAP 2000). According to this statement, the GMP supports a co-management model for the overlap areas and provides a starting point for negotiations with the SMT.

From the perspective of the SMT, the land defines their cultural and ethnic identity. It is their source of life and well-being. In their management plan, the SMT propose to classify the forests as strict protection zone. Although "strict protection" is used, the interpretation varies. They interpret this to mean that no timber will be cut except for subsistence use and no trees will be cut near water bodies. The harvest of honey, vines, and other nontimber forest products would still be allowed in this forest block.

In a workshop to discuss the merits of a co-management agreement with the park, the SMT developed a set of guidelines that would be proposed in succeeding negotiations: (1) full recognition and respect for the rights of the SMT; (2) adherence to ADSDPP and the SMT's customary laws; (3) free and prior informed consent of the community; (4) right to select members to the co-management board; (5) funds and income sharing from activities within the protected area; (6) transparency; (7) official designation of the SMT as forest guards; (8) indigenous justice system to be applied; (9) transfer of knowledge, skills, and technology; (10) disposition of equipment and facilities after expiration of co-management agreement; and (11) right to revoke agreement in the event of violation or deviation from the plan.

The SMT view the co-management agreement as an interim arrangement, with their ultimate objective being to excise the overlap area from the protected area and have management turned over to them. The IPRA also guarantees that the indigenous people's decisions pertaining to the use of the resources within their domains shall at all times prevail. The application to segregate the overlapping portion of the CADT from the protected area has been filed by the SMT and is now in Congress. Not surprisingly, such declarations of independence and self-governance faced concerted resistance from the PAMB.

From the perspective of the local government units of San Fernando and Cajidiocan, there is also concern over the impact of indigenous activities on the watershed. San Fernando is concerned about the availability of freshwater that comes from the Cantingas watershed, which forms part of the domain (Rios, personal communication 2001). A river that flows from this watershed feeds into an existing irrigation canal that waters the extensive rice fields in San Fernando. All three local government units are also endorsing the construction of a proposed 900 kilowatt hydroelectric power facility to be located downstream from this watershed. From these municipalities' points of view, this critical watershed should not be left under the exclusive management control of the SMT.

As a result of failed projects, there is also a history of distrust between indigenous peoples and the local government units that aggravates their conflict. A reforestation project that was funded by the DENR and implemented by the local officials encroached on swidden farms of the SMT that were under fallow. With no farms to plant on, the indigenous people had no recourse but to cut the narra trees that were planted on their plots and clear the land to plant root crops. The DENR and local government accused the SMT of illegal logging, leading to strained relations and distrust.

The lowlanders have stereotyped the indigenous people of Sibuyan Island as illegal loggers since the time of Spanish rule. According to retrieved Spanish documents, the indigenous people were contracted and exploited by lowland merchants to cut and haul wood. Timber cutting was halted only when the park was established in 1996, but because of the long history, the activity has defined the island's social relations and politics ever since.

Resistance to Ancestral Claim

As in the cases of many indigenous groups throughout the country, the SMT struggle for community title has been fraught with obstacles toward securing tenure over their ancestral domains. The full impact of the IPRA law and the claim by the SMT for ancestral domain title on resource use inside their ancestral domain, local governance, and on social relations was underestimated. Resolute opposition to the delineation of the ancestral domain by "alleged" indigenous peoples came from local officials and the DENR. Ironically, it was the very same authorities who initially recognized the presence and identity of the indigenous peoples on the island who later questioned the legitimacy of these people's indigenous origins and the validity of their ancestral claim.

To make sense of this resistance, it is worth noting that the opposition of other stakeholders to the indigenous claim stems largely from the fact that a significant proportion of the island's population does not own land. To have this small and marginalized group claim and have title to a third of the protected area shifts the balance in the local political economy.[4]

Some lowlanders fear that the SMT may mismanage the resources inside the domain or hold hostage the water resources that feed the agricultural lands and support the households below. Likewise, many who have vested interests in the uplands through land claims and ties to illegal logging are threatened by controls enforced by the SMT over forest resources. Isolated incidents of harassment and physical violence related to the ancestral claim have been reported.

SUSTAINABILITY ISSUES

It is held in conservation circles that indigenous peoples, by occupying ancestral territories since time immemorial, have developed sustainable forms of livelihoods that are intrinsically linked with nature. The recognition of indigenous rights to ancestral domains, it is believed, will allow these sustainable practices to continue indefinitely. Both the NIPAS and IPRA provisions recognizing indigenous rights and claims for ancestral domain are premised on this view.

However, the assumption of harmonious coexistence may turn out to be false when viewed in a longer temporal scale. A number of factors conspire against this assumption, such as: (1) indigenous populations becoming dense; (2) a high rate of land use that causes changes to the surrounding protected areas; (3) assimilation by non-natives; and (4) erosion of sustainable forms of livelihood owing to market forces, cash economy, etc. Indigenous people's entry into modern ways of farming and harvesting is accelerated by their introduction to cash economies, commercialization, and access to technology to harvest resources or/and increase farm yields. Studies reveal that the commercialization of nontimber forest products in forested areas has led to shifts from traditional to more unsustainable extraction practices (Neumann and

Hirsch 2000). The use of labor-saving technologies, export-driven production of nontimber forest products, debt peonage, and weak regulatory mechanisms and monitoring have led to overexploitation of such resources and degradation of the forest (Neumann and Hirsch 2000). There are cases of co-management strategies that are planned with indigenous people based on their current commitment to a stable ecosystem but that have failed because of assimilation and changes in native lifestyles (Burrel 1987).

Studies of swidden agriculture report evidence that upland communities tend to shorten their farming fallow periods to meet increasing demands for food. Unable to meet their food and cash requirements, families clear forests to open new areas for new farms. Although swidden cultivation per se has been found to be a sustainable form of resource management, under short fallows it can be unsustainable and unproductive (Warner 1991).

The prohibition in the ADSDPP on new swidden plots and on intensified use of existing ones poses a long-term dilemma for the SMT. By limiting the plots for cultivation, they in effect *enforce* shortened fallow. The length of fallow is crucial to the sustainability of swiddens. With the customary practice of subdividing the family's plot among the children when the latter start their own family, there will be fewer plots from which to feed an increasing number of family members. With limited arable land, increasing population, hunger, and poverty, indigenous people have little recourse but to clear the adjacent forest for swidden farms (see also Shepherd this volume). The failure of land reform programs in Latin America to address forest clearing is well documented (FoE and GTA n.d.; Utting 1993; Barraclough and Ghimire 1995). Experience shows that giving away lands to farmers is not enough to address poverty issues that, in turn, drive destructive activities in the forest.

Lack of Government Support

The NCIP, as the government arm mandated to implement IPRA, is charged with monitoring the implementation of all management plans in ancestral domains. But since its inception, it has lacked the technical competence, financial resources, and political authority to defend indigenous lands. For three and a half years, it languished under excess bureaucracy, unable to make any progress in advancing indigenous people's rights. Conflicts of interests were also evident. Not a single CADT was awarded during its first three years of operation. The record of government implementation of the IPRA has been dismal and disconcerting.

The NCIP bureaucracy inherited the old structure and personnel that dealt with indigenous affairs during the Marcos years. The Office of Southern Cultural Communities (OSCC) and the Office of Northern Cultural Communities (ONCC) were responsible for delivering basic services to and implementing projects with the indigenous cultural communities. Large amounts of aid passed through the office, and there were charges of corruption and diversion of assistance to nonindigenous groups. The same set of officials is found in the NCIP hierarchy.

The state's reluctance to recognize tenurial claims by indigenous people can be traced to land laws that are largely premised on the Regalian doctrine inherited from the Spanish colonial times. Under this doctrine, or theory, all public domains belong to the state (see Gartlan this volume), and forestlands,[5] where the majority of the country's indigenous peoples live, are part and parcel of the public domain. Legal documentation stating ownership of land has never been part of indigenous customary ways, since their concept of land is based on communal ownership. Moreover, current land laws favor individual ownership, as opposed to communal ownership as practiced by indigenous peoples even before the creation of the state. Currently, the Regalian theory finds legal expression in the 1987 constitution as well as in other pieces of legislation. Even the National Integrated Protected Areas System Act of 1992 (RA 7586), enacted for the establishment and management of protected areas, was premised on this theory. A recent resolution of the supreme court has upheld the right of ownership of indigenous Filipinos to their ancestral lands and domains and recognized their role as managers or stewards of all natural resources within their ancestral domains and ancestral lands, in accordance with the dictates of the IPRA and their respective customary laws (Supreme Court G.R. No. 135385).

Lack of Meaningful Participation by Indigenous Peoples

Although the SMT have a seat on the Protected Area Management Board of the MGGNP, their meaningful participation in the processes of park management is yet to be felt. The three indigenous representatives serving on the board have all been declared inactive since September 2001. The SMT attribute their absences to unfavorable PAMB "politics," lack of support, and increased frustration over the slow decision-making processes of the board.

Co-management denotes equal voice in decision making (see also Brown this volume). Although the SMT are accorded membership in the PAMB, the PAMB is still not the appropriate management structure to promote co-management. The PAMB is dominated by the DENR and by local executives who, by their sheer number, can easily outvote the SMT in decisions when their interests run counter to that of the indigenous community. On several occasions, the PAMB quashed motions by the SMT to increase tribal representation on the PAMB and to excise the domain from the protected area.

Lessons Learned

Building Institutional Capacity for Conservation and Development

It has been argued that awarding rights over large tracts of lands and underlying resources therein unduly burdens the indigenous peoples, who may not have the technical capability to manage them sustainably. Indeed there is some merit to this argument. Therefore, beyond securing tenure for indigenous peoples over their ancestral domains, due recognition should be given to building the capacities of indigenous peoples to achieve economic well-being, social development, and ecological integrity. The DGIS-funded KKP project targeted building on the indigenous capacities to sustainably manage their resources and implement their management plans.

The immediate and evident results of the claim-making process has been the revival of SMT traditional political structures wherein the power to formulate and enforce policies was restored to their council of elders. While the structure maintained the leadership of these elders, it likewise recognized the leadership of the much younger and educated members of the community, especially in dealing with matters that involve external agents. An ancestral domain committee in each tribal cluster was formed to implement the programs contained in the ADSDPP. A special committee was formed to safeguard the forest and rivers. The project provided radio communications equipment and field gear to the forest stewards.

The SMT have been observed to be more assertive in their rights and are able to negotiate for their interests with local authorities (i.e., provision of mini-hydroelectric and other facilities to be constructed inside the domain, bioresearch by foreign scientists, active participation in the selection of local leaders, etc).

The development of this discerning attitude among the SMT was supported through the many study tours and seminars they attended wherein they had the opportunity to hear and learn from the lessons of other tribal groups. Their confidence was further reinforced through paralegal training to deepen their understanding of the IPRA and its relation to other laws that directly impact their tenurial security and access rights (i.e., NIPAS law, forestry code, Mining Act, etc.). This paralegal training has produced a cadre of paralegals within the community who have increased knowledge about rights, assisted in awareness campaigns, initiated community dialogues, mobilized community members, written petition letters, gathered evidences for violations, and advocated on behalf of indigenous concerns in various gatherings and meetings.

Customary laws and traditions on conflict resolution have likewise been revived and codified. The communal stewardship that the CADT prescribes also sets the venue for federating the previously fragmented and nonfunctioning seven tribal clusters inside the ancestral domain. More importantly, the CADT has become a powerful instrument for unifying the SMT to pursue their common objective of securing land rights, self-determination, and the right to representation through their own institutions (see also Child and Dalal-Clayton this volume regarding governance issues).

Use of GIS/GPS Technologies and Indigenous 3-D Mapping

In the Philippines, "more lands are lost to maps than bullets." The lack of accurate maps has resulted in incursions into indigenous lands by more powerful interest groups. Without maps to indicate the extent of their lands, indigenous communities are powerless to oppose the claims of logging, mining, and other companies on their lands. In Sibuyan, the use of GIS/GPS technologies was important in helping the communities generate, define, and articulate spatial information through maps, as well as translate these maps to become understandable to policy makers and acceptable to mapping agencies. The process of 3-D mapping involved community gatherings and training that provided community members an opportunity to chronicle their culture, economy, history, and struggle as a distinct community. A map that uses local dialect and traditional place-names speaks volumes and advocates the community's knowledge and predominant role as the steward of the area.

People First and the Environment Will Follow

This project employed an ICDP approach to address the two-pronged goal of protecting the forest and providing for the human needs of those traditionally dependent on it. Without addressing the need of local people to gain access to the natural resource base, the conservation message would have been lost. The perception of local communities that a protected area exists for their benefit and not simply as a goal in itself is important. If indigenous people are to support protected areas, park authorities must address their livelihood concerns, especially their property rights. Only then will indigenous peoples make decisions that benefit the long-term interest of the park. It should be appreciated that the long-term objective of preserving the island's threatened ecosystems cannot start anywhere but in the short-term cross-sectoral approach of addressing poverty—the root cause of many of the threats (see Shepherd this volume).

The Value of Education and Raising Awareness

The tenurial assistance and capacity-building components of the project have resulted in the increased awareness and capability of the SMT to take responsibility for their resources. Communal ownership of resources has now been revived. Vigilance in protection efforts has been heightened. Each community has devised its own scheme to step up security within its ancestral domain. Through community efforts, illegal loggers have been apprehended. This has earned the SMT the respect of concerned lowlanders and even some local officials. They have likewise started a communal fund as an initial step toward making themselves self-sustaining. The federation of the seven tribal clusters and the reorientation of their goals and objectives toward their customary laws and traditions are enabling conditions that favor the implementation of the SMT management plan.

Self-Determination and Participatory Development

KKP's assistance to the SMT has always operated on the premise that the SMT have the full right to determine development priorities for the territories they traditionally occupy and use. In all processes undertaken, community consultation and validation were strictly observed. The KKP's dilemma lies in performing dual roles as facilitator and an advocate in community processes. The fine line between a partner organization's facilitative role and that of pushing an agenda must be clear at all times. It has been difficult to make a clear distinction between, on the one hand, the need to ensure a balanced process that fairly addresses the interests of stakeholders, and, on the other, the interest of partner organizations in the outcome of this process. An NGO must constantly remind the community (and itself) of the hat it is wearing when shifting roles between facilitator and advocate (see also Franks and Blomley this volume).

Many development initiatives directed not only toward indigenous people but also to other marginalized sectors of the community have failed because of the lack of observance of these two distinct and separate roles. Because they will ultimately have an impact on the indigenous way of life, development initiatives should carefully take into account the sociocultural and economic realities under which these communities operate. In this way, initiatives can be tailored to address community needs and therefore minimize failures and negative impacts.

Recommendations

Five years after its establishment and operation, the MGGNP is faced with both challenges and opportunities for collaboration between the park and the indigenous community. It is becoming increasingly apparent that fundamental changes in how the protected area is managed and designated, in so far as the overlap is concerned, must be addressed. It should be noted that the park's ultimate objective of protecting the forest's biodiversity and resources does not run counter to the SMT objectives as expressed in their management plan.

CO-MANAGEMENT AS THE WAY FORWARD

A co-management framework looks to be a workable strategy for the overlap areas in Mount Guiting-Guiting, as evidenced by the common objectives of the park and the indigenous people with respect to the remaining forest block. There remains a need to elaborate on the non-timber forest products that may be extracted, the harvestable levels, and the monitoring systems.

As of now, there is no ongoing enforcement by DENR in the upland areas. The lack of resources hinders the DENR from maintaining the rangers to patrol the hinterlands. Devolving the daily surveillance and patrol function to indigenous tribal councils will not only save park enforcement costs but allow the SMT to exercise their rights to their domain. While de facto enforcement by the SMT is ongoing, an official recognition of their role through a co-management agreement could strengthen the basis of their enforcement actions.

MONITORING AND EVALUATION SYSTEM

As a corollary to the co-management agreement, the PAMB and the SMT need to jointly develop a monitoring and evaluation system that would be applicable to the overlap areas. This system should include environmental monitoring of biodiversity, resource use, and threats using both scientific and social methods. The monitoring results should be used by both the PAMB and the tribal council for adaptive management. The monitoring and evaluation system should also include a component on capacity building to improve the competence of park rangers and community officers to monitor the domain. In this way direct links between SMT management and biodiversity conservation could be better established (see also Salafsky and Margoluis this volume).

Harmonization Between Indigenous Plans and Municipal Plans and Policies

There is a need for constructive collaboration between the local government units and the SMT in instituting governance systems in the domain areas (see Singh and Sharma this volume). At the local policy level, this entails recognition and integration of ancestral domain land and resource-use plans into the municipal development plans, including future land-use and zoning plans. Policies may also focus on establishing a better link between the traditional rights of local and indigenous populations and the "formal" legal system of forestland ownership and access. This goes back to the need to establish a better and more realistic agreement between traditional rights and those of the national legislation (Brandon and O'Herron this volume).

Basic Services

Health and education are important interventions that should be provided by government agencies. Access to primary health care, including immunization programs and reproductive health services, will go a long way in reducing the high infant mortality rate, maternal morbidity and mortality rates, and the common occurrence of easily preventable diseases. Assistance in the construction of sanitation facilities and water systems is much needed not only to prevent diseases but also for environmental protection. These interventions will likewise help in overcoming indigenous community distrust of local government.

Conclusion

Conservationists and development planners worldwide have been learning that the conservation of biological diversity in the developing world will not succeed in the long term unless local people perceive those efforts as beneficial to their economic and cultural interests

(Brown and Wyckoff-Baird 1995). ICDPs have struggled to effectively involve indigenous peoples in natural resource management. The examples presented in this case study highlight many of the issues and challenges of linking indigenous peoples and protected areas. By recognizing the traditional rights of indigenous peoples over their resources and building their capacities to manage the same, indigenous peoples can be powerful allies in the fight to protect biodiversity inside their territories and promote sustainable resource use.

Central to the issue of conservation and development and indigenous peoples is tenurial security. Contrary to the common understanding of land as something physical and therefore finite—a possession on which shelter may be built or where food or cash might be had—land is often what defines indigenous people. The political, cultural, economic, and social fabric that binds them together is inextricably linked to the territories and resources they occupy and utilize. When governments ignore traditional rights, this creates not only intense conflicts between indigenous populations and the state and other resource users, but also incentives for illegal schemes. Uncertain property rights do not encourage long-term investments in sustainable management. Thus, policy reforms should recognize traditional rights and devolve responsibility for the management of forestlands to indigenous communities, with the state concentrating its attention on ensuring that the interests of society in general are protected. In the case of the Philippines, where many of the laws are in place, serious attention must be given to law enforcement and development of compliance mechanisms at the national, regional, provincial, and municipal levels.

In this context, indigenous peoples should be logical allies in the conservation of natural resources both within and outside of protected areas. There will be conflicts, as described in the Sibuyan example, but negotiation, compromise, and trade-offs by all involved are necessary if conservation on these lands is ultimately going to be successful in the long term.

Endnotes

1. In 1996, these upland and forest dwellers were formally recognized by the Office for Southern Cultural Communities (OSCC) as an "indigenous cultural community" (ICC).

2. Integrated conservation and development projects, or ICDPs. For over a decade, the experience with ICDPs as a strategy in protected area management has shown mixed results (Brandon and Wells 1992; Barrett and Arcese 1995; Larsen, Freudenberger, and Wyckoff-Baird 1998; Hughes and Flintan 2001).

3. Upon assumption of the new administration on January 17, 2001, President Macapagal-Arroyo suspended the awarding of all approved CADTs by the NCIP. The Sibuyan CADT is awaiting resolution from the presidential task force that was created to review all applications.

4. In Peru, the minimum size of the territory needed to accommodate land-use practices (e.g., hunting, fishing, gathering based on distances traveled by male hunters, shifting cultivation based on female work) and ethnic perceptions of space was calculated. For one community, the study calculated 164,950 hectares for a population of 1,198 people (Uriarte 1999).

5. Under Presidential Decree 705 (otherwise known as the Revised Forestry Code of the Philippines 1975), public lands are classified as those having a slope of 18 percent and more.

References

Agenda 21. 1992. *Program of Action for Sustainable Development: Rio Declaration on Environment and Development*. New York: UN Department of Public Information.

Barraclough, S. and K. Ghimire. 1995. *Forests and Livelihoods: The Social Dynamics of Deforestation in Developing Countries*. London: Macmillan Press.

Barrett, C. S. and P. Arcese. 1995. Are integrated conservation and development projects sustainable? On the conservation of large mammals in sub-Saharan Africa. *World Development* 23:1073–1084.

Beltran, J., ed. 2000. *Indigenous and Traditional Peoples and Protected Areas: Principles, Guidelines, and Case Studies*. Gland, Switzerland: IUCN–The World Conservation Union.

Brandon, K. and M. Wells. 1992. Planning for people and parks: Design dilemmas. *World Development* 20:557–570.

Brown, M. and B. Wyckoff-Baird. 1995. *Designing Integrated Conservation and Development Projects*. Washington, D.C.: Biodiversity Support Project.

Burrel, T. 1987. *Parks, Plans, and People: Protected Areas and Socio-Economic Development*. Environmental Encounters Series. Strasbourg: Council of Europe.

Colchester, M. 1994. *Salvaging Nature: Indigenous Peoples, Protected Areas, and Biodiversity Conservation*. Discussion Paper no. 55. Geneva: UN Research Institute for Social Development.

East, K. M. 1991. Joint management of Canada's northern national parks. In P. C. West and S. R. Brechin, eds., *Resident Peoples and National Parks: Social Dilemmas and Strategies in International Conservation*, 333–345. Tucson: University of Arizona Press.

FoE and GTA. n.d. *Coherent Public Policies for a Sustainable Amazon*. São Paulo and Brasília: Friends of the Earth International and NGO Amazon Working Group (GTA).

Goodman, S. M. and N. R. Ingle. 1992. Sibuyan Island in the Philippines—Threatened and in need of conservation. *Oryx* 27:174–180.

Griffith, T. and M. Colchester. 2000. *Indigenous Peoples, Forests, and the World Bank: Policies and Practice.* Report on a workshop hosted by the Forest Peoples Program and the Bank Information Center, 9–10 May, Washington, D.C. http://www.wrm.org.uy/actors/WB/IPreport.html (December 2001).

Herlihy, P. 1997. Indigenous peoples and biosphere reserve conservation in the Mosquitia rain forest corridor, Honduras. In S. Steven, ed., *Conservation Through Cultural Survival: Indigenous Peoples and Protected Areas,* 99–130. Washington, D.C.: Island Press.

Hughes, R. F. and F. Flintan. 2001. *Integrated Conservation and Development Experience: A Review and Bibliography of the ICDP Literature.* London: International Institute for Environment and Development (IIED).

ILO (International Labor Organization). 1989. *ILO Convention 169: Convention Concerning Indigenous and Tribal Peoples.* Geneva: International Labor Conference.

Indigenous Peoples Rights Act (IPRA). 1998. Implementing rules and regulations. Administrative Order No. 1. Manila: National Commission on Indigenous Peoples, Republic of the Philippines.

IUCN. 1993. *Parks for Life: Report of the IVth Word Congress on National Parks and Protected Areas.* Gland, Switzerland: IUCN–The World Conservation Union.

Larsen P. 2000. *Co-Managing Protected Areas with Indigenous People: A Global Overview for IUCN/WCPA and WWF International.* Gland, Switzerland: World Wildlife Fund.

Larsen, P. S., M. Freudenberger, and B. Wyckoff-Baird. 1998. *WWF Integrated Conservation and Development Projects: Ten Lessons from the Field 1985–1996.* Washington, D.C.: World Wildlife Fund.

NAMRIA (National Mapping and Resource Information Authority). 1987. Romblon, sheet 2518, 1:250,000 series. Manila, Philippines: Fort Bonifacio.

Neumann, R. P. and E. Hirsch. 2000. *Commercialization of Non-Timber Forest Products: Review and Analysis of Research.* Bogor, Indonesia: Center for International Forestry Research (CIFOR).

NIPAP (National Integrated Protected Areas Program). 2000. *General Management Plan for Mt. Guiting-Guiting Natural Park, Sibuyan Island.* Manila, Philippines: Department of Environment and Natural Resources, Protected Areas and Wildlife Bureau. (DENR, PAWB).

NIPAP (National Integrated Protected Areas Program). 2002. *Final Report 1995–2001.* Manila, Philippines: Department of Environment and Natural Resources, Protected Areas and Wildlife Bureau (DENR, PAWB).

NIPAS (National Integrated Protected Areas System). n.d. Implementing rules and regulations. Administrative Order 25. Manila, Philippines: Department of Environment and Natural Resources, Protected Areas and Wildlife Bureau (DENR, PAWB).

Supreme Court G.R. No. 135385. 2001. Isagani Cruz and Cesar Europa v. Secretary of Environment and Natural Resources, et al. Resolution of Court En Banc, 18 September. Manila, Philippines.

Uriarte, L. 1985. Los nativos y su territorio: El caso de los jivaro achuar en la Amazonía Peruana. *Amazonía Peruana* 6 (11):39–64. Cited in Davis, S. and A. Wali. 1993. *Indigenous Territories and Tropical Forest Management in Latin*

America. Policy Research Working Papers no. I 100. Washington D.C.: The World Bank.

Utting, P. 1993. *Trees, People, and Power*. London: Earthscan.

Warner, K. 1991. *Shifting Cultivators: Local Technical Knowledge and Natural Resource Management in the Humid Tropics.* Community Forestry Note 8. Rome: Food and Agriculture Organization of the United Nations (FAO).

Weaver, S. M. 1991. The role of Aboriginals in the management of Australia's Cobourg (Gurig) and Kakadu National Parks. In P. C. West and S. R. Brechin, eds., *Resident Peoples and National Parks: Social Dilemmas and Strategies in International Conservation,* 311–332. Tucson: University of Arizona Press.

Wells, M. 1995. *Biodiversity Conservation and Local People's Development Aspirations: New Priorities for the 1990s.* Rural Development Forestry Network Paper 18a (Winter–Spring). London: Overseas Development Institute (ODI).

Wells, M., S. Guggenheim, A. Khan, W. Wardojo, and P. Jepson. 1999. *Investing in Biodiversity: A Review of Indonesia's Integrated Conservation and Development Projects*. Washington, D.C.: The World Bank.

WWF (World Wildlife Fund). 1996. *Indigenous Peoples and Conservation: WWF Statement of Principles.* Gland, Switzerland: World Wildlife Fund.

10

Land Tenure and State Property: A Comparison of the Korup and Kilum ICDPs in Cameroon

Steve Gartlan

Introduction

This paper compares and contrasts two projects in western Cameroon that have specifically adopted the integrated conservation and development project (ICDP) philosophy (figure 10.1). Both of the projects have been in existence for some fifteen years. While the two have elements in common (both were initiated by international conservation NGOs, and both are located in the forest zone), there are also major differences, including scale, biological scope, legal status of the project areas, social structure of the people of the area, and ecological status of the local environment. This paper examines these differences together with the historical factors that have been important in creating the current situation. The current status of each project is assessed, a prognosis is given for each, and the salient points are discussed.

The Korup Project Area covers about 660,953 hectares (of which 130,837 ha, or 19.7 percent, is Korup National Park), with 40,000 people dependent on the project area. It has a relatively low human population density—in many areas less than one person per square kilometer (Draft Korup Project master plan 1989). The principal economic activities are hunting, fishing, and collecting of nontimber forest products (bush mango, chewing stick); there is also a little cultivation of oil palm and cacao. Resources such as wildlife, fish, water, and firewood are

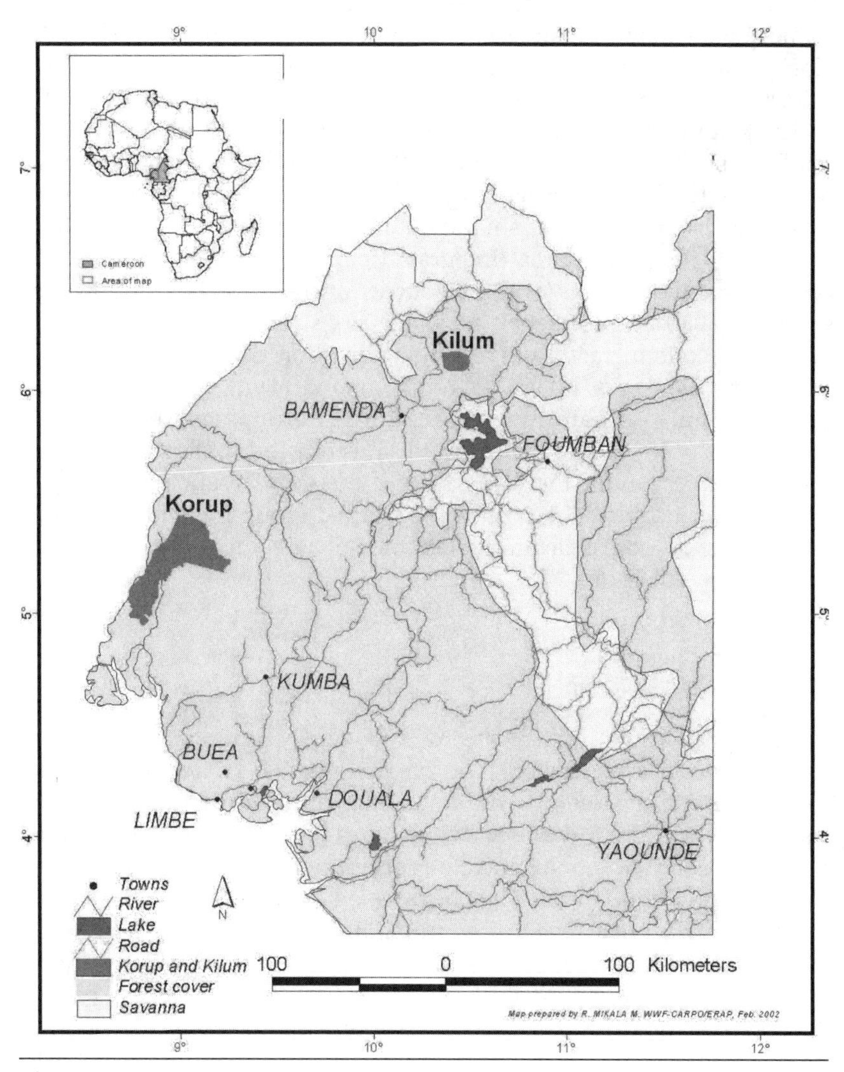

Figure 10.1.
Korup and Kilum, Cameroon, Africa.

abundant and nonlimiting. Korup lies on important smuggling routes from Nigeria, and villagers derive income from this. The population of Korup is ethnically complex, and lines of authority within the tribal structure are unclear and occult; it is hard to determine who has authority and how robust this authority is. A secret society, Ekpe (the leopard cult), is a major factor in tribal decision making; negotiation and agreement are highly complex issues. The main ethnic groups are the

Bantoid Ekoi, including the Ejagham and the Ibibio tribes, as well as the Korup people. Cameroon-Congo Bantu in the east of the area include the Oroko and the Mbo tribes (Laird, Cunningham, and Lisinge 1999).

The Kilum/Ijim Project Area covers some 50,000 hectares (less than a tenth the size of Korup), with about 20,000 hectares being the core project area (which is not formally protected). It has a population density that exceeds 65 people per square kilometer, and some 100,000 people depend on the project area (Integrated natural resource management programme n.d.). Some of the resources of the area (land, fuelwood, grazing space, sites for beehives, water) are stretched to the limit. The principal economic activities are the cultivation of cash crops (coffee, Irish potatoes, beans, honey, cola nuts), woodcarving, and papermaking. At Kilum there are two principal tribal groupings, the Oku and the Kom peoples, living on opposite sides of Mount Oku. Strong, linear hierarchies under the leadership of a strong chief, or *fon,* whose authority is absolute, characterize both tribal groupings. If negotiation with the fon is successful, implementation of agreements is then a fairly straightforward issue.

Historical Factors

The two decades between 1930 and 1950 were exceptionally important in Central Africa for the creation of reserves and protected areas. In Southern British Cameroon the total amount of forestland put under reserves in the 1930s was 390,782 hectares (or 14 percent of the land area). Some of Cameroon's most important wildlife reserves were also created at this time in the French-speaking zone: Benoué (1932), Faro (1932), Douala Edea (1932), Campo (1932), Kalfou (1933), and Waza (1934). The creation of these reserves was consistent with the mandate of the London Convention of 1933 (*Convention Relative to the Preservation of Fauna and Flora* 1933) that entered into force on 1 January, 1936. The contracting parties to this convention (the Union of South Africa, the United Kingdom of Great Britain and Northern Ireland, Egypt, Spain, France, Italy, Portugal, Anglo-Egyptian Sudan) were committed to creating national parks and reserves throughout their African colonial territories. The cause behind this activity was essentially public disquiet in Europe at the "orgy of wildlife slaughter," particularly of elephants, that was taking place throughout Africa. Article 3, paragraph 1 of the convention states: "The Contracting Government will explore forthwith the possibility of establishing in their territories national parks and strict natural reserves." Article 7, paragraph 5 states: "Take, so far as in their power lies, all necessary measures to ensure in each of their territories a sufficient degree of forest country and the preservation of the best native indigenous forest species" (*Convention* 1933).

The classification of the forest reserves between 1930 and 1950 engendered strong local opposition. When approached (in 1936) regarding the demarcation of forest reserves, the villagers of Korup were "universally suspicious" (Sharpe 1998), and in the montane forest regions of the Northwest Province where Kilum is located, opposition was very strong. Numerous well-documented cases of disputes shortly after World War II were linked to the actions of the Forestry Service in creating forest reserves. In the mid-1950s, continual protests in the newly created Cameroun legislative assembly led the French colonial government to suspend the creation of new forest reserves for fear of negative publicity reaching the UN Mandate Commission and so endangering their mandate (Burnham 2000).

In Cameroon, no national parks were created in the colonial era. This pattern prevailed throughout Africa; areas were designated as reserves in the precolonial period and were raised to the status of national parks in the postcolonial era. It was at this time the rules and restrictions began to be strongly enforced; this is the situation that affects as much as three-quarters of Africa's national parks. (The situation differed in South Africa, which was not a colonial entity. There the rules and restrictions were enforced from the start [Gartlan in prep.], and in consequence the problems are now somewhat different, with land claims over protected areas being made by many communities who were displaced from the protected areas.)

The Korup Project

Korup National Park (figure 10.2) is located in the South-West Province of Cameroon, along the Nigerian border between latitudes of 4°53′ and 5°28′ north and longitudes 8°43′ and 9°16′ east. The Korup Native Authority Forest Reserve was created by the British colonial administration in 1937, and was reduced in area in 1962 by the West Cameroon government to cover an area of 88,370 hectares. In 1986 a further area of almost 42,000 hectares was added, and the resulting area of 130,837 hectares became Korup National Park by Presidential Decree No. 86/1283 of October 30, 1986.[1] The official gazettement of the national park was a requirement for the release of funds by the British Overseas Development Agency (ODA) for the initiation of the Korup Project. Other than the legal gazettement, there were few specific further commitments by the Government of Cameroon to Korup at that time.

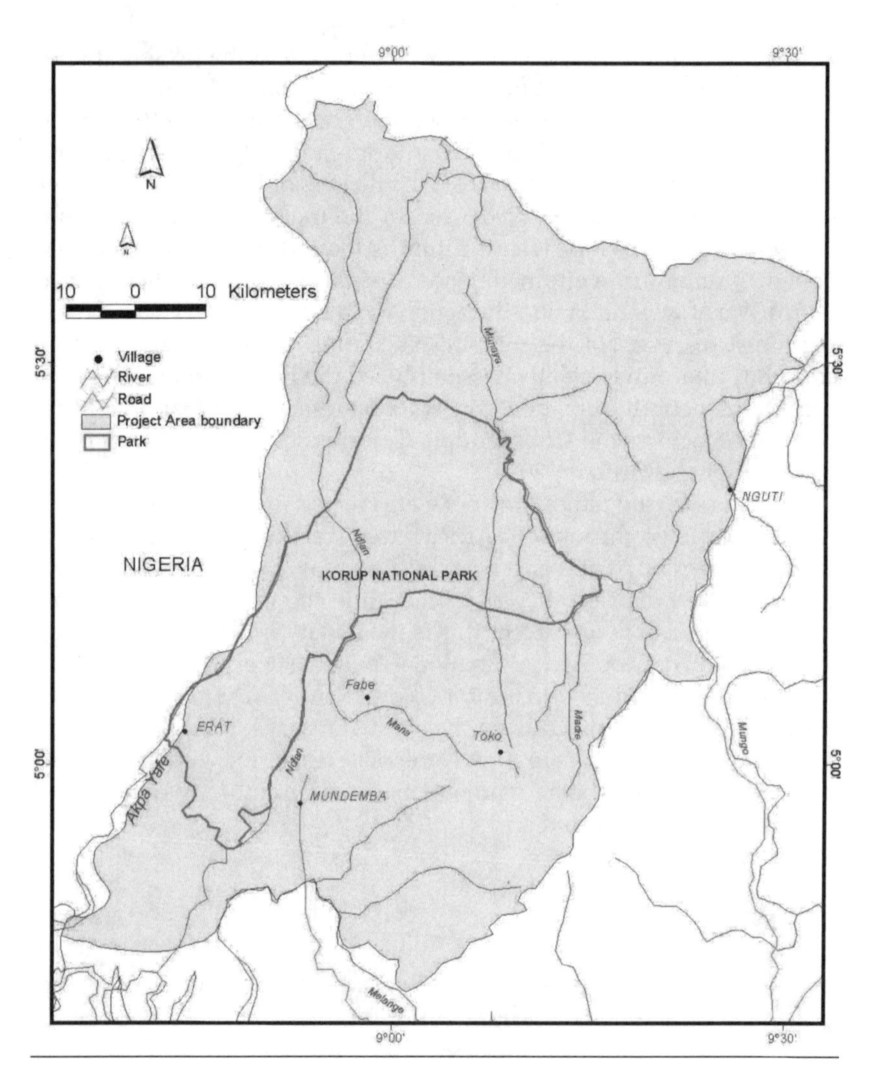

Figure 10.2.
Korup National Park and Korup Project Area.

The Korup Forest lies in what Letouzey has called the Biafran forest formation (Letouzey 1985). This is coastal, humid, evergreen lowland forest dominated by the Caesalpiniaceae, with a high rainfall of up to 6,000 millimeters (240 in.) per year. The species richness and high degree of endemism at Korup indicate a community that has persisted and evolved over a very long period (Maley 1988; Newbury, Songwe, and Chuyong 1998). Korup is part of a forest refugium, extending as far south as Gabon, that has sheltered forest during periods of dry climate, not only during the recent Quaternary and Pleistocene, but possibly also far back into the Miocene[2] (Hamilton 1989).

Korup contains 400 known tree species of some 68 families (Thomas 1993). The total known plant species list of 1,700 species will certainly eventually exceed 2,000. There are 400 fish species, 174 species of reptile and amphibian, including all three of Africa's species of crocodile. There are 407 bird species and a mammal fauna of 119 species of 30 families.[3] The primate fauna (fifteen species) includes four that are endangered—the drill *(Mandrillus leucophaeus)*, Preuss's red colobus monkey (*Piliocolobus badius preussi*), Preuss's guenon *(Cercopithecus preussi)*, and the chimpanzee *(Pan troglodytes)*—and one vulnerable species, the red-eared guenon *(Cercopithecus erythrotis)*. Other species not listed by IUCN as critical or endangered, but which have not been seen in Korup for the past two decades, include the African golden cat *(Felis aurata)* and the leopard *(Panthera pardus)*. Forest elephant *(Loxodonta cyclotis)* populations have been much reduced in recent years, as have those of the yellow-backed duiker *(Cephalophus silvicultor)*. The biological value of Korup is very high indeed, but that value is being degraded through unsustainable exploitation particularly of the large mammal fauna and of some plant species, principally for commercial purposes. Creation of Korup National Park has not halted the loss of large mammals, but the forest has been protected against logging and the boundaries have been respected (apart from an incursion for military purposes as part of the ongoing conflict with Nigeria over the Bakassi Peninsula).

The motivation for the development of the Korup Project was the perceived loss of tropical forest throughout the world and the need to preserve a large, functioning forest ecosystem intact. This was to be achieved through creation of a protected area and development activities outside the protected area to attract people away from it; it was one of the first examples of the ICDP philosophy.

At the time of Korup's classification as a Native Authority Forest Reserve in 1937 and until independence in 1961, Southern British Cameroon was administered as part of Nigeria. Korup was located in the Kumba Division near the village of Ndian. The only economic activity there was a palm oil plantation (Ndian Oil Palm plantation), managed and supplied from Calabar in Nigeria. Korup was remote and inaccessible by road from Cameroon. After its classification as a forest reserve, it was little visited by the authorities over the next twenty-five years. At independence, construction of a bridge and dirt road somewhat simplified access. It became possible to take a river boat or barge from the end of the road to Ndian. This trip could take from as little as eight hours to as long as three days depending on tides, sandbanks, and the height of the rivers; in those days, Ndian was known as "Cameroon overseas."

In the late 1960s, the biological significance of Korup was recognized first by primatologists.[4] Research to document its biological value in more detail was interrupted by the outbreak of the Biafran War in adjacent areas of Nigeria in July 1967, and the research focus moved elsewhere, as it was impossible to continue to work there. Prior to this, however, the idea of creating a Korup National Park had been suggested to the Biology Department of Federal University of Cameroon, who had in turn taken up the matter with the presidency. Eventually President Ahidjo agreed to the idea as long as the park was associated with Federal University. For a variety of reasons, principally the Biafran War, this initiative was not followed through. It had also proved impossible to interest the international community in funding Korup at that time.

The Biafran War ended in January 1970, and the Ndian Division was created in 1974 with a prefect and a divisional administration. A road from Lobe to Ndian was finally opened in 1985, and it was finally possible to drive from Kumba to Ndian by road, but the unsurfaced road was often impassable in the rainy season. In 1982 a film, *Korup: An African Rainforest,* made by the young British filmmaker Phil Agland, was released in the United Kingdom to great critical acclaim. Agland's specific goal in making the film was to raise consciousness in the developed world about the worldwide crisis of the rainforest biome. Public concern about Korup as a result of this film resulted in agreement from the British Overseas Development Administration to fund the Korup Project through the charity Earthlife and later through the World Wildlife Fund (WWF). The then permanent secretary of the ODA, Sir Crispin Tickell, visited Korup to announce this agreement.

The philosophy behind the Korup Project was explicitly stated in 1990 (Wicks 1992):

> The Korup project takes the view that no protected area can survive in isolation without the active support of the community that live in or around it. Every effort has been made to find out through detailed socio-economic surveys the effect the creation of the park would have on the local population and vice versa.
>
> The main threat to the park comes from the expanding population of Nigeria, many of whom cross into Cameroon to farm and hunt. The park is also threatened by the hunters in the 6 small villages inside the park and the 27 villages inhabited by 12,000 people that live within 3 Kms of it.
>
> The villages inside the park are inhabited by approximately 750 people. These 6 villages depend on hunting and slash and burn agriculture. They kill some 12,000 animals weighing over 140 tonnes each year, which

> they sell in the towns in Cameroon and Nigeria. Every effort was made to find an alternative sustainable source of income for these villages. As they are situated on very poor acidic soils, and the villages are isolated both from each other and any potential markets for other less valuable crops, this proved unsuccessful.
>
> In addition the villages want the same level of development as other villages outside the park, including road access to towns, hospitals and schools. As this cannot be provided inside the park they have agreed to move to more fertile soils outside the park. They will stay within their tribal area and within the Korup forest which has been their home for many years. All resettlement is voluntary and the villagers are being helped to build their own villages on sites of their choice....
>
> The objective of the rural development program is to replace the income and the protein obtained from unsustainable hunting by other sources of income. It is also aimed at helping the local community develop sustainable land use systems including agro-forestry....
>
> The project will only be judged a complete success when the villagers who live in the park are happily settled on good soils outside the park, and when the local people provide the main protection for the park because they are convinced of its importance to them and their successors.

Despite the humane and entirely laudable goals of the project, the legal land-use model used, the national park, was derived directly from the colonial framework of the 1933 convention—as is much current African conservation legislation. Furthermore, the intervention logic chosen by the project, focusing on development activities outside the protected area to "attract" people out from the park did not work because the Government of Cameroon failed to provide the legally required compensation, which had the effect of further antagonizing the already alienated inhabitants of the park area. The absence of financial compensation raises an interesting legal question on the legality of the gazettement of Korup, as the Cameroon legislation (Régime Foncier et Domainial) requires that compensation be paid in order for the expropriation to become effective. Moorehead and Hammond (1994) confirmed that the people who lost the most from the creation of Korup were not the primary beneficiaries of rural development initiatives carried out by the project.

Despite the fact that hunting and trapping are formally forbidden inside Korup, they are carried out intensively in the park and surrounding areas because bushmeat is such an important source of income. The sale of bushmeat is estimated to account for half of household income (Amadi 1993). Similar figures are reported from other forest areas of Cameroon; among the Mvae, far from urban influences, one domestic group (family) during a thirteen-month period captured 169 animals, or 13 per month, for a total weight of 130 kilograms. Of this total, 62 kilograms were sold, 44 kilograms consumed, and 24 kilograms were the object of various transactions. This translated to an income of 17,310 CFA per month (U.S.$23), when the minimum monthly wage is 19,000 CFA (U.S.$25) (de Garine 1998).

In 1990, WWF–UK stated that in the four years since the project started, WWF and various aid agencies (including the ODA and the German aid agency GTZ) had spent over U.S.$2 million (*Project Proposal: The Korup Project, Cameroon* 1987). At the same time, WWF–UK produced a management plan in which it estimated that the total cost of project implementation would be U.S.$30 million. The European Union (EU) took over the funding of the project in 1993, and GTZ also made a long-term commitment. They were still the principal funders at the end of 2001. The EU budget (1993–2000) for Korup has been 16.6 million Euros (U.S.$14.2 million); that of GTZ (1993–2002) has been approximately DM 9.7 million, or U.S.$4.2 million over the same period.[5] Thus in the last decade and a half, a total of some U.S.$20.4 million has been spent on the Korup Project by the WWF, EU, and GTZ, when the annual budget for the entire national protected area system has sometimes been as little as U.S.$31,200. Other significant inputs have come from the U.S. Department of Defense, Exxon, Operation Raleigh (all three involved with the design and construction of bridges), and the Wildlife Conservation Society (principally involved in research). It is difficult to be precise about budget allocation, but the vast majority of funds (over 90 percent) have been spent on development and infrastructure projects and relatively little on park management and protection (see also Brandon and O'Herron this volume). Biodiversity conservation has been a conspicuously poor relation in terms of total allocation of financial resources to the Korup Project. Conservation has not been an equal, or even significant, partner in this particular ICDP.

It is clear that by the terms of its own ambitions, the Korup Project has not succeeded. The villagers are not "happily settled on good soils outside the park" (Wicks 1992). Most of the population is still living inside the park; indeed, the population there may have doubled since the project began. Unsustainable levels of hunting and trapping continue, and there has been little progress in replacing the income and the protein obtained from them. The major focus of the development interventions of the project has not been those people who were most affected by the creation of the park, but the populations living outside in the "support zone." It is clear that problems and unjustifiable assumptions were made at the design stage of the project and that these have greatly hindered the implementation of the project (see also McShane and Newby this volume). These should have been detected at an early stage and the project reformulated, but even if this had been done, it is doubtful that the Korup Project could have achieved its objectives. Effectively, the Korup Project continued a process that had begun in the colonial era of land and resource alienation and brought these latent problems to a head. The problems were exacerbated because necessary government guarantees on resettlement, compensation, law enforcement, counterpart staffing, and funding were not obtained before the project began. If Korup fails, this will not be because of an inherent flaw in the ICDP approach, but because the appropriateness of the land-use paradigm in the particular local context was not adequately considered in historical and social terms. Later palliative measures did not address this fundamental issue and consequently did not work, and a strategic decision was made to focus on development in the support zone. This further exacerbated the relations with those isolated inside the park. The inception of the project crystallized the fact to the local communities that what they had been treating as their communal land and their communal resources no longer belonged to them. Although this had been true for almost fifty years when the project started, there was little local knowledge of, or information on, the situation or acceptance of it. As far as the local communities were concerned, the Korup Forest and all it contained was theirs.

Once it became evident that they might make gains in acceding to resettlement plans, villagers played on the ambitions of the project by employing a stop-start bargaining methodology: the resettlement program was thus virtually doomed. This was the only way in which park inhabitants could extract some kind of financial benefit in the absence of official compensation.

Ekundukundu, on the periphery of the park, the first village to be resettled, was moved in 1999, thirteen years after the project was initiated and five years after the WWF estimated that the whole resettlement process would be completed. The total cost of resettling this village of 47 houses was 336,855,200 francs CFA (U.S.$445,300) (Vabi 1999). The total cost of resettling the 375 families in Korup, at the level of cost incurred at Ekundukundu, would be some U.S.$3.63 million. This excludes the costs of compensation for permanent structures and tree crops (principally cocoa and oil palm) that is required by law. In 1987 these costs were estimated at about U.S.$900,000, and they have since certainly greatly increased with the increased number of plantations of both crops that have been established in anticipation of compensation. There has been little success in providing alternative economic activities and protein. Unsustainable habits continue and have even multiplied, e.g., harvesting of eru *(Gnetum bucholzianum)*, toothbrush sticks (*Garcinia mannii* and *Massularia acuminata*), and Hausa sticks (*Carpolobia lutea* and *C. alba*). There has been no attempt to set up managed communal hunting zones outside the forest, nor any success in obtaining agreements not to hunt the most endangered species (drill or red colobus).

A midterm review of the Korup Project was carried out in 1996 (EDG 1996). According to this review, Korup National Park had not been secured, its status as a Government of Cameroon institution had not been strengthened, and its institutional integrity remained insecure ten years after initiation of the project. Surveillance, management, and infrastructure were all weak. The review recommended resolution of the resettlement issue—either resettle or redesign the strategy.

Essentially, the park had become a satellite of a more vigorous development project. The conservator of the national park was not seconded to the project and in fact had no formal affiliation with it. The project was directed by a series of expatriates with a development background. There was an overoptimistic assumption that a complex, donor-driven, multifaceted ICDP could succeed in transferring technology and training to the Government of Cameroon in four years of EU funding. There was an unrealistic assumption that the rural development component could operate in an area as large as the support zone (WWF–UK 1997) and that it could be effective when the principal human targets of the activity had not been resettled.

The park has never operated from a position of political strength. The political pressures brought to bear in 1986 for gazettement basically derived from the interests of two powerful elites of the region. Their principal goal was not forest conservation but to bring development to the remote, undeveloped Ndian Division; the WWF became a proxy for the Government of Cameroon in bringing rural development to the division. The francophone central government, sensitive to land issues in the anglophone region, was complaisant but lacked a fundamental political commitment to the project. The two elites both suffered politically for their involvement in the Korup Project and were widely perceived to have "sold" their traditional forests and failed to bring adequate development in return. Furthermore, for personnel of the local administration, a posting to the newly created and remote Division of Ndian ("Cameroon overseas") was difficult enough. With almost no local markets and food in very short supply, they had traditionally obtained their meat from the forest. The Korup Project interfered with that tradition and helped create a hostile perception of the project among the local administrators. It became difficult to arrest and pursue poachers through the courts; they would "escape" from custody or would not be convicted. Relations eventually became so bad that guard patrols ran the risk of being beaten or taken hostage. In an effort to improve the technical skills of guards, the WWF initiated military training for guards in late 1997 from a former British Army major (WWF–UK 1997). This so incensed the local authorities that the prefect closed the national park to all project personnel and ordered the deposition of all guard weapons and ammunition with the local Gendarmerie Brigade. Strong law enforcement within the park was not and is not politically possible, and this too helped shift the focus from biodiversity protection to development outside the park.

While the Korup Project has a vision of "people happily settled on good soils outside the park," it does not have an exit strategy. The multiple actors in the project have tended to pursue their own agendas, synchronized neither with each other nor with the project. At the beginning of ODA intervention (in 1986), "environment" was a major theme of this aid agency; by 1993 it was "poverty alleviation," and the exigencies of the donor agency deformed the goals and the execution of the project.

Mount Kilum Forest Project

The Kilum-Ijim Forest of Mount Oku lies between latitudes 6°05′ and 6°20′ north and longitudes 10°20′ and 10°36′ east (figure 10.3). Mount Oku, at 3,011 meters, is the second highest mountain in West Africa after Mount Cameroon (Collar and Stuart 1988). Rainfall at Kilum is over 3,000 millimeters at the summit of Mount Oku and around 2,000 millimeters lower down. The forest has never been classified or gazetted, and traditional rights are exercised over it. Attempts had been made as far back as the 1930s to create a reserve in the area by the colonial authorities, but without success, the local opposition proving too strong. Some forty years later, with the montane forest greatly reduced, traditional authorities in both the Oku and Kom areas attempted to demarcate a forest zone in the 1970s in an attempt to preserve the traditional forest for their peoples. It was unsuccessful, but the attempt does indicate a growing interest in forest conservation as the resource dwindles.

The area within the project boundary is about 20,000 hectares, of which some 7,000 hectares are montane forest, the rest being montane grassland and various types of recolonization formation: this represents a major part of Cameroon's remaining montane forest resource. As recently as 1963, forest covered 17,500 hectares in Kilum; this was reduced to 8,700 hectares by 1983, with 1,700 hectares of this being lost by fire and another 2,700 hectares badly damaged in 1986 (Integrated natural resource management programme n.d.). The Kilum Project of ICBP (International Council for Bird Preservation—now BirdLife

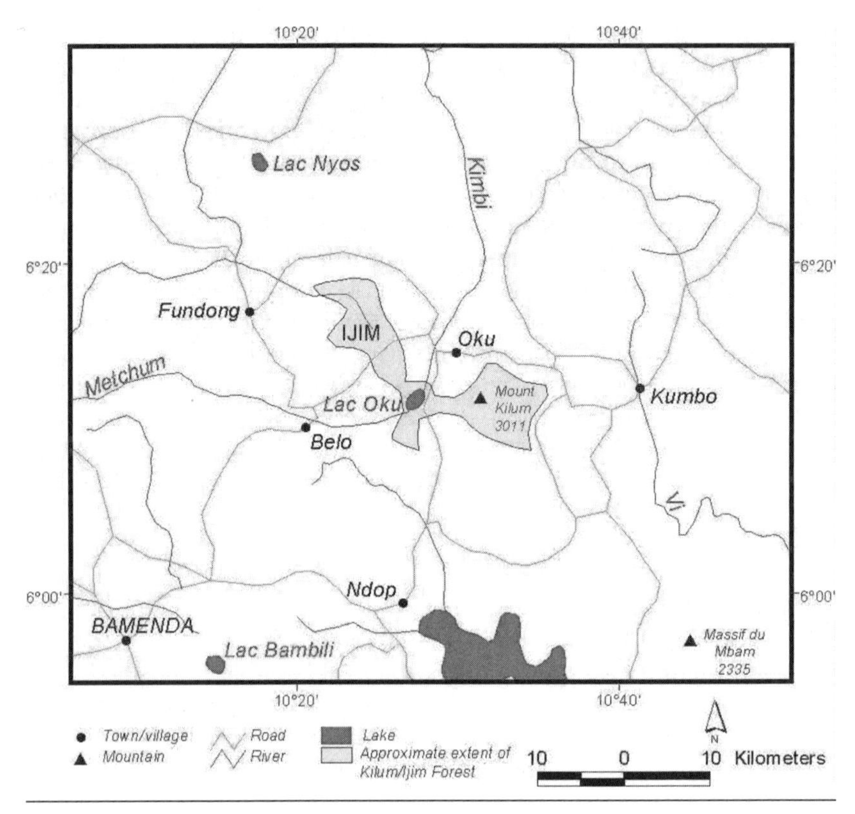

Figure 10.3.
Kilum/Ijim Project Area.

International) was initiated in 1986 to stop the loss of forest and help provide sustainable alternatives to forest clearance. The forest of Mount Oku is very seriously threatened by agricultural encroachment, illegal grazing, fire damage, and overexploitation of *Prunus africanum* bark (Maisels 1997). Kilum has suffered and continues to suffer serious ecological problems (water shortage, soil erosion, loss of large mammals, massive deforestation). The forest is still highly vulnerable to fire; the use of the forest for grazing animals, especially goats, which browse on regenerating forest and thus halt the successional process, poses another serious threat. Other more subtle but harmful influences are the selective removal of plant and animal species for food, medicine, carving, paper production, and firewood. However, the Kilum-Ijim Forest now has an agreed-on boundary, and conversion to farmland has essentially ceased within the project area since 1992.

The Kilum/Ijim Forest lies in the montane forest formation largely above 2,000 meters altitude, a formation that is relatively rare in West Africa. The high-altitude forest is relatively species-poor in terms of trees. The paleontological events responsible for the distinctive biological character of Kilum are much more recent than those of Korup, dating back to perhaps 18,000 to 25,000 years ago. The principal biological affinities of Kilum (and other montane forests in Cameroon) are with the montane forests of eastern Africa.

The montane forests of Cameroon contain twenty-one endemic bird species. Two of these occur at Kilum: Bannerman's turaco *(Tauraco bannermani)* and the banded wattle-eye *(Platysteira laticincta)*. Two other threatened birds—the green-breasted bush-shrike *(Malaconotus gladiator)* and Bannerman's weaver *(Ploceus bannermani)*—and the near-threatened Cameroon mountain greenbul *(Andropadus montanus)* also occur on Mount Oku. The existence of these birds was the primary reason for the ICBP's initiation of the Mount Kilum Forest Project (Maisels 1997).

The large mammal fauna of Mount Oku is virtually extinct. Elephant, leopard, and chimpanzee occurred in the recent past, with probably up to four species of *Cercopithecus* monkeys and three or four species of duiker. Bushbuck also existed there. The elimination of these large mammals must certainly have had fundamental effects on the dynamics of the forest ecosystem, but these are presently undocumented. However, in practical terms, their absence also means that trespassing into the forest for hunting does not occur.

Kilum is protected by the power of the traditional authorities and also by Prefectorial Order No. E26/38/RPB/RS/93 of 15 November, 1993, forbidding and regulating certain activities within the forest (Maisels 1997). Article 1 of this order prohibits the setting of bush fires, the felling of all categories of trees, the cutting of alpine bamboos, farming within the reserve, rearing and grazing of domestic animals, disrespect for forest patrollers, and hunting and trapping of animals within the forest reserve. However, Article 2 stipulates that the inhabitants of villages surrounding the Mountain Forest Reserve may harvest material from the forest for traditional medicinal use, collect dry wood for domestic use only, practice beekeeping, catch rats for consumption, and collect mature Alpine bamboo for traditional building and other allied uses.

The primary reason for the Kilum Project was to prevent the imminent disappearance of the remaining montane forest and with it the endemic bird species that it contained.

The Kilum Project is an ICDP without the complications of state land tenure. The land remains in the "domaine national," and traditional rights may be (and are) exercised over it. Kilum is located in an area where traditional structures are visible, explicit, and very powerful. The fon determines the traditional structures in Kilum, and the fon of Oku is one of the more powerful of the traditional rulers. Partly because he perceived the destruction of the traditional forests, with their socioreligious significance, as a threat to his traditional authority, he became convinced of the value of the project's approach. The local and influential power structure has actually been seeking on its own initiative the conservation of the remaining fragment of montane forest. Unlike the situation in Korup, there was strong local political support at all levels of society for the project at its inception, which was important for the local acceptance of the project but also contains the seeds of danger. Many of the sons and daughters of the region work in the major cities of Cameroon (and abroad). They tend to be well educated and are often also well off. When they retire, they may return to the village of their birth. They are not likely to submit easily and uncritically to authoritarian traditional rule. The ultimate weakening of traditional authority will almost certainly occur, and if the forest program is too closely associated with it, it too may suffer in consequence (see also Child and Dalal-Clayton this volume).

While Kilum has been a much smaller project than Korup, project funding has also gone to development rather than conservation activities. Forest nurseries, soil erosion control and contour planning, development of honey marketing, wood carving and paper-making cooperatives, etc. have used most of the funds. Although over the fifteen-year life of the project, expenditures have been little over 15 percent those of Korup, in terms of the size of the project area, the two projects are comparable, with Kilum perhaps being more expensive in terms of expenditure per hectare. Both are also similar in the relatively small percentage spent on biodiversity monitoring and protection; Korup and Kilum are alike in the fact that the bulk of project funding has been spent on development.

The Kilum forests are widely used by densely distributed local communities. The forests provide the water catchment for the irrigation of the village fields (very important in this dry area), they are a source of firewood, and they are an area for trapping rodents (very little remains in Kilum that is larger than a squirrel). Although a biological disaster, this is a social advantage, as there is no pressure on the forest from unsustainable commercial hunting. (However, the unsustainable exploitation of the montane forest tree *Prunus africanum* poses a very real threat to the future of this forest.) The Kilum forest provides habitats for beehives, an important source of honey for cash income. The forests were being cleared for agriculture (Irish potatoes and coffee) and for subsistence in this very densely populated area. Burning and encroachment were the major problems, as well as grazing by nomadic Fulani in the upper grasslands—and sometimes fires were set, with serious consequences.

The situation of the forest guards at Kilum provides a strong contrast to the situation at Korup. At Kilum, community-selected forest wardens patrol the forests, principally to detect fires. When transgressors are arrested, they are dealt with in customary courts, quickly and effectively.

The principal office of the Kilum Project is located in the center of the town, whereas in Korup the park headquarters is some two kilometers out of town and physically and socially isolated. The Kilum Project administration worked within existing administrative structures—the rural development agencies of the Ministries of Agriculture and the Forestry Department, and the Northwest Development Agency (MIDENO). In contrast, as a national park, Korup required a whole new administrative infrastructure that was installed in a new, custom-built headquarters away from the local community. Kilum worked within the status quo, did not require resettlement, and did not require from the government large amounts in compensation. Existing staff complements of divisional representations of the Ministries of Agriculture and Forests were expanded, rather than requiring new agencies, new offices, and new staff as at Korup.

The Kilum/Ijim Project, working from a basis in which the community desired the conservation of the remaining forest fragment, has been able to work from a position of strength and cooperation with the local communities. The project has provided inputs that have helped raise the standard of living of the local community, improved quality of honey, and increased income from paper making and wood carving. Needs such as firewood and medicinal plant harvesting are met. The local community respects the visible boundary of the reserve, planted with distinctive trees.

Discussion

A comparison of the Korup and Kilum projects provides insights in two main areas: first, the nature of preconditions necessary for the eventual success of such ICDPs, and second, their capacity to deliver biodiversity conservation.

It is clear that neither in Korup or Kilum is there any interest in biodiversity conservation as such. "Contrary to certain ideas, rural people have a relatively negative traditional perception of nature: it is not a friend. The bush is a frightening place and villagers are only at ease in a profoundly human environment, symbolically protected by the ancestors. Farmers, especially crop farmers, consider wild animals harmful" (de Garine 1998:84). If the drill were to become extinct tomorrow, it would be a catastrophe for the local villagers, not because it is a unique and interesting species, but because it is large, easily shot, and very much in demand as bushmeat. Similarly, if Bannerman's turaco were to become extinct, probably not a single villager in Kilum would shed a tear; only the more farsighted would perhaps wonder what effects this would have on the number of bird-watchers visiting the project. The focus of the project is the integrity of the forest boundary. The value is in the biomass, not the biodiversity. In both these cases, the fundamental local concern is the resource base and its continued availability. Resources have value; biodiversity is a characteristic or aspect of the resources with no added value and therefore no economic or social significance. A successful ICDP may provide stable use of the resource base that is compatible with the development of a conservation ethic but will not itself produce that ethic. Such an ethic is likely to develop only if people benefit from the presence of drills or of Bannerman's turaco. For example, if villagers at Korup were to receive an annual bonus based on numbers of drills and red colobus enumerated in an annual census, this would provide an immediate incentive for biodiversity conservation; these primates would begin to have a value greater than that of the other thirteen primate species that occur in that forest.

Both project sites have suffered a marked loss of biodiversity and local extinctions of large mammals since the colonial powers began the reservation process. At Kilum, project intervention has probably slowed the process, but only because most mammals are already extinct—large mammal biomass value is essentially zero. Kilum is a tiny forest barely on the cusp of existence. A single forest fire or sudden soil erosion on the steep slopes in heavy rain are events that could in a few days easily wipe out the whole fragile system. Korup has seen its large mammal populations decline, but has been successful in preserving the forest from logging. One may draw the conclusion that where the process of ecological degradation has advanced to the point where people are beginning to suffer the consequences, and where biomass resources of economic value such as large mammals have been eliminated, it may be possible to halt further ecological decline when the social conditions are appropriate and provided that such areas are not too large and there is local control over land use. It is noteworthy that colonial attempts to create a forest reserve at Oku (Kilum) in the 1930s, when the forest was much more extensive, met with opposition; not until the forest was almost gone did conservation become an acceptable option.

While varying a great deal in the details, the land-tenure systems of precolonial Africa possessed many common characteristics (Okoth-Ogendo 1987). The first of these relates to the notion of tenure itself. Precolonial tenure was essentially regulation of land use and a function of the relationships between people in a given socioeconomic and political context; ownership of land as such did not exist. Control of land is an expression of sovereignty (in its jurisdictional nonproprietary sense) and therefore resides in the political authority of a given society. Access to land is a function of membership in society and is determined by the network of social relations in which a member participates. There is no general access; access rights vary with the perceived value to the community of a given individual (Okoth-Ogendo 1987). In any given community a number of members could each hold access rights, expressing a different range or variety of functions (cultivation, grazing, transit, fruit gathering, etc.) over the same parcel of land. Colonization introduced the concept of private property and of exclusion.

The legal conception of land in English property law is summarized in the maxim *cujus est solum, ejus est usque ad coelum et ad inferos* (whoever owns the land owns whatever is above and whatever is below it). In colonially imposed English law, not only is absolute ownership of land admitted, it is conceived in basically individual terms rather than a feature or product of communities or societies. To "own," therefore, is to exclude the possibility of others having simultaneous rights of the same or a different quality over particular resources. The introduction of the concept of private property weakened and destroyed finely tuned traditional systems, systems that were, ironically, amenable in principle to incorporating into their norms and practices the principles of biodiversity conservation and sustainable management in the long-term interests of society. The concept was also a potent force in the weakening of traditional society through elimination of a major function of that society.

In the case of protected areas, the exclusive individual private owner was the distant and often nebulous state or government. When the Korup Forest Reserve was established in 1937, traditional footpaths and rights of way within the reserve were maintained as well as "the rights to hunt and fish. The rights to collect snails, tortoises, land crabs, honey, cola, bush fruits and nuts, and any other food materials. The right to collect palm produce ..." ("Other Rights" 1962). These usufruct rights were retained until they were extinguished by the national park legislation in 1986.

In most such cases the changes were imposed with little consultation and a minimum of information or explanation. It was often not clear to local communities what the purpose of the changes was, as there was often no perceptible "use" of the resources by the new state landlord. Thus while the benefits of forest preservation and its ecological riches accrue at the national and international level, the costs accrue at the local level (Egbe 1997). The resulting disaffection, confusion, and political destabilization (which continue) are major causes of environmental degradation and destruction (Sikod et al. 2000). In French Equatorial Africa (AEF) territories, nationalization of land occurred as early as 1934, and the immediate result of the loss of local control was extensive deforestation (Riddell 1987). It is a historical irony that having destroyed what was essentially a continent-wide, fully functional community system of land allocation according to social need and replaced it with a system of exclusive private property, donor agencies of the developed world are now trying in an ad hoc way and on a piecemeal basis to restore some elements of community land management.

It may be asked whether the situation at Korup is salvageable. The Government of Cameroon is in the process of revising land-use plans in the southwest of Cameroon.[6] These include proposed changes to the boundary of Korup National Park and an extension of 30,149 hectares. If, as part of this process, the park boundary around the village of Erat (figure 10.2) were modified to exclude it from the national park, this would effectively remove about half of the population living within the park who depend on the cross-border trade with Nigeria and are reluctant to move. It is clear that people stranded inside the park need to be able to hunt, fish, and farm to survive, and the usufruct rights that they obtained in the colonial era should be restored and moratoriums imposed on the hunting of endangered species (and incentives to reward communities for the preservation of these species). It is essential from a legal, moral, and practical point of view that compensation owed for cash crops and permanent buildings should finally be paid. People who received compensation would then probably move voluntarily toward the new roads. The expensive resettlement program would probably thus become redundant and could be abandoned. A return to a traditional land tenure system at Korup is probably not politically feasible, nor indeed would such a change be likely to lead to conservation and long-term protection. But more local control and community management of the resource is necessary.

The Kilum Project has been successful in developing, through participatory mapping, an agreed-on forest-use plan for the local communities that reconciles the needs of the beekeepers, the firewood gatherers, the grazers, the wood carvers, and the rodent trappers. This participatory mapping is a process in which forest use is defined and usage allocated in respect of membership in the society and in terms of specific need. It regulates use; it does not determine ownership (see also Brown this volume). In this crucial respect, Kilum has essentially returned to the precolonial land tenure system.

Kilum has been able to maintain its focus and achieve qualified success as a project because the project addressed the sustainable use of the resource by the community—the focus was the resource and its use by the people. At Korup, on the other hand, the focus of development was outside the park in the support zone and diverted the attention of the project away from the park. This led to the development of a "soft center" at the core, with villagers in the park feeling neglected and hostile toward the project. While both projects have been focused on development rather than conservation, there is no doubt that the much smaller and more intimate management style of Kilum helped it; Korup became highly bureaucratized. Kilum also has the benefit of being small enough to be subsumed under the hegemony of the local tribal structure. (The Kilum-Ijim area is less than twenty kilometers across at its widest point; Korup is seventy-five kilometers north to south and fifty kilometers east to west.) Furthermore, at Kilum the population has suffered the first consequences of ecological catastrophe and have had their eyes opened; they are more ready to practice careful resource management than those in a large, resource-rich area that has not suffered the results of environmental damage.

Neither of the two projects has been successful in the sense of achieving conservation of a viable, intact ecosystem with informed community involvement in its management. Neither has a realistic biodiversity conservation goal, although Kilum has achieved some agreement over resource management. It is clear that biodiversity conservation projects cannot succeed where there is ambiguity and antagonism over land tenure. Throughout Cameroon, most communities who are neighbors of national parks are overwhelmingly hostile to them, and the continuing depletion of their resources is highly alarming. None of the protected areas of Cameroon covers its costs; all are negative drains on the national economy. ICDPs are not an efficient route to conservation in the national park context; levels of antagonism and hostility are too high (e.g., see Brandon and O'Herron this volume). Cameroon needs to move to fully devolved community management of protected areas and face the possible (even likely) loss of valuable biodiversity, or it needs to set up effective protection regimes that include compensation for lost rights. If the biodiversity is of global significance, as it often is in Cameroon, a major share of the costs of conservation should be borne by the international community. Mixed measures or compromises will not work.

Conclusion

At their most efficient, ICDPs may deliver sound resource management. The use of this approach to develop an ethic and a demand for biodiversity conservation in rural communities has not been demonstrated but might be feasible. But it will not be feasible when the land-tenure system embodies the arbitrary and unfair legal underpinnings of a state land-tenure system, as is the case with most national parks. Unless there is a willingness to fully devolve resource management to local communities, it is probably best not to attempt ICDPs in the context of state private property.

Sustainable resource management, highly important in itself, is not biodiversity conservation and does not automatically lead to it. Biodiversity conservation needs to be addressed through its own institutions and address specifically biological issues—while taking into account demographic trends and the development context. If the world values tropical biodiversity, the developed world must pay the bulk of the costs. Most of the problems in African conservation are not failures of systems in themselves, but result from a lack of resources; all African countries pay a higher relative proportion of their GNP into protected area management than do North America and Europe, but in absolute terms the amounts are insufficient. It is politically unrealistic to expect Africa to increase financial inputs for biodiversity conservation when there is no African constituency for it. There is an international constituency for biodiversity conservation, and it is this constituency that should pay. The Western world is currently spending more than U.S.$33 million per day on the war in Afghanistan; one month's costs of this war could secure a major part of Africa's biodiversity heritage in the long term. If the costs of biological conservation continue to be borne by poor and disaffected local communities and African states, there is a very bleak outlook indeed for the future of Africa's forest biodiversity.

Acknowledgment

I thank Rufin Mikala Mussavu, WWF–CARPO/ERAP, Libreville, for providing the maps.

Endnotes

1. The Korup Forest Reserve was established by Order 25 of the chief commissioner of the Southern Provinces (of Cameroon) in 1937. It was modified by the Kumba Western Council (Korup) Forest Reserve Order of 1962, reducing

the reserve area to 343 square miles (88,837 ha). In 1986, by presidential decree, an area of some 42,000 hectares was added, and the area became the Korup National Park.

2. There has been much recent discussion on the extent of anthropogenic influences on these forests and the extent to which the forests are "pristine." Discussions by Vansina (1990) and Hamilton (1989) and evidence provided by Maley (1988) indicate that while these forests may have been influenced by humans, especially after the inventions of agriculture and of iron, there is no evidence that the forest cover was ever removed in the coastal region. Even during climatic events such as the El Niño Southern Oscillation (ENSO) that occurred between 1765 and 1799, when rainfall was severely reduced, altering the floristic structure of Korup, the forest cover evidently remained.

3. There are some anomalies in mammal distribution at Korup. The western lowland gorilla *(Gorilla gorilla gorilla)*, which occurs both to the north and to the southeast of Korup, is absent from the Korup Forest, as is the bongo (*Tragelaphus euryceros;* also absent from Nigeria but present in Upper Guinea) and the giant forest hog (*Hylochoerus meinertzhageni;* also absent from Nigeria).

4. The two primatologists in question were Dr. T. T. Struhsaker and the present writer. The writer must take some responsibility for suggesting the idea of Korup as a national park and being involved in the early stages of its implementation. Korup was in fact one of the first projects to explicitly espouse the ICDP philosophy.

5. EU budget data from *Rapport Annuel 1999–2000,* Délégation de la Commission Européenne en République du Cameroun. Montant de la Convention: 7,340,000 Euro (FED), 1,981,837 (Stabex). Montant engagements secondaires: 7,208,820 (FED), 101,722 (Stabex). Details of GTZ budget from GTZ office, Yaoundé. Preliminary phase, DM 1,850,000; first phase (1993–1996), DM 3,500,000; second phase (1997–1999), DM 3,500,000; third phase (2000–2002), DM 2,713,615. Exchange rates as of July 15, 2001.

6. The current plans with Canadian Technical Assistance include for Korup a proposal to de-gazette a small section in the extreme south of the park that has been impacted by a military road and by agricultural incursions. In compensation, a much larger area, the "extension ouest" and the southern part of Ejagham Forest Reserve, would add 30,149 hectares to Korup National Park.

References

Amadi, R. M. 1993. *Harmony and Conflict Between Non-Timber Forest Product Use and Conservation in Korup National Park.* Rural Development Forestry Network Paper. no. 15. London: Overseas Development Institute (ODI).

Burnham, P. 2000. Whose forest? Whose myth?—Conceptualisations of community forests in Cameron. In A. Abramson and D. Theodossopoulis, eds., *Land, Law, and Environment: Mythical Land, Legal Boundaries,* 31–58. London: Pluto Press.

Collar, N. J. and S. N. Stuart. 1988. *Key Forests for Threatened Birds in Africa.* International Council for Bird Preservation, monograph no. 3. Cambridge, U.K.: International Union for Conservation of Nature and Natural Resources.

Convention Relative to the Preservation of Fauna and Flora in Their Natural State. 1933. London: Government of the United Kingdom.

De Garine, I. 1998. Traditional attitudes towards wildlife: A case study of Cameroon. In: N. Christoffersen, B. Campbell and J. du Toit, eds. *Communities and Sustainable Use: Pan African Perspectives.* Harare, Zimbabwe: IUCN-the World Conservation Union.

Draft Korup Project master plan. Godalming, U.K: World Wildlife Fund. Unpublished manuscript.

EDG (Environment and Development Group). 1996. *Mid-Term Review of the Korup Project.* Produced on behalf of the Ministry of Environment and Forests. Oxford, U.K.: Environment and Development Group.

Egbe, S. 1997. *Forest Tenure and Access to Forest Resources in Cameroon: An Overview.* Forest Participation Series no. 6. London: International Institute for Environment and Development (IIED).

Gartlan, S. In prep. *Creed, Greed, and Conquest: An Ecological and Social History of Africa and Its Peoples.*

Hamilton, A. 1989. African forests. In H. Leith and M. J. A. Werger, eds., *Tropical Forest Ecosystems,* 155–182. Amsterdam: Elsevier.

Integrated natural resource management programme for the Kilum/Ijim Mountain Forests, North-West Province, Cameroon. n.d. Cambridge, U.K.: International Council for Bird Preservation. Unpublished manuscript.

Laird, S. A., A. B. Cunningham, and E. Lisinge. 1999. One in ten thousand? The Cameroon case of *Ancistrocladus korupensis.* In C. Zerner, ed., *People, Plants, and Justice: The Politics of Nature Conservation,* 345–372. New York: Columbia University Press.

Letouzey, R. 1985. *Notice de la Carte Phytogéographique du Cameroun au 1:500.000.* Toulouse, France: Institut de la Carte Intérnationale de la Végétation.

Maisels, F., ed. 1997. Kilum-Ijim Forest Project: Conservation objectives. Cambridge, U.K.: Birdlife International. Unpublished manuscript.

Maley, J. 1988. *Synthèse sur l'histoire de la végétation et du climat des forêts de l'ouest Cameroun au quaternaire récent.* Actes du Colloque "Peuplement Ancients et Actuels des Forêts Tropicales." Orléans, France: Ecofit.

Moorehead, R. and T. Hammond. 1994. *An Assessment of the Rural Development Program of the Korup National Park Project.* London: CARE–U.K.

Newbery, D. M., N. C. Songwe, and G. B. Chuyong. 1998. Phenology and dynamics of an African rainforest at Korup, Cameroon. In D. M. Newbery, H. H. T. Prins, and N. D. Brown, eds., *Dynamics of Tropical Communities,* 267–308. London: Blackwell.

Okoth-Ogendo, H. W. O. 1987. Tenure of trees or tenure of lands? In J. B. Raintree, ed., *Land, Trees, and Tenure,* 225–229. Madison, Wis.: Land Tenure Center, University of Wisconsin; Nairobi, Kenya: International Council for Research in Agroforestry.

"Other Rights." 1962. Article 2 of the Second Schedule (Rights Within the Reserve) of the Kumba Western Council (Korup) Forest Reserve Order of 1962. *West Cameroon Gazette,* 27 January (suppl.), p. B16.

Project Proposal: The Korup Project, Cameroon. 1987. Godalming, U.K.: WWF. Project proposal presented to Overseas Development Administration.

Riddell, J. C. 1987. Land tenure and agroforestry: A regional overview. In J. B. Raintree, ed., *Land, Trees and Tenure,* 1–16. Madison, Wis.: Land Tenure Center, University of Wisconsin; Nairobi, Kenya: International Council for Research in Agroforestry.

Sharpe, B. 1998. First the forest: Conservation, "community," and "participation" in South-West Cameroon. *Africa* 68:25–45.

Sikod, F., E. Lisinge, J. Mope-Simo, and S. Gartlan. 2000. Cameroon: Bushmeat and wildlife trade. In P. Stedman-Edwards and J. Mang, eds., *Socio-Economic Root Causes of Biodiversity Loss,* 126–152. London: Earthscan.

Thomas, D. W. 1993. Korup project plant list (all species). Mundemba, Republic of Cameroon: WWF Korup Project. Unpublished manuscript.

Vabi, M. B. 1999. Development of a management plan for the Korup National Park. Final draft: WWF CPO. Yaoundé, Republic of Cameroon: WWF–Cameroon. Unpublished manuscript.

Vansina, J. 1990. *Paths in the Rainforest: Toward a History of Political Tradition in Equatorial Africa.* Madison: University of Wisconsin Press.

Wicks, C. 1992. Korup Project. In J. A. Sayer, C. S. Harcourt, and N. M. Collins, eds., *The Conservation Atlas of Tropical Forests: Africa,* 110–118. Gland, Switzerland: IUCN–The World Conservation Union.

WWF–UK. 1997. Contractor's closing report for the period 17/12/1993–16/12/1997. The Korup Project: Republic of Cameroon. Mundemba, Republic of Cameroon: WWF Korup Project. Unpublished manuscript.

11

Trade-off Analysis for Integrated Conservation and Development

Katrina Brown

The Limits of Win-Win Solutions to Integrating Conservation and Development

The experience of integrated conservation and development projects (ICDPs) in the last two decades has demonstrated that there are very few win-win solutions. In integrating conservation and development, there are winners *and* losers. The relationship between conservation and development is clearly more complex than either the direct conflicts perceived by the protectionist conservation lobby or the overoptimistic assumptions of synergy or win-win held by early proponents of ICDPs and community-based conservation. As reviews of ICDPs have shown, the interactions between conservation and development are more complicated and more fragile than often assumed (see McShane and Newby this volume). This relationship is context and culture specific, dynamic, and nuanced. Therefore, ICDPs need to be carefully considered and evaluated for who will be the winners and losers within the same society and for the impacts on different aspects of biodiversity (genetic, species, and ecosystem). This leads us to an appreciation of *trade-offs* as a necessary part of the design, planning, and implementation of ICDPs. Rather than expecting either conflicts or complementarities, we need to understand that both conflicts and complementarities exist in tandem. They need to be evaluated, negotiated, and managed through inclusionary processes, and the costs and benefits for different stakeholders need to be carefully weighed.

What then is the nature of these trade-offs in ICDPs? Trade-offs exist between different stakeholders or users, and between different geographical and social scales, as exemplified by Wells's (1992) analysis that distinguishes between local, regional/national, and transnational/global spatial scales. Trade-offs exist between different interests and priorities, particularly between economic development, social welfare, and environmental goals. Trade-offs also exist between long-term and short-term time horizons, where typically biodiversity conservation as a long-term objective is traded off against short-term economic benefits. A large literature is emerging that details the trade-offs between different sections of society and biodiversity conservation, often suggesting that rich people benefit from conservation while the poor bear the brunt of the costs, with resulting social injustice (Wells 1992; Ghimire and Pimbert 1997). Franks and Blomley (this volume), for example, illustrate the trade-offs between stakeholders in the Uluguru Mountains ICDP in Tanzania. There may also be trade-offs between different aspects of biodiversity. Redford and Richter (1999) show the range of impacts according to different uses and the numerous components or attributes of biodiversity. For example, they use the example of the Pantanal, Brazil, to explore the impacts of different management options on these dimensions of biodiversity, thus exposing the potential trade-offs when choosing management strategies.

There is no single method for evaluating integrated conservation and development trade-offs. Indeed, a range of methods will be necessary depending on the context, the nature, and the scale of the trade-offs, and on whether such an analysis is undertaken as part of an appraisal procedure—for example, choosing between different management options or different sites for intervention or investment—or as part of an ongoing management process. One approach to evaluating the trade-offs between different users and uses of natural resources, and to supporting decisions on management strategies, has been developed by Brown and colleagues (Brown, Tompkins, and Adger 2001; Brown et al. 2001). Termed *trade-off analysis*, this suite of techniques was designed and applied in devising a co-management strategy for a marine protected area, the Buccoo Reef Marine Park, in southwest Tobago in the West Indies. The approach is used not only for evaluating the trade-offs inherent in attempts to integrate conservation and development objectives, but for facilitating deliberation by, and the participation of, key stakeholders in decision making and management. These goals of deliberation and inclusion are increasingly important in many conservation and development initiatives. The next section describes the application of trade-off analysis at Buccoo Reef. Subsequent sections discuss the application of this approach in other situations, including forest ICDPs, and within the wider context of participatory and adaptive management strategies for integrating conservation and development.

Evaluating Trade-Offs Between Economy, Society, and Ecosystem Health: The Case of Buccoo Reef Marine Park

Buccoo Reef Marine Park is located in the southwest of the small island of Tobago in the eastern Caribbean and consists of the Buccoo Reef and Bon Accord Lagoon Complex. The marine protected area itself comprises a reef system that protects an extensive shallow lagoon bordered by a fringing mangrove wetland. It covers an area of 150 hectares, with a terrestrial area of 300 hectares, shown in figure 11.1. The management of the conservation area, the range of uses of complex biological systems, and the collaboration with different stakeholder groups create some classic conservation and development dilemmas. The economy of Tobago is dependent on tourism and fishing, which generate national revenue and support myriad livelihood activities in both the formal and informal sectors. As Tobago is an island of only sixty-two square kilometers, the management of the coastal margins is critical to the whole island ecosystem. Buccoo Reef is one of the most visited recreational sites in Tobago. Both international and local visitors enjoy the beauty of the coral reefs, the clear waters, and the abundant marine life that can be found there. Tourism makes an important contribution to local incomes, yet it degrades the natural resource base on which many islanders directly depend for their livelihoods. For instance, tourists cause physical damage to the coral reef itself, hotel and infrastructure construction results in clearance of mangroves and other natural vegetation, beach developments may undermine natural storm protection and enhance vulnerability to damage, and increased runoff and nutrient loading bring pressure on water treatment plants and discharge of polluted effluent into the lagoon system. Thus tourism brings benefits to Tobago, but not for everyone. The challenge therefore is to find ways of managing the Buccoo Reef that are acceptable to stakeholders while maintaining environmental quality and conserving the important biodiversity of the reef and associated systems. It is clear that trade-offs between these objectives are necessary, but how are decisions made, by whom, and on the basis of what information?

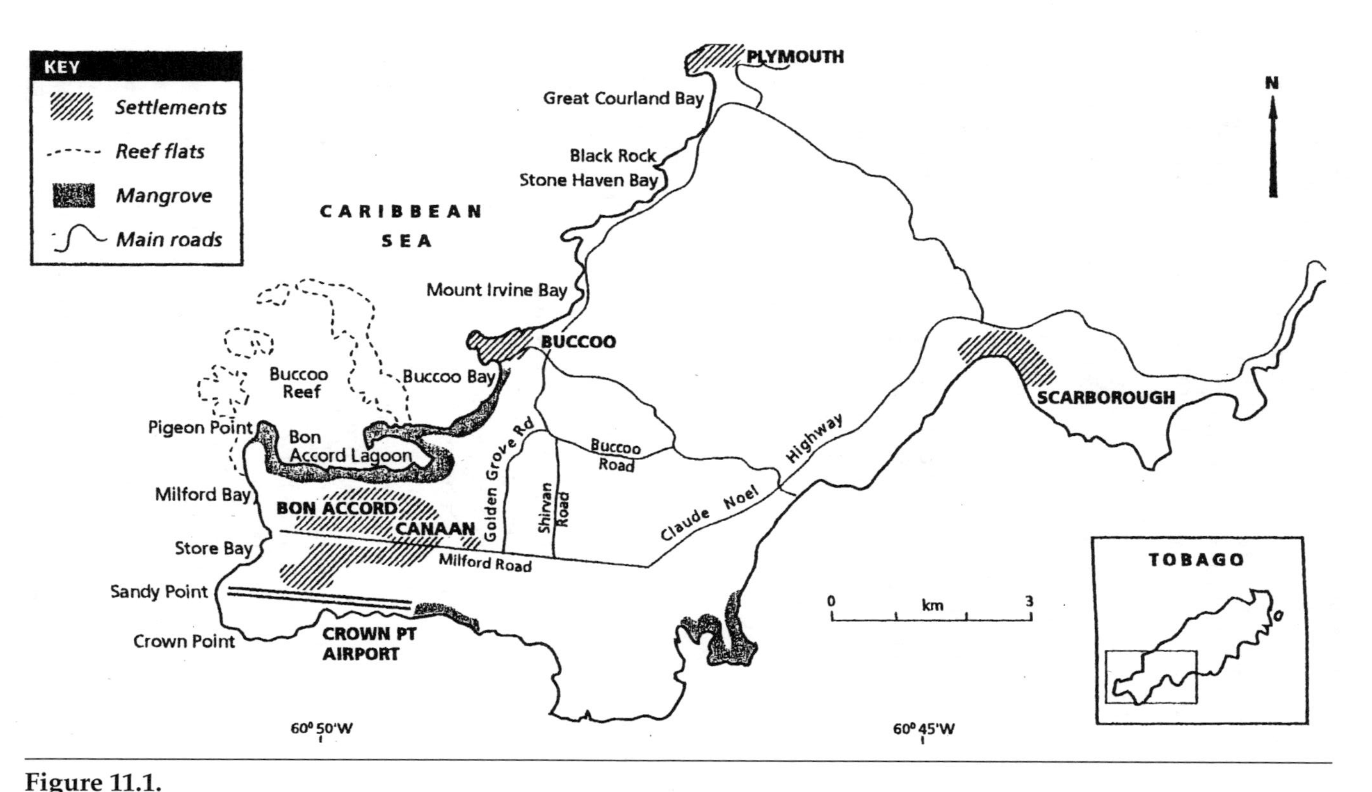

Figure 11.1.
Buccoo Reef Marine Protected Area. Tobago, West Indies.

How best to manage Buccoo Reef has been debated since the 1960s when attempts to stop mangrove clearance for tourism development were first made. The impacts of development have been a major issue for Buccoo Reef from then to the present day. Different forms of protection have been enacted to conserve the biological resources of the reef and surrounding systems, with legislation restricting use, access, and extraction of resources. However, there has been a history of conflicts over and resistance to these restrictions by local stakeholders, absentee landowners, and international tourists alike. Poor enforcement of regulations, lack of communication about rules and boundaries, and active infringement and lawbreaking have resulted in resentment and loss of trust among various stakeholders, including the authorities responsible for protected area management and local environmental non-governmental organizations (NGOs) and community groups. Trade-off analysis was initiated in partnership with these organizations in an attempt to support decision making that would be inclusive and deliberate, facilitate flows of information, and enhance compliance and co-management of the various resources within the marine park and around it in order to better integrate the conservation and development objectives of the protected area.

Trade-off analysis as developed for this marine protected area consists of a suite of techniques, as shown in figure 11.2. The method is described in greater detail in Brown, Tompkins, and Adger (2001). This approach uses techniques such as focus groups, Participatory Rural Appraisal, formal and informal surveys, and consensus building to engage with the diverse stakeholders. It uses envisioning techniques to help stakeholders discuss their visions of sustainable futures and their priorities for the protected area. It specifically asked stakeholders about their views on the trade-offs between ecological, social, and economic criteria, and used information and measures that the stakeholders had defined as important to show the impacts of different development options on these criteria. The diagram presents a rather simplified and linear model of what is, as practiced, an iterative process with simultaneous repeats of, and feedbacks between, the different stages and techniques. The analysis may be used as part of an adaptive management strategy (see also Salafsky and Margoluis this volume).

Multicriteria analysis is used as an overarching framework for the organization of information about different management strategies and for the deliberative and inclusionary processes. Multicriteria analysis conventionally proceeds by generating information on a decision or management problem from available data and then evaluating alternative solutions to provide a transparent understanding of the decision. Additionally, multicriteria analysis is often used by a team of experts to

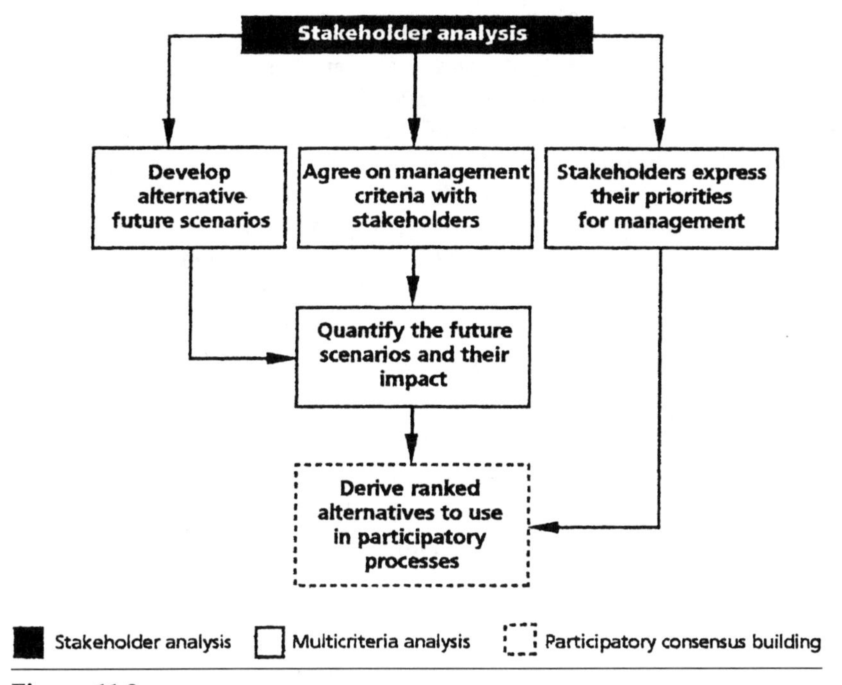

Figure 11.2.
Trade-off analysis.

assess the pros and cons of alternative courses of action. It has advantages over alternative decision-support tools, such as extended cost-benefit analysis, because it enables evaluation based on both qualitative and quantitative indicators of change. All the alternatives can be compared without having to reduce indicators to a common unit of measurement, such as monetary values. It therefore enables quantitative impacts of different options on the environment (levels of pollution, vegetation cover) with changes in income to be compared with qualitative changes in social or cultural or even aesthetic aspects. Experts weight different criteria in order to assess the impacts. For example, they may decide that negative environmental impacts have to be avoided and so weight environmental criteria above economic or social criteria. Multicriteria analysis has been applied to environmental management issues such as waste management strategies or the siting of wind energy farms.

Trade-off analysis uses the multicriteria analysis framework in a different way, supporting a process by which stakeholders can examine information on criteria used to make decisions and explore the outcomes and impacts of decisions made based on different priorities.

They do this through applying different weights to economic, social, and ecological criteria. It is therefore a process-oriented rather than outcome-oriented use—the multicriteria analysis is used as a tool to facilitate the deliberations of stakeholders. Multicriteria analysis offers opportunities to present trade-offs and to rank different priorities and criteria in a systematic manner that does not specify an overall single value framework. Used in this way, it allows the sensitivity of both social and physical data to be tested, and it makes explicit the trade-offs between competing objectives, impacts, and stakeholders.

IDENTIFYING AND ENGAGING STAKEHOLDERS

In the management sciences literature, a stakeholder is a person, organization, or group with an interest in an issue or resource. Stakeholders are both the people with power to control the use and management of resources and those with no power or influence but whose livelihoods are affected by changing use or management. Stakeholder analysis is a system for collecting information about groups or individuals who are affected by decisions, for categorizing that information, and for explaining the conflicts that may exist between important groups and identifying areas where trade-offs may be possible. It can be undertaken simply to identify stakeholders, to manage stakeholders and their interests, or to explore opportunities for getting groups or individuals to work together. In a stakeholder analysis, the stakeholder groups are often described by socioeconomic classifications such as income level, occupational group, and employment status; by degree of formal involvement in the decision-making processes; or by how the group is organized, such as whether it has formal legal standing (a parish council, for example) or is informal (an ad hoc self-help group). Stakeholder identification is often complicated by the fact that stakeholders tend to fall into more than one category. For example, in the Tobago case, a fisher might also be a member of the parish council.

One method for identifying stakeholders is to use a continuum of stakeholders from the macro to the micro level (see Grimble, Aglionby, and Quan 1994, for application to tree resources). Categorizing stakeholders means that key stakeholders can be engaged in further stages of trade-off analysis and other management processes. Stakeholders are commonly categorized according to their level of influence and their importance. *Importance* refers to the degree to which the stakeholder is considered a focus of decisions or management. For example, if an ICDP is targeted to improve the livelihoods of the poor, the poor who directly and indirectly use natural resources within or around the ICDP may be the most important stakeholders. If, on the other hand, an ICDP is primarily aimed at conserving endangered species, then resource managers, resource owners, and conservation agencies may be the most important stakeholders. Importance varies according to the objectives of the project itself. *Influence* refers to the degree of power a stakeholder has to control decisions and management and is dictated by the stakeholders' control of, or access to, power and resources. Influential stakeholders, such as well-established lobbying groups, wealthy landowners, or respected local leaders, often are already engaged in decision-making and management processes.

The relative levels of influence and importance determine whether someone is a primary, secondary, or external stakeholder. *Primary stakeholders* have low influence over the outcome of management decisions, but their welfare is important. Often, the primary stakeholders are those who stand to lose most from decisions—although this is not always the case. *Secondary stakeholders* are often managers or decision makers and those charged with implementing decisions, although their welfare is not a priority. *External stakeholders* are those individuals or groups who can exert significant influence through lobbying, and might include NGOs.

Table 11.1 shows the classification of Buccoo Reef stakeholders according to their importance and influence and identifies the primary, secondary, and external stakeholders in managing the Buccoo Reef Marine Park. This analysis was used to identify key stakeholder groups for inclusion in the trade-off analysis process.

Table 11.1 *Diverse stakeholders in Buccoo Reef Marine Park*

	Stakeholder groups
Primary stakeholders	Fishers
	Diver tour operators
	Reef tour operators
	Tourists—local, national, and international
	Park managers
Secondary stakeholders	Hoteliers
	Informal businesses—hawkers, taxi drivers
	Environmental planners and personnel from other government departments
External stakeholders	Conservation groups—local, national, and international

Alternative approaches also have been used to categorize stakeholders. For example, according to Mikalsen and Jentoft (2001) in their analysis of the management of a Norwegian fishery, the priority given to various stakeholders depends on the extent to which stakeholders possess each of three attributes: *power,* or the stakeholder's ability to influence the decision; *legitimacy,* defined as the degree to which the stakeholder's relationship with the decision maker has been established; and *urgency,* or the degree to which the stakeholder's needs require speedy action. Instead of primary, secondary, and external stakeholders, Mikalsen and Jentoft describe definitive, expectant, and latent stakeholders. *Definitive stakeholders* are those stakeholders possessing all three qualities of power, legitimacy, and urgency, so that they have an unequivocal claim on the attention of management and are key to successful implementation. *Expectant stakeholders* possess combinations of two out of the three attributes. They include those who are powerful, with urgent needs, but who are not legitimate and may be very dangerous to management stability; or those who are legitimate, with urgent claims, but who lack power and have to rely on advocacy or building alliances. *Latent stakeholders* possess only one of the three attributes, and include powerful groups who have no motivation to use their power, legitimate groups without power or any demand for urgent action, and groups with urgent needs who have no power and legitimacy.

More complex analyses can be undertaken. In analyzing the relevance of different stakeholders for sustainable forest management in Trinidad, Gunter (2001) has applied the approach developed by Colfer (1995) that assesses stakeholders according to six dimensions. These are proximity, preexisting rights, dependency, indigenous knowledge, culture-forest integration, and power deficit. These dimensions were scored 1–3 (high to low) to identify the key stakeholders. This approach shows that, depending on the objectives (identifying stakeholders for inclusion in poverty alleviation, participatory processes, impact mapping), different attributes may be more appropriate or significant. However, the scoring of numerous attributes, especially those difficult to judge, may introduce biases and add complexity to the process.

DEFINING SCENARIOS AND CRITERIA

A set of scenarios, which represent the options for future development, and criteria, which are the priorities for decision making, is required to frame the multicriteria analysis within our trade-off analysis. Both the scenarios and the criteria are developed in consultation with the relevant stakeholders, which in the case of Buccoo Reef Marine Park, involved a series of interviews, discussions, and public meetings. The

Table 11.2 *Scenarios for Buccoo Reef Marine Park*

Scenario	*New tourist beds in BRMP area*	*Population in BRMP area*	*% waste treated*
A: Limited tourism development without complementary environmental management	240	6,900	9
B: Limited tourism development with complementary environmental management	240	6,900	49
C: Expansive tourism development without complementary environmental management	1,580	7,400	18
D: Expansive tourism development with complementary environmental management	1,580	7,400	69

scenarios were derived by asking people the question: What will Tobago look like in ten years' time? The scenarios of future change are based on existing development plans and options, such as proposals for hotel development and national parks legislation. The key drivers of change are the number of new tourism developments in southwest Tobago, changes in subregional population growth, and the standard of waste treatment. These then are the factors that will determine what Tobago will look like in ten years' time: whether tourism is allowed to expand unchecked or only very limited and select development is allowed; and whether there is investment in a strict enforcement of regulations on waste and water treatment and effluent. These are what define and quantify the different scenarios, as shown in table 11.2.

The four scenarios represent feasible and believable futures for Tobago. The advantage of stakeholder involvement at this stage of the process is that the scenarios are not perceived as being unrealistic but are credible descriptions of the options facing the stakeholders in southwest Tobago. The important attributes of the scenarios for trade-off analysis are that they are understandable to all participants and are distinct from one another.

A similar participatory process, using key informants and public discussion groups, was used to develop criteria for the multicriteria analysis. The resulting economic, social, and ecological criteria are related to the impacts on community, social development, and cultural integrity as well as on ecosystem health. The subcriteria represent indicators or

Table 11.3 *Criteria for assessing management options for Buccoo Reef Marine Park*

Criteria	*Subcriteria*	*Measure/basis of calculation*
Economic criteria	1. Macroeconomic benefits of tourism to Trinidad and Tobago	Tourism revenue × economic multiplier × (1-marginal propensity to import)
	2. Tourist benefits	Consumer surplus of recreational users of BRMP
Social criteria	3. Local employment in tourism	Additional full-time "quality" jobs × proportion of jobs to Tobagonians
	4. Informal sector benefits	Changes in informal sector benefits
	5. Costs of local access to BRMP	Change in costs of accessing BRMP for recreation and subsistence extractive purposes
Ecological criteria	6. Water quality	Nutrient concentration—nitrate loading and concentration
	7. Productivity of sea grasses	Unit productivity
	8. Coral reef health	% live coral cover
	9. Mangrove habitat	Change in area of mangrove (ha)

measures of these impacts. They do not necessarily describe the whole system or attempt to model its complexity; rather they are usable and widely understood indicators of aspects of the overall picture as perceived by the major stakeholders. The subcriteria must be measurable across the different scenarios. The basis of their calculation is shown in table 11.3. Operationalizing the multicriteria analysis involves estimating the effects of the scenarios on each subcriterion, or how each indicator changes under each scenario. The data were collected using diverse techniques, discussed in Brown, Tompkins, and Adger (2001).

The key feature of the criteria is that they each change across the different scenarios and that they change in different directions. In other words, for each scenario, some criteria are enhanced and others worsen. This captures the essence of the trade-off problem. In a very common situation, a new development may simultaneously result in increased economic revenue and a decline in ecosystem health. This is shown in table 11.4. The two scenarios with expansive tourism development, C and D, show the highest economic revenues from tourism. However, visitors' enjoyment of the marine park, which depends on the extent of environmental management, differs between C and D. But each of the indicators of ecological health is worse for scenarios C and D than for scenarios A and B, which have strictly controlled tourism. This table, which is called an "effects table" (because it shows the effects of different options or courses of action on the criteria) in multicriteria analysis, illustrates the trade-offs between the economic gains, the social and environmental costs of tourism development, the mixed economic and social impacts, and ecological benefits of enhanced environmental management. This provides the framework and information with

Table 11.4 *Trade-offs between economic, social, and ecological aspects of four development scenarios for Buccoo Reef*

	Scenarios			
Criteria	*A*	*B*	*C*	*D*
Economic				
1. Economic revenues to Tobago (U.S.$million)	9	11	17	19
2. Visitor enjoyment of BRMP (U.S.$million)	1.2	2.5	0.9	1.7
Social				
3. Local employment (# jobs)	2,500	2,600	6,400	6,500
4. Informal sector benefits (score)	5	4	3	2
5. Local access (score)	6	5	6	7
Ecological				
6. Water quality (μg N/l)	1.5	1.4	2.2	1.9
7. Sea grass health (g dry weight/m^2)	18	19	12	15
8. Coral reef viability (% live stony coral)	19	20	17	18
9. Mangrove health (ha)	65	73	41	65

Scenarios (as explained in table 11.2):
A: Limited tourism development without complementary environmental management
B: Limited tourism development with complementary environmental management
C: Expansive tourism development without complementary environmental management
D: Expansive tourism development with complementary environmental management

which stakeholders can evaluate these trade-offs and make informed choices and decisions about the management of Buccoo Reef Marine Park and the area of southwest Tobago around it.

Weighting Criteria and Evaluating Scenarios

The development and estimation of scenarios and criteria is an entry point into stakeholder-led negotiations on priorities for management of the protected area. The set of systematic information for the multicriteria analysis is used to engage with stakeholder groups to explore their priorities in terms of decision-making criteria and development scenarios and outcomes. In the case of Buccoo Reef Marine Park, an iterative three-stage process was utilized. First, each of the stakeholder groups met separately to discuss the issues. Second, each of the groups was presented with the outcome of their own deliberations and the outcomes of the other stakeholder groups, thereby challenging their preconceptions of how others perceived the management issues. Third, the stakeholders were brought together in a series of consensus-building workshops. Researcher meetings with individual stakeholders built trust around the concepts and procedures of negotiation and validated local knowledge of the reef system and the ecological linkages in coastal areas, enabling the final discussions.

The stakeholder groups included fishers, local communities, local businesses and entrepreneurs, reef tour and water sports operators, recreational users, and technical personnel from various departments of the Tobago House of Assembly (THA). Some of the stakeholder groups were not well represented in focus groups. The priorities of the THA and park regulators were derived through a series of informant interviews, as well as participatory exercises that included ranking exercises. The interests of the tourists and recreational users were important, as this stakeholder group is crucial to the financial sustainability of the island economy. However, because they are highly transient, they could not very easily be brought into focus groups. Therefore, a formal survey was used to reveal their priorities and preferences. Thus appropriate techniques need to be employed, depending on the stakeholders.

In the first and second stages of the process, stakeholders were asked to weight their priorities among criteria for making decisions about future development options. The weights were derived through focused and structured discussions of the implications of the scenarios and the options for management of the protected area and development in southwest Tobago. Participants were asked to allocate weights between the three priorities for management—economic growth, social issues, and ecosystem health—according to the importance these issues should be given in decision making. The technical details of these procedures can be found in Brown, Tompkins, and Adger (2001). This process revealed that each group of stakeholders prioritized ecosystem health, and each group ranked ecosystem health highest, above both economic and social criteria. To an extent this was unexpected, as the conflicts surrounding management of the park and the comments made by stakeholders about others' actions had led us to expect much more contrasting views. We discussed the rankings and the reasons for them at length with stakeholders. Each of the stakeholder groups expressed the view that, in the long term, their and their children's welfare and economic prosperity depended on the maintenance of the Buccoo Reef ecosystem. Thus reef tour operators acknowledged that tourists would not continue to visit the marine park if the reef became further degraded; fishers understood that the lagoon and sea grass beds provided a nursery for many fish species. The views expressed were not all motivated by self-interest, however; many local people value the environment for its intrinsic and aesthetic qualities and insisted that an environmental conservation ethic was a shared value of Tobagonians (see also Glick and Freese this volume). They wanted their children and grandchildren to "enjoy the reef." The discussions revealed considerable agreement on the long-term priorities for managing the park and pointed the way to a basis for consensus. Eventually a series of consensus-building meetings were convened to reach agreements on management strategies and development options for the management of the marine park.

Evaluation of the scenarios through their impacts on the criteria as shown in the effects table (table 11.4) is the first step in the multicriteria analysis. Applying weights to the criteria generates an ordered ranking of the development scenarios. The highest scoring scenario can be considered the most desirable scenario. Figure 11.3 shows the rank ordering of scenarios for a range of stakeholder weights. These are compared to a base case of equal weighting of economic, social, and ecological criteria. Through their weighting of ecological criteria, all the stakeholders are in effect demonstrating concern for proactive management and limitation of development in southwest Tobago. Scenario B is ranked highest across the range of weightings (other than the base case of equal weighting) (figure 11.3). However, stakeholder weightings differ in the subsequent ordering of scenarios. A higher emphasis is placed by both the regulators and recreational users on ecosystem health and a lower emphasis on economic criteria, which would mean limitations on tourism development (scenario A following scenario B). For local stakeholders, the implications of their prioritizing ecological criteria are that they favor enhanced environmental management (scenario C following

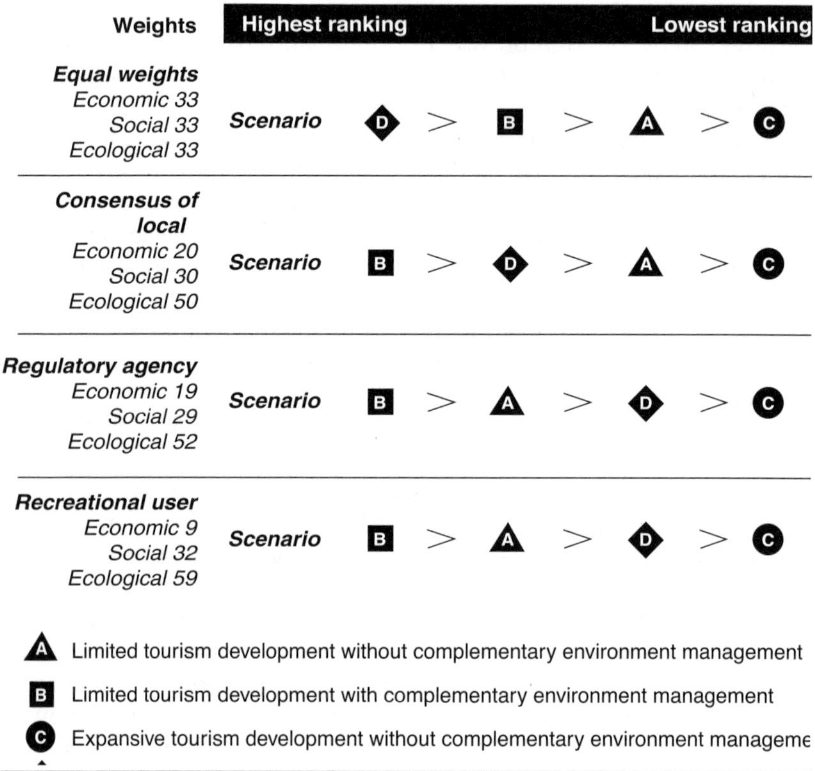

Weights	Highest ranking						Lowest ranking
Equal weights *Economic 33* *Social 33* *Ecological 33*	***Scenario*** D	>	B	>	A	>	C
Consensus of local *Economic 20* *Social 30* *Ecological 50*	***Scenario*** B	>	D	>	A	>	C
Regulatory agency *Economic 19* *Social 29* *Ecological 52*	***Scenario*** B	>	A	>	D	>	C
Recreational user *Economic 9* *Social 32* *Ecological 59*	***Scenario*** B	>	A	>	D	>	C

A Limited tourism development without complementary environment management

B Limited tourism development with complementary environment management

C Expansive tourism development without complementary environment manageme

Figure 11.3.
Ranking of the development scenarios for Buccoo Reef and southwest Tobago.

scenario A in rankings). This difference in priorities is substantiated by discussions in stakeholder meetings and at consensus-building workshops. These rankings therefore reveal the types and extent of trade-offs different stakeholders are prepared to accept or make.

TRADE-OFF ANALYSIS AND CONSENSUS BUILDING

Trade-off analysis means that stakeholders can not only be explicit about their priorities for management and decision making, but they can also see the potential outcomes and impacts. In this way they can be informed about the trade-offs inherent in decisions on resource management, conservation, and development. In the case of Buccoo Reef, the use of these techniques made a key contribution toward deliberative inclusionary decision making and consensus-based approaches to managing the protected area and toward the evaluation of the conservation and development trade-offs inherent in the management options and scenarios.

In Tobago the application of trade-off analysis opened up spaces for more consensual discussions of management of the marine park, where stakeholders could express their concerns about the impacts of development on such things as the reef, the lagoon, the types of jobs and livelihoods available for local people, and access to beach and mangroves. The basis of agreement, expressed through their ranking of the criteria, enabled a consensus-building workshop to concentrate in the first instance on long-term development plans. All stakeholders shared a similar vision of what kind of development they wanted for Tobago. They wanted their children and grandchildren to have the same options in the future that current stakeholders enjoyed; the visions they articulated were remarkably close to the Brundtland definition of sustainable development. Maintaining the health and integrity of the ecosystem was critical to this. Working from the future, a series of medium-term planning needs were identified. These were then discussed and prioritized, using various ranking and voting exercises. A very large majority of stakeholders identified water treatment and control of effluent as the most important medium-term planning issue.

In the case of Buccoo Reef, as in many other situations, most of the conflicts between stakeholders revolve around quite short-term actions. People are aggrieved because they perceive others to be flaunting rules and regulations; they are threatened by uncertainties about enforcement and boundaries; and, importantly, they feel disempowered because they are not included in decisions—their views are not valued and they are not informed about plans. We discussed these grievances at length but in the context of what people—all different stakeholders, including government agencies—could do about it. A series of action plans were agreed on, which included clear marking of boundaries and sign-boarding of regulations; establishment of mooring buoys; and information evenings and events at local schools. Commitments were made by local communities, resource users, government officers, and politicians. The most significant of these actions was the establishment of a stakeholder group, which meets regularly and has entered into a co-management arrangement with the government department responsible for managing the marine park. A process of social learning has begun, and a new institutional arrangement has been established. Trade-off analysis was an important first step toward co-management of the protected area. In the next section the discussion turns to further applications and the broader institutional and management opportunities for trade-off analysis in ICDPs.

Trade-Offs in Conservation and Development

The trade-off analysis approach described above offers a set of techniques and a methodology that can be applied in a range of contexts and in different ways. Similar approaches have been applied in other situations that use some but not all aspects of trade-off analysis. For example, stakeholder analysis, multicriteria approaches, and use of scenarios and envisioning are becoming more widely applied in resource management and in conservation and development interventions (see also Franks and Blomley this volume).

Multicriteria analysis in particular is increasingly used in natural resource management, notably for forest management (Mendoza and Macoun 1999), in public assessment of environmental risk (Stirling and Mayer 2001), and in regional-scale biodiversity conservation assessments (Faith and Walker 1996). The advantages of using multicriteria approaches in assessing multiple-objective resource management issues, where outcomes are uncertain, where there are multiple and potentially conflicting stakeholders, and where measurement can be difficult, are fivefold. First, of course, the approach has the capability to accommodate multiple criteria into the analysis, rather than being limited to the single objectives of most decision-support or management analytical tools. This is important for ICDPs that have both development and conservation objectives. Second, multicriteria can work with mixed data, and the results need not be data intensive. Critically, it allows the incorporation of both quantitative and qualitative information. There is no need to reduce all data to a common unit of measurement, as, for example, in extended cost-benefit analysis. Third, it allows the direct involvement of multiple experts, interest groups, and stakeholders. Fourth, the analysis is transparent to participants. Fifth, the multicriteria analysis includes mechanisms for feedback concerning the consistency of judgments made.

Multicriteria analysis can be used at different scales of analysis in integrating conservation and development. For example, Faith and Walker (1996) develop "trade-off curves" to inform investment in regional-scale biodiversity protection. Such an approach extends a traditional cost-benefit approach to prioritize areas for conservation, taking into account the different components of biodiversity and possible complementarities in protected area strategies, as well as incorporating sensitivity analysis and safe minimum standards rules. However, it does not get inputs from different stakeholders—although it could conceivably be used in a more interactive way. In a later development of their approach, Faith and Walker (2001) apply their trade-off framework to national conservation planning for Papua New Guinea, and also suggest using trade-offs to investigate international priorities for conservation. They use the examples of the Millennium Ecosystem Assessment[1] and the Critical Ecosystems or "Hotspots" program[2] to illustrate the global analysis. The approach they develop recognizes that a range of environmental dimensions of conservation options—including biodiversity and environmental services or functions—need to be taken into account, although it places emphasis on the opportunity costs, or the development opportunities foregone by investing in conservation. The authors and their collaborators have developed specific software for the analysis, called TARGET. They argue that the trade-off framework can provide a linkage between local, regional, and global levels for conservation planning. The analysis of these different scales can be integrated so that, for example, the trade-offs in individual protected areas feed into the assessments of trade-offs at regional levels.

The analysis of the trade-offs between different aspects of biodiversity has been developed in studies by Redford and Richter (1999) and Putz et al. (2000). Redford and Richter develop a quasi-effects table that outlines in a qualitative fashion the impacts of different management options on different components and attributes of biodiversity. In the example of the Roanoke River in North Carolina, six different water management options (equivalent to scenarios in our trade-off analysis) are developed, ranging from channeling the river to removing all dams and restoring the "natural" flood regime. Continuing current conditions ("business as usual") is another option. The impacts of these different options are then assessed on three community or ecosystem-level biodiversity attributes, three population or species-level attributes, and three genetic attributes. These correspond to criteria and subcriteria in our trade-off analysis multicriteria framework. In the example of the Pantanal wetland in Brazil, five management options are defined. The feasibility of the different management options is incorporated into the analysis by assessing the "option space" of each scenario (e.g., in the case of Roanoke River, removal of all dams is not currently feasible). The matrices produced by the analysis demonstrate the sensitivity of different biodiversity components under different management options. Putz et al. (2000) similarly consider the impacts of different logging regimes and silvicultural treatments on different attributes of biodiversity. For example, management for nontimber forest products and "reduced impact logging" plus reserve designation are two options that conserve landscape, ecosystem, and community biodiversity components and have limited impacts on genetic attributes. Conventional logging and enrichment planting, on the other hand, have greater impacts on genetic, species, and community components. But each strategy will have different cost implications, and these will be different for different stakeholders. Putz et al. (2000:28–29) demonstrate the patterns and potential trade-offs of the different management options when financial profitability and commercial timber production are measured against biodiversity impacts. This captures some of the dimensions of the trade-offs in selecting logging and silviculture techniques.

These analyses can be used to inform regional investment and priority setting, or to help form management strategies for a given protected area or set of natural resources. None of these examples, however, specifically sets out to include or explicitly recognize the different weights and values given to the different impacts by different stakeholders. They each imply that, so long as accurate information is provided, rational choices can be made to optimize management of the trade-offs.

The Center for International Forestry Research (CIFOR) has used multicriteria analysis in their research program to develop criteria and indicators for sustainable forest management. Their approach aims to reconcile the maintenance and enhancement of ecosystem integrity with human well-being—another articulation of integrating conservation and development objectives. Their manual on multicriteria analysis (Mendoza and Macoun 1999) suggests that the approach can be used to support decision making in sustainable forest management at a range of scales and contexts: for example, to develop certification schemes, in defining forest policy, in deciding funding priorities, in management of particular forest areas, or in project design and implementation. CIFOR applies multicriteria analysis within its criteria and indicator assessment process, where the methods are used to evaluate the relative importance of each criterion and indicator; to ascertain different participants' perspectives; and to help to aggregate individual values and find consensus or group-based evaluations. The manual they have produced outlines both top-down and bottom-up applications and argues that the two approaches are not mutually exclusive, and can be used in combination.

Multicriteria analysis is used in a more constructivist or stakeholder-driven way by Stirling and Mayer (2001), who aim to recognize and incorporate the plurality of values and knowledges, and attempt to integrate different cultural and social perceptions of risk to inform policy making in regard to the use of genetically modified crops. This approach is termed "multicriteria mapping" and is used as a heuristic rather than a prescriptive tool—in other words, to explore different options and to develop learning-based approaches. Participants appraise a series of options—ranging from organic agriculture to use of genetically modified crops with voluntary controls—and a set of criteria, including biodiversity, agriculture, health, economy, and social and ethical issues. Participants assign weights to these criteria, examine the implications of different weightings, and change their weights to explore the sensitivity of different outcomes. They use the multicriteria in an interactive way and as a result "map" the criteria and options, deriving "optimistic" and "pessimistic" outcomes for each stakeholder group based on individual participant's weightings. This differs from the usual applications of multicriteria analysis that aggregate weightings. Thus the approach aims to account for individual subjective values, fears, and uncertainties, and, rather than arriving at definitive right or wrong prescriptions, is a more open-ended process. This application exposes how different criteria and factors are involved in evaluative procedures by different participants and helps to secure trust and confidence in decision making.

The CARE ICDPs described by Franks and Blomley (this volume) use some techniques similar to trade-off analysis in applying vision-driven approaches to project design. Using these techniques in more open-ended and discursive forums supports deliberative and inclusionary processes in decision making and management. The term *deliberative inclusionary processes* (or DIPs) is used to cover a range of participatory policy-making processes, management practices, and community empowering actions. These processes are variously applied to more effectively implement policy, to redistribute power and benefits, and as part of efforts to bring about deliberative democratic approaches to environmental decision making. Deliberative inclusionary processes have been applied in many different economic, political, and cultural contexts worldwide, and are reviewed by Holmes and Scoones (2000). Their objectives will clearly change depending on context, site, and issue. Two key features distinguish these approaches: *deliberation* and *inclusion*. *Deliberation* means careful consideration or discussion. Decision making and planning thus require social interaction and debate. Deliberation implies that different positions of stakeholders are recognized and respected. Participants are expected to reflect on and evaluate and reevaluate their and others' positions. The process of deliberation aims to bring about some kind of transformation of values or preferences and foster negotiation between participants in order to reach a decision. *Inclusion* is the action of including different participants in these processes.

A number of procedural techniques, mechanisms, and methods have developed, including citizens juries, focus groups, issue forums, Participatory Rural Appraisal, visioning exercises, and various types of workshops and working groups. Associated with these methods are principles such as working with small groups of people, focusing on the future and on common ground, urging full attendance and participation, incorporating the widest possible range of interests, and seeking public commitments to action (Holmes and Scoones 2000). These different deliberative inclusionary processes reflect the range of forms of participation in conservation projects described by Pimbert and Pretty (1997). In developing countries, deliberative inclusionary processes have been associated with the rhetoric of participation and empowerment, and in the developed countries, with democracy and representation (see also Singh and Sharma and Child and Dalal-Clayton, both this volume). The recognition of diverse values and local knowledge is also of fundamental importance to how these processes operate. Deliberative inclusionary processes are used to link government agencies and civil society groups in co-management strategies for natural resource management. Such co-management strategies are often a key feature of ICDPs (e.g., Tongson and Dino this volume). Trade-off analysis as developed in the Tobago case study can be viewed as a technique to support deliberative and inclusionary approaches to resource management or to participatory conservation and development.

Conclusions

This paper has presented an approach to assessing the trade-offs between conservation and development and has illustrated how this has been applied to developing co-management strategies for a protected area in Tobago in the Caribbean. Similar approaches can be used to assess conservation priorities at local, regional, and global scales and to make decisions about management strategies at specific sites. This approach to trade-off analysis is used as a bottom-up, participatory means to support a deliberative inclusionary process in implementing ICDPs. The trade-off analysis and the multicriteria framework are used as tools to encourage deliberation by, and inclusion of, a wide range of stakeholders in deciding management priorities, choosing between different management options, and selecting ways of implementing management objectives. The trade-off analysis process undertaken in Tobago culminated in a series of consensus-building workshops (see Brown, Tompkins, and Adger 2001) that helped to initiate a new, discursive, and participatory approach to co-management of the protected area. Trade-off analysis is a set of techniques that support social learning approaches or adaptive co-management (cf. Ruitenbeek and Cartier 2001; Fortmann, Roe, and Van Eeten 2001; and Salafsky and Margoluis this volume, for perspectives on adaptive management). However, trade-off analysis is not a blueprint but rather a suite of techniques that can be adapted and adopted as appropriate.

Early applications of trade-off analysis offer a number of lessons for understanding trade-offs in ICDPs. First, it is critically important to conduct a rigorous and thorough analysis of stakeholders at the outset. Experience has shown that homogeneous stakeholder groups with clearly defined interests do not exist in ICDPs and that the social, economic, and political complexity and dynamics of stakeholders and their interests and relationships must be appreciated if participatory or inclusionary processes are to be successful. Stakeholders are not just "local people" but include policy makers, governments, and NGOs.

Second, developing trust between stakeholders is a prerequisite to the process, but is by no means a straightforward task. Developing trust by all major interests in the decision-making processes, in the institutions and individuals making and implementing decisions, and in the institutions that define ICDPs takes time and effort. To be successful, the process of defining management objectives has to be transparent—trade-off analysis helps to achieve this.

Third, designing appropriate institutions for integrating conservation and development can be supported by deliberative inclusionary approaches and social learning. *Social learning* means sharing and legitimizing knowledge, acknowledging that different forms of knowledge and different sets of values and beliefs exist and are valid and are to be included in evaluating trade-offs. One must show sensitivity to the dynamics of ecosystems and the impacts of human interventions, and to the ecological and social resilience that can be maximized by integrated conservation and development, and accept that uncertainty and complexity are part of an adaptive strategy and need to be dealt with explicitly.

Effective integration of conservation and development demands new ways of working and new means of evaluating and managing interventions. Old, "scientific," expert-centered and top-down approaches are no longer acceptable. This paper's arguments support the involvement of stakeholders not only in implementing ICDPs but in defining policy priorities. The plurality of values and perspectives and the priorities of multiple stakeholders—ranging from local forest dwellers to national forest planners to international conservation organizations—must be recognized. Trade-off analysis, as outlined in this paper, enables evaluation of the trade-offs between different policy options, different priorities for conservation and development, and different stakeholders. Used in conjunction with other deliberative and inclusionary processes, it can be part of a process of empowering and sensitizing stakeholders to facilitate social learning. This is a vital component of adaptive co-management strategies and necessary for the long-term success and environmental, social, and economic viability of integrated conservation and development initiatives worldwide.

Acknowledgments

Research in Tobago was funded by the U.K. Department for International Development (DFID) Natural Resources Systems Program; and research in Brazil by the DFID Social and Economic Research Program for the benefit of developing countries. Views expressed are not those of DFID. I am extremely grateful to colleagues, including Neil Adger, Emma Tompkins, Emily Boyd, and Esteve Corbera Elizalde, who collaborated on parts of the research discussed here, and who have helped in the development of ideas in the paper. I also acknowledge the kind support of collaborating institutions in Tobago, including the Tobago House of Assembly and University of West Indies in Trinidad. All errors are the sole responsibility of the author.

Endnotes

1. See www.millenniumassessment.org for further details.
2. Details at www.cepf.net (Critical Ecosystem Partnership Fund).

References

Brown, K., E. Tompkins, and W. N. Adger. 2001. *Trade-off Analysis for Participatory Coastal Zone Decision Making*. Overseas Development Group (ODG), in collaboration with the Center for Social and Economic Research on the Global Environment (CSERGE), both at the University of East Anglia.

Brown, K., W. N. Adger, E. Tompkins, P. Bacon, D. Shim, and K. Young. 2001. Trade-off analysis for marine protected area management. *Ecological Economics* 37 (3):417–434.

Colfer, C. 1995. *Who Counts Most in Sustainable Forest Management?* CIFOR Working Papers no 7. Bogor, Indonesia: Center for International Forestry Research (CIFOR).

Faith, D. P. and P. A. Walker. 1996. Integrating conservation and development: Effective trade-offs between biodiversity and cost in selection of protected areas. *Biodiversity and Conservation* 5 (4):417–429.

Faith, D. P. and P. A. Walker. 2001. The role of trade-offs in biodiversity conservation planning: Linking local management, regional planning, and global conservation efforts. Sydney: Biodiversity and Systematics Department of the Australian Museum. www.amonline.net.au/systematics/faith4.htm (26 October 2001).

Fortmann, L., E. Roe, and M. van Eeten. 2001. At the threshold between governance and management: Community-based natural resource management in southern Africa. *Public Administration and Development* 21:171–185.

Ghimire, K. B. and M. P. Pimbert, eds. 1997. *Social Change and Conservation: Environmental Politics and Impacts of National Parks and Protected Areas.* London: Earthscan.

Grimble, R. J., J. Aglionby, and J. Quan. 1994. *Tree Resources and Environmental Policy: A Stakeholder Approach.* Chatham, U.K.: Natural Resources Institute (NRI).

Gunter, M. 2001. Intergenerational equity and sharing of benefits in a developing island state. In C. J. P. Colfer and Y. Byron, eds., *People Managing Forests: The Links Between Human Wellbeing and Sustainability,* 171–189. Washington, D.C.: RFF Press; Bogor, Indonesia: Center for International Forestry Research (CIFOR).

Holmes, T. and I. Scoones. 2000. *Participatory Environmental Policy Processes: Experiences from North and South.* IDS Working Paper 113. Brighton, U.K.: Institute of Development Studies, University of Sussex.

Mendoza, A. and P. Macoun. 1999. *Guidelines for Applying Multi-criteria Analysis to the Assessment of Criteria and Indicators.* The Criteria and Indicators Toolbox Series no. 9. Bogor, Indonesia: Center for International Forestry Research (CIFOR).

Mikalsen, N.H. and S. Jentoft. 2001. From user-groups to stakeholders? The public interest in fisheries management. *Marine Policy* 25:281–292.

Pimbert, M. P. and J. Pretty. 1997. Parks, people, and professionals: Putting "participation" into protected area management. In K. Ghimire and M. P. Pimbert, eds., *Social Change and Conservation,* 297–332. London: Earthscan.

Putz, F. E., K. H. Redford, J. G. Robinson, R. Fimbel, and G. M. Blate. 2000. *Biodiversity Conservation in the Context of Tropical Forest Management.* Environment Department Paper no. 75. Washington, D.C.: The World Bank.

Redford, K. H. and B. D. Richter. 1999. Conservation of biodiversity in a world of use. *Conservation Biology* 13 (6):1246–1256.

Ruitenbeek, J. and C. Cartier. 2001. *The Invisible Wand: Adaptive Co-management as an Emergent Strategy in Complex Bio-Economic Systems.* CIFOR Occasional Paper no. 34. Bogor, Indonesia: CIFOR.

Stirling, A. and S. Mayer. 2001. A novel approach to the appraisal of environmental risk: A multi-criteria mapping of genetically modified crop. *Environment and Planning* 19 (4):475–632.

Wells, M. 1992. Biodiversity conservation, affluence, and poverty: Mismatched costs and benefits and efforts to remedy them. *Ambio* 12 (3):237–243.

12

Transforming Approaches to CBNRM: Learning from the Luangwa Experience in Zambia

Brian Child and Barry Dalal-Clayton

Introduction

This paper provides a brief description of the Luangwa Integrated Resource Development Project (LIRDP). It uses background and data from the project to emphasize several principles that make for effective community-based natural resource management (CBNRM) (cf. Dalal-Clayton and Child 2003). It also discusses some of the linkages between conservation, development, donor financing, and political economics.

The LIRDP was one of the first programs to recognize the linkages between poverty and wildlife conservation. Initiated in the early 1980s, it tackled the serious elephant and rhino poaching in the southern Luangwa Valley, Zambia, directly through improved law enforcement, but it was primarily designed to make large investments in rural development. The project was built on the concept that addressing poverty and the long neglect of the Kunda people was necessary to provide the socioeconomic environment for sustainable wildlife conservation (Dalal-Clayton and Lewis 1984). If people remained desperately poor, they would continue to poach. It was also assumed that a sustainable project and local economy could be built by promoting the growth of the wildlife sector. This NORAD (Norwegian Agency for Development Cooperation)-funded integrated conservation and development project (ICDP) effectively became a "mini-government" for Mambwe District, assuming the responsibilities of ineffective line ministries to provide a range of services to the district. It was simultaneously the park management authority.

The project context has always been extremely difficult. The economy was chronically mismanaged, and the Zambian National Parks and Wildlife Service disliked the project. Project reviews consistently emphasize the interinstitutional tensions and personal rivalries that dominated the launch and implementation of LIRDP and conspired against effective coordination, implementation, and integrated planning in the project area.

The project area (figure 12.1) covers some 15,000 square kilometers in eastern Zambia, incorporating the South Luangwa National Park (SLNP) and the adjacent Lupande Game Management Area (GMA), with an estimated human population of 45,000. Administratively, the GMA is also the Mambwe District, but the district council is largely ineffective. It has no money and, importantly, no subdistrict structures. Elected councilors, the number of which recently increased from four to twelve, represent all 45,000 people. Far more important to local people are the hereditary chiefs, who exert much power because they control access to land. The six chiefdoms (Jumbe, Kakumbi, Malama, Mnkhanya, Msoro, and Nsefu) of the matrilineal Kunda ethnic group are shown in figure 12.2.

The SLNP is Zambia's premier wildlife tourism attraction, known internationally for abundant wildlife such as elephants, leopard, lions, hippos, buffalo, giraffes, and antelope. Soils are generally infertile, and the majority of people live on fertile alluvial deposits along the Luangwa's major tributaries. Agricultural productivity is low (compared to the adjacent plateau lands) and highly variable. Tsetse flies make it impossible to keep livestock for ploughing, so farms are small. People rely mainly on subsistence and semicommercial agriculture, with maize and sorghum grown on about 70 percent of the cultivated land and cotton, sunflower, and rice on the rest. Poverty and malnutrition are widespread, and people get some 30 percent of their nutrition from wild foods (Sana Chipeta, personal communication 2001; see also Shepherd this volume). There is a tradition of hunting that continues to the present. Access is difficult, particularly in the rains. Government services are notable by their absence, with few development inputs, extension services, credit sources, and opportunities to market agricultural produce. Schools and clinics are poorly serviced and equipped. Beginning in 1975, commercial poaching escalated rapidly in the Luangwa Valley. The few scouts employed by the under-resourced and ineffective Zambian National Parks and Wildlife Service (NPWS) were little deterrent. The black rhino population in the Luangwa Valley was reduced from about

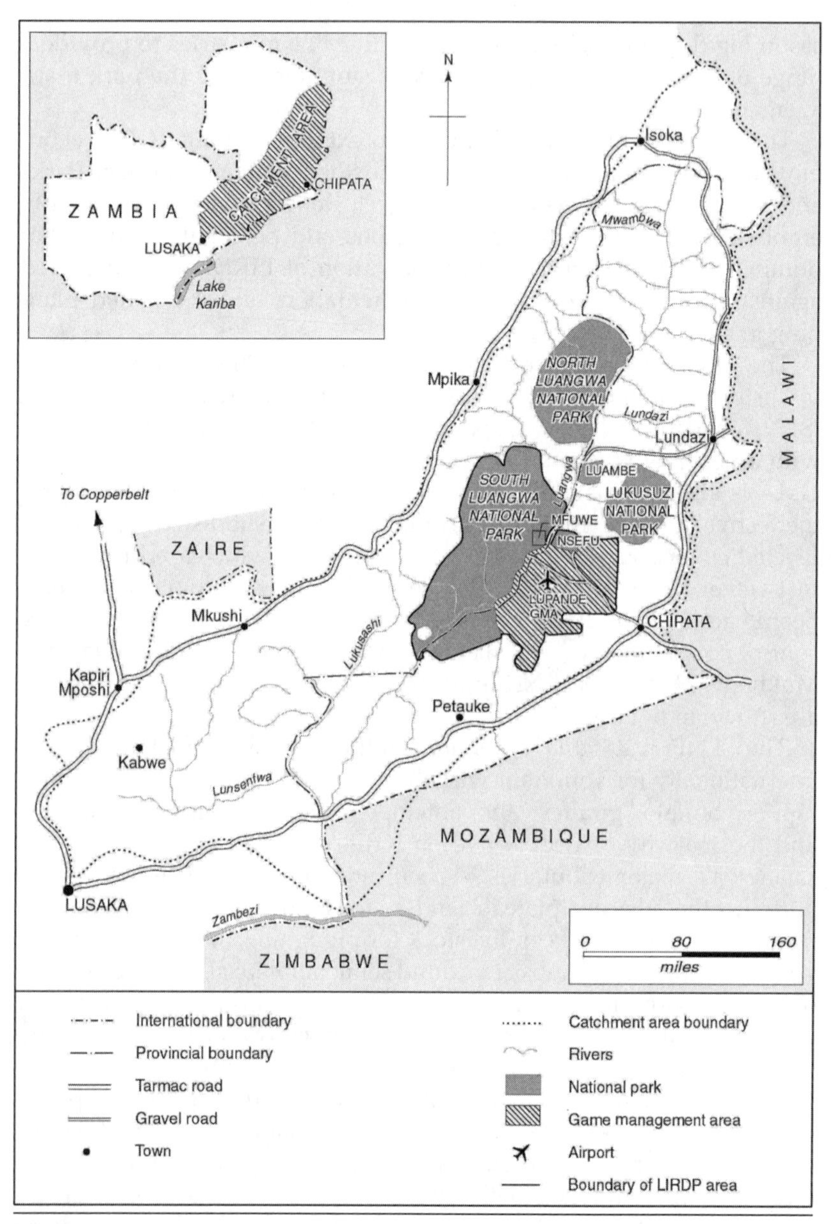

Figure 12.1.
The Luangwa Valley, Zambia.

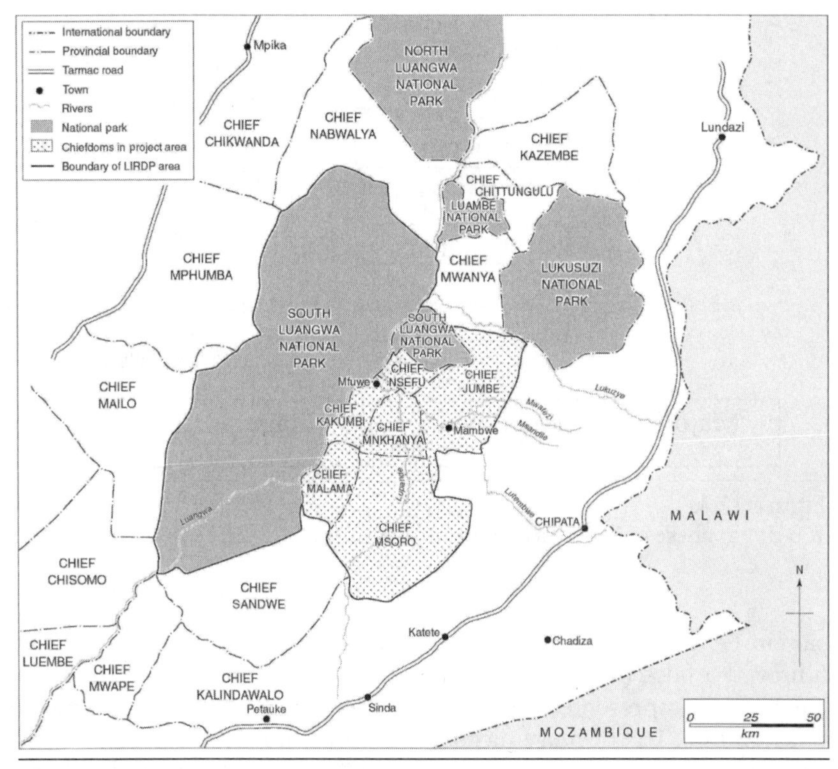

Figure 12.2.
Chiefdoms in the Luangwa Valley, Zambia.

8,000 in the early 1970s to fewer than 100 by the mid-1980s, and disappeared soon thereafter. Over the same period, elephant numbers were cut from 90,000 to 15,000. In the project area (SLNP and Lupande GMA), the population fell to as low as 2,500 in 1989.

There was little incentive for local people to resist commercial poachers or inform on their activities. Revenues from wildlife, such as hunting license fees, park entrance fees, and safari earnings, went to the central government or businessmen living outside the area. Communities living in the Luangwa Valley gained little direct legal benefit from local wildlife resources, and wildlife was seen mainly as meat on the hoof, a serious agricultural pest, and the plaything of rich white westerners.

The Early CBNRM Program

The LIRDP's community-based natural resource management (CBNRM) program had two distinct phases, separated by a major policy deci-

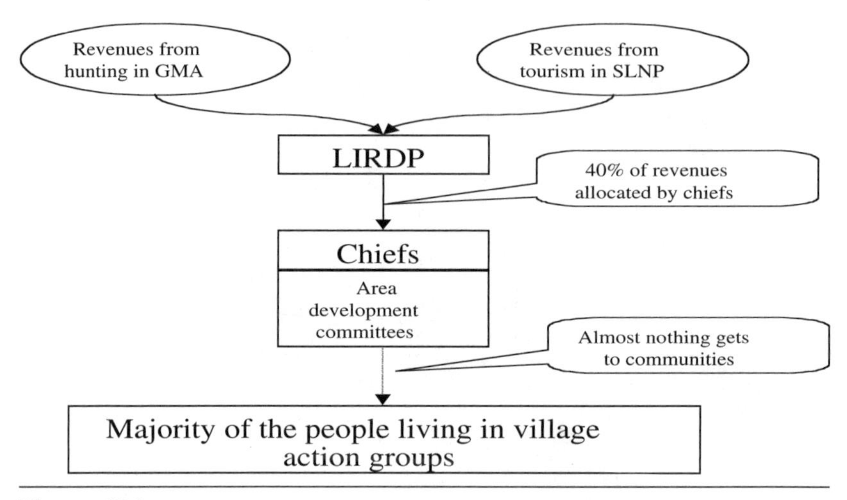

Figure 12.3.
Top-down phase of LIRDP's community program.

sion in 1996. The early program focused primarily at the area level and followed a false (but logistically simpler) assumption that traditional chiefs fairly represented their people (see also McShane and Newby this volume). The primary mechanism for community interaction was a regular formal meeting with the six chiefs. At this meeting, the chiefs instructed the project managers on what to do with wildlife revenues. As a result, between 1987 and 1992, LIRDP policy returned 40 percent of all wildlife income from tourism and hunting in the project area to communities (figure 12.3). This took the form of community projects chosen by chiefs.

Participatory mapping exercises in 1996 showed the community had little idea of any projects built over the previous seven years. Moreover, they mistrusted both the project implementers and chiefs, inferring irregularities in finances and project construction (such as theft of cement, wastage, etc.). For example, the chiefs persuaded the project to establish a bus service to service the community. When this proved unprofitable, the bus was shifted to routes outside the project area. Revenues were not accounted for, and when the bus was sold the income was given to the chiefs. Again, no records of this money exist, though one chief's new house is commonly linked to the bus project. The results from culling programs were similar. While the "community" (read chief) was given carcasses, much meat was sold outside the area, and financial accounts were never forthcoming. Several years of highly frustrating, conflict- and stamina-intense work were necessary to undo the initial empowerment of nonaccountable, nondemocratic leadership.

Introduction of Devolution: The 1996 Policy Shift

In 1996, the LIRDP's Review and Policy Committees (comprising senior civil servants) approved a new CBNRM policy for the project. The turnaround in the community program and all later progress in the CBNRM program can be traced back to this four-page document.

The new policy defined village action groups (VAGs) as the primary organizational building blocks (figure 12.4). Each VAG developed a written constitution. VAGs were constituted at a scale that ensured the face-to-face participation necessary for grassroots democracy in largely illiterate communities (roughly 200 households). The revenue distribution process became the driving mechanism. This returned 80 percent of wildlife money to VAGs and greatly empowered them. Procedures were put in place so that all villagers participated in making decisions, so that all activities were reported to them no less than quarterly, and to ensure financial accountability. These procedures reversed the previous status quo by incorporating citizen-centric principles of democratic, accountable, transparent, and equitable governance, and by making leaders answerable to their people (citizens) and not people to their leaders (subjects). Management was improved by incorporating clear procedures.

This refocusing between 1995 and 1998 returned all GMA revenues directly to communities, while retaining park revenues for park management. Income was allocated as follows (figure 12.4):

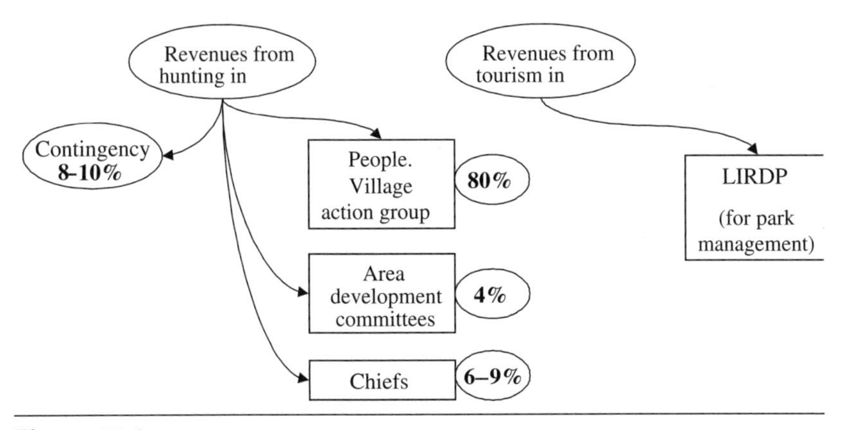

Figure 12.4.
Bottom-up phase of LIRDP's community program (post-1996 policy shift).

- VAGs — 80 percent
- ADCs — 4 percent
- Chiefs — 6 percent (later they received an additional ±2 percent from contingency funds)
- Contingency — 10 percent (reserved to employ ADC executive secretaries)

The revenue from the two hunting concessions in the project area is immediately banked in an account with joint community and project signatories. In April of each year, the money is divided equally between the six areas (not necessarily the optimal arrangement—see the discussion below). Areas divide money among their VAGs using different formulas, sometimes equally and sometimes according to the number of members. An annual general meeting is then held. The first session (day) is used to inform the community—to present information such as members' rights and responsibilities, the number of animals killed and their values, etc. The second session (day) is a formal meeting. Previous activities are reported, and a new VAG committee is elected. Once in place, the community then decides what to do with their revenue; for example, allocating it toward projects (40 percent), take-home cash (40 percent), VAG administration (10 percent), and wildlife management (10 percent). The percentages in parentheses are representative, and vary from area to area. The percentages are also a good indication of the health of the community's institutions. Where there have been problems of transparency, or where the chief has imposed projects or taken money, the amount allocated to cash dividends has increased relative to the other uses. Once the allocation of revenues has been decided, the VAG committee collects cash from the bank and pays each member his or her dividend, usually on the following day. Each VAG has a bank account. Their books are audited two to four times annually, and the committee (usually the treasurer) is trained and obligated to present a quarterly financial (and progress) report to the community.

The Village Action Group

This is the "action" level (figure 12.4). A committee of ten elected people implements the decisions made by the community. This committee reports quarterly to the community on all activities, especially finances. Thus, each of the forty-three communities has its own income, projects, bank account, double-entry bookkeeping system, is audited twice annually, and is required to provide an acceptable financial report (verbal and flip chart) to the community. This is validated (or not) through an independent audit report prepared by project staff. Initially, project staff administered the distribution of revenue, but within four years communities were quite able to do this for themselves. VAGs have received revenues provided they have accounted for money and reported back to communities properly. Where there have been problems, revenue distribution has been delayed until they have been resolved.

The Area Development Committee

Each chief's area has an area development committee (ADC) comprising between three and ten VAGs (figure 12.4). They receive 4 percent of revenues. Their primary role is coordination and oversight and, more recently, wildlife management. Against strong advice, chiefs have assumed the chair of most ADCs, with the concomitant problem that chiefs are "above" audits and accountability. Customarily, chiefs cannot be held accountable, a situation that is not suited to modern financial management. Predictably, ADCs have proved far less able to implement projects, and financial misappropriation has been an order of magnitude higher than VAGs (though this is an improvement on the earlier situation when financial records were simply not required). Management at the area level has proved unavoidably problematic because of weak accountability at this scale (how can a committee report to, and be scrutinized by, 1,000 households?) and the dominance of chiefs at this level.

Opposition to Change from Beneficiaries of the Status Quo

As would be expected, devolution, which changes the structure of the political economy, creates conflicts. Serious tensions developed between the project and the chiefs. Chiefs in Zambia enjoy significant power over their communities, and particularly so in the Luangwa area (see also Gartlan this volume). Chiefs' opinions and actions customarily cannot be questioned, and they can throw people off "their" land. The new system was criticized by chiefs, who said it undermined their authority and caused division within the communities.

Revenue distribution was delayed by many months in 1997 when the six traditional chiefs demanded an additional 9 percent (1.5 percent per chief) over and above the 6 percent in the new formula. In a tense period, it was resolved that chiefs would receive an additional ZK 1.5 million but could claim no revenue from community funds meant for development projects and household cash dividends. This agreement was broken when some chiefs continued to expropriate money from their communities in 1997 and 1998. The project implementers and chiefs argued over these matters, and the project's access to communities was blocked for months at a time. The project did not intervene directly in this conflict, but ensured that all finances were transparent. Consequently, the communities knew accurately how much money had been misused. When the newly inaugurated Senior Chief Nsefu gained confidence, he replaced Chief Kakumbi's informal and insidious leadership, agreed that chiefs should repay monies illegally obtained, and began to support the new system. No doubt this was a response to internal, if largely invisible, community pressure. By 1999, relations with five of the chiefs were good, with some beginning to play a more positive role. The new system is unanimously supported by individuals in the community and has been the crucial factor in the recent turnabout of attitudes toward wildlife.

RESULTS OF NEW CBNRM POLICY

Financial Accountability

Implementation of the new CBNRM policy has provided evidence that devolution to communities works, but that midlevel institutions are prone to accountability problems. By 1999, the community program was working and delivering good institutional progress. Only ZK 3.35 million out of ZK 400 million allocated to VAGs was unaccounted for (0.8 percent) (table 12.1). Two ADCs had serious problems, mostly related to two chiefs. In total, some 40 percent of revenues managed by ADCs was mismanaged, a level of misappropriation common to many development projects that fail to work transparently at the village level.

Table 12.1 *Summary of money unaccounted for from 1999 CBNRM disbursements (Zambian Kwacha)*

Total VAG income (42 VAGs)	400,000,000
Msoro	400,000
Malama	1,000,000
Jumbe	600,000
Mnkhanya	200,000
Nsefu	1,150,000
Kakumbi	–
Total money missing from VAGs	**3,350,000**[a]
Money unaccounted for by ADCs	
Nsefu ADC	10,000,000
Senior Chief Nsefu (recorded loan)	10,500,000
Chief Kakumbi (no records)	24,000,000
Total missing in ADCs/chiefs	**44,500,000**[b]

[a] Village level: 0.8% leakage
[b] Area level: 40% leakage

Table 12.2 *Summary of projects (1996 to 1999)*

Type of project	*Number*
Teachers' houses	16
School block renovation or construction	36
Clinic or health projects	14
VAGs doing wells	26 (about 100 wells)
Other projects (maize, electric fence, sport, women's clubs, chief's vehicle, road maintenance, local court, ADC office, bus shelter, toilets)	60
Total projects	**152**

Community Projects

By mid-1999, some 150 community-funded projects were in progress or completed (table 12.2). By 2001, this figure exceeded 300.

Community Attitudes

The impact of revenue distribution on community attitudes is striking, with the combination of tangible and transformational benefits leading to a metamorphosis of attitudes toward wildlife and the project (see table 12.3).

In 1996, the great majority of people in Lupande GMA did not understand LIRDP policy. They believed community development came from donor aid channeled through the LIRDP and were unaware that funding for projects derived from 40 percent of revenues from the park and GMA. They were also critical that projects were few, incomplete, and inappropriate and tended to benefit the chiefs alone (unpublished LIRDP PRA assessments).

By May 1998, local people knew that wildlife provided money each year, and many knew the prices of hunting licenses. In the villages, people were often found talking about *tyolela* (household dividends) and community projects. Their general view was that, if they looked after their wildlife, then the money they received each year as tyolela and projects would increase. Indeed, there was a sense of ownership of wildlife by the people. It was now seen as "their wildlife." This shift in attitude toward wildlife conservation was a direct consequence of the change in LIRDP policy in 1996. This attitude shift was measurable by comparing the results of a series of master's thesis studies conducted between 1990 and 1999 (Balakrishnan and Ndhlovu 1992; Wainwright

Table 12.3 *Comparison of community attitudes before and after revenue distribution*

Before (1996)			*After (1999)*
"Do you think that your living standard has been improved because of LIRDP activities?"		*"What is the importance of wildlife cash in your household economy?"*	
Not at all	38.8%		
Not very much	19.0%	37.2%	Not important
Neutral	15.0%		
Somewhat	16.0%	37.8%	Minor
Very much	12.0%	25.0%	Important
Survey done in 1996, *N* = 200; Wainright and Wehrmeyer 1998.		Survey done in 1999, *N* = 185; Phiri 1999.	

1996; Butler 1996, 1998; Phiri 1999). In the early 1990s, few people in communities believed they were benefiting from wildlife: Balakrishnan and Ndhlovu (1992) found that 88.2 percent of respondents did not support safari hunting.

By the late 1990s, there was wide understanding and support for the community program and its benefits (revealed by questionnaire surveys conducted by Elias Phiri and Nasson Tembo during master's thesis field studies). Butler (1996, 1998) also found local people's attitude toward tourism and the protection of national parks to be very positive. In 1996 she cited employment as the key benefit, but also mentioned markets for provisions and handicrafts. By 1998, direct benefits had usurped employment, and people had a better understanding of the economic potential of wildlife, which featured prominently in their responses. People closer to Mfuwe, the center of tourism, perceived a wider range of benefits. This attitudinal shift laid the social foundation for later progress in the uptake of responsibility for resource management by communities (cf. North 1980; Olsen 2000).

Uptake of Responsibility for Wildlife Management

Within a year of the first revenue distribution under the new policy, communities began to invest in wildlife (table 12.4), even before they had worked out what exactly was required. Progress was spearheaded by Chief Msoro's area, which built four small dams to provide water for wildlife and employed the first fourteen community scouts. By 2001, all six areas had community-based scouts, some seventy-seven in all,

Table 12.4 *Allocation of VAG revenue to wildlife management (Zambian Kwacha)*

	1996	*1997*	*1998*	*1999*
Amount (ZK)	nil	3,082,516	10,413,995	53,243,707
%		1.4	4.4	13.3

and they were patrolling fairly regularly, with the main limiting factor being the ability to train them and provide them with sufficient authority: in short, the communities' willingness to pay, act, and learn was not the limiting factor.

There is little doubt that community wildlife management provides a sound social basis for conservation where wildlife has a comparative economic advantage. This leaves the question of whether communities can protect their own wildlife. Getting benefits to communities revolutionizes their attitudes toward wildlife and provides the social platform for wildlife management. Changing attitudes have been enough to reduce subsistence poaching in many communities. But in some (known) communities, there remains a clear tendency to poach. Regardless of social forces, active law enforcement will always be necessary to control the criminal element that continues to poach. Communities, or at least their leaders, have employed community-based scouts partly as a response to this. While it is too early to know how effective these efforts will be, it is likely that with proper training and clearly defined responsibilities and rights, communities will control poaching. The real question is whether the authorities have sufficient confidence to allow this to happen (see also Tongson and Dino this volume).

Improved Community Organization

While VAGs were developed initially to support CBNRM, they represented the only formal subdistrict governance structures in the area. Consequently, other programs began to use them—for instance, the Zambia Social Investment Fund, which funds community projects if communities are organized and provides 15–20 percent of the project cost.

Another significant effect of the CBNRM program is the increasing number of community initiatives evolving in the area. The most notable are the establishment of community-run cultural tourism businesses (Kawaza Village Tourism Project, Nsendamila Cultural Village) and efforts to control or charge for natural resources, including the Kakumbi Natural Resources Business, which charges for timber, firewood, grass, sand, etc.; Nsefu Fishing Association, which is beginning to regulate fishing in an area of conflict; and other communities that are addressing charcoal burning and tree cutting (see also Kiss this volume and Brandon and O'Herron this volume).

WEAKNESSES OF THE NEW CBNRM POLICY

Principle of "Producer Communities" Is Not Applied

A key principle in the conceptually powerful CAMPFIRE program in Zimbabwe was the "producer community." This states that the income from an animal should be returned to the community in which it was produced/shot (see also Kiss this volume). In Lupande GMA, this gave rise to a serious conflict. Chiefs adjoining the national park (Kakumbi, Malama, and Nsefu) argued that their areas had more wildlife, and, as a consequence, they should get more money. Nevertheless, as a body, the six chiefs ruled that, for the sake of the unity of the Kunda people, they would continue to divide revenues equally. This issue reemerges periodically. While the project supports the concept of this principle, it also sees the benefit of a phased approach to allow the other areas to develop their wildlife. All six communities have considerable potential. Transparency and dialogue are seen as the solution.

A similar criticism is that, under the LIRDP, the South Luangwa National Park has only benefited communities in the Lupande GMA and that other communities on the southern, western, and northern flanks of the park have been alienated. It is true that the latter have not benefited; the government has retained much of their hunting revenue. However, the Lupande communities benefit only from wildlife revenues earned within the Lupande GMA, not from those earned in the park (which loses money). Even then, the Lupande communities have been heavily disadvantaged. Between 1992 and 1996, they received only 53 percent of the revenue earned from Lupande GMA (ZK 564 million from ZK 1,056 million). After 1996, they received all GMA revenues.

Communities Do Not Control Their Income Stream

Most of the hunting revenues in Zambia do not reach the communities producing them (Lupande is the only exception); instead, they are absorbed by the Ministry of Finance into central coffers. In addition, the centralized and questionable management of hunting concessions reduces the amount of revenues. Quota management is also poor. A large proportion is allocated to low value uses (meat hunting for political favors), while the allocation of hunting concessions is less than transparent. As a result, Zambia earns less than 10 percent of the income per unit area that its neighbors generate, e.g., Zimbabwe, South Africa, Namibia, and Botswana (Safari Club International hunting flyer).

A serious deficiency is that communities are not privy to the real negotiations with concessionaires, and so lose large potential empowerment and financial benefits. Consequently, the LIRDP has for several years provided evidence that revenues could be quadrupled and communities significantly empowered by redrawing concession boundaries and improving the quota system so that they are set by communities, and by giving communities rights to market hunting concessions themselves. Not surprisingly, the response to these suggestions has been silence. Gibson (2000) provides a fascinating account of the role of wildlife as a tool of political patronage in Zambia, confirming the common knowledge that senior politicians are heavily involved in hunting concessions, while free special licenses provide ample opportunity for reward (the meat hunting for political favors referred to above). A single buffalo, for instance, equates to several months' salary for a civil servant.

Eternal Challenges: Problem Animals

The age-old conflicts between people and wildlife, especially elephants, are a constant challenge. Resolving human-elephant conflicts is especially difficult because this most destructive of animals is the only species that has no financial value to local people. While local wildlife officials support carefully planned use of wildlife, pressure from a few international groups at the political level means elephant hunting is banned in Zambia. In the Luangwa area, this is counterproductive because of expanding elephant populations on shrinking range. Elephants damage crops, destroy houses, or kill people. So villagers want them shot. But this clashes with the reluctance of project staff to kill a large number of "problem elephants." Consequently, problem elephants remain a bitter topic in meetings, and are the biggest negative factor in an otherwise increasingly positive relationship between the project and the communities.

Although killing elephants makes little practical difference to crop raiding, people still demand it vocally, perhaps because of a desire for meat or to punish the perpetrator. A number of measures have been tried in response to this demand. Providing cash dividends to communities is a partial compensation, though people are keenly aware that they get nothing from elephants, regularly bringing up the fact that elephants are the only animal they are not allowed to sell at meetings. Three small-scale two-strand electric fences (2–4 km long, each encircling up to twenty-five households) were constructed and maintained by the affected people, and have shown good success. The careful introduction of high-fee hunting of elephants would almost certainly help. Some ten elephants are killed each year to protect crops, and not one cent is paid to communities for this—an opportunity cost approaching 30 percent of the entire wildlife dividend. In Zimbabwe, for example,

Table 12.5. *Investment in CBNRM in Lupande GMA (U.S.$thousand)*

	1995	*1996*	*1997*	*1998*	*1999*	*2000*	*2001*
CBNRM	35	134	192	145	86	80	38
Technical assistance		60	60	30	15	15	10
Total	35	194	252	175	101	95	48

safari hunting clients are used to deal with crop raiders. This covers the costs of investigating and following up crop raiding elephants, and generates close to U.S. $10,000 where an animal is shot. By implication, Lupande is forgoing U.S. $100,000 in lost revenue.

BENCHMARKING: WHY DOES IT COST TO IMPLEMENT CBNRM?

Table 12.5 shows that the CBNRM program costs somewhere between two and four dollars per person per year to administer, cheaper than many programs, and less than a quarter of the cost of Namibia's program in the Caprivi area (Taylor et al. 2001).

In 2000, the budget for managing the community program was streamlined to U.S. $80,000. However, in 2001 the budget was cut again to under U.S. $40,000, too little to support the program. The project was rightly subjected to pressure by NORAD to become financially self-sustaining (nothing concentrates the minds of managers so strongly), and this was an important factor in recent improvements in the efficiency of park management. However, the design flaw is trying to fund CBNRM, a non-income-generating social service, out of the same budget as the park, which is being forced to become a viable business operation. The promising conservation education program (providing support to twenty-eight Chongololo Clubs in schools) and the community newspaper (*Malco News,* targeted at the 45,000 people in the GMA) were also cut out for the same reasons.

Lessons and Insights for CBNRM

COMMUNITY ORGANIZATION AND ACCOUNTABILITY

For ease of explanation, and to allow comparisons across countries and community programs, we define three levels of organization: the village, representing some 150 households; the area, representing several to ten villages; and the district (figure 12.5).

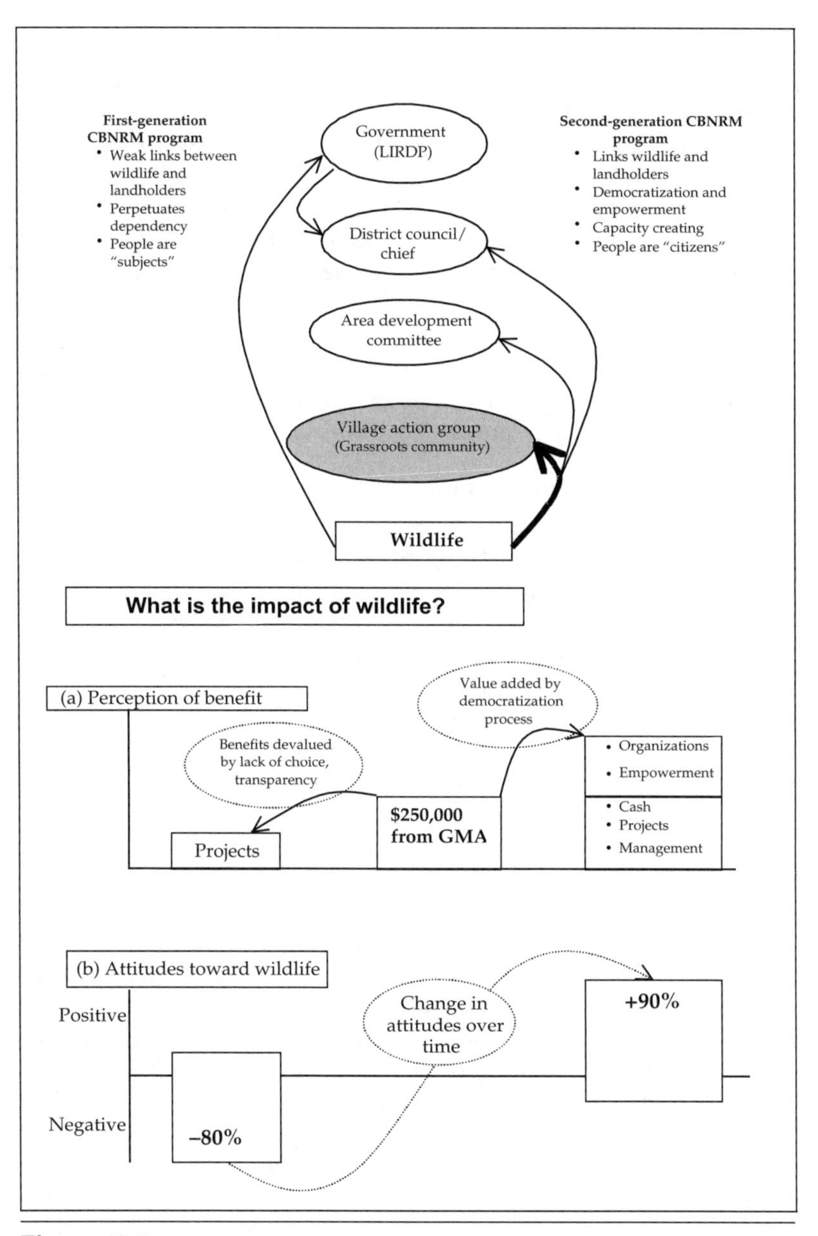

Figure 12.5.
Characteristics of first- and second-generation CBNRMs.

Superiority of Participatory Over Representational Democracy

To fully understand how LIRDP's communities are organized, we need to define two kinds of rural democracy. Far more common is "representational democracy," where a group of people elects a single individual to an area committee that makes decisions on behalf of the communities. More effective, but rare, is "participatory democracy," where the community itself makes decisions in general meetings (village level) and elects a committee to execute these decisions. For the purposes of coordination, the chairperson of such committees may also sit on an area committee. The LIRDP's community policy guidelines emphasize participatory democracy, and reinforce it by channeling 80 percent of wildlife revenues to the village, maintaining the village as the decision-and-doing level. The area merely has a coordinating role.

Upward and Downward Accountability

Similarly, there are two kinds of accountability. Upward accountability is more common. In CBNRM programs, the local committee (generally at an area level) is usually accountable to government agencies or to an NGO that funds it. Downward accountability, linked as it is to participatory democracy, is rare. Here the local committee is primarily accountable to its constituency.

The case for bottom-up accountability is made powerfully by the experience of LIRDP: it is remarkable just how little money has been misappropriated by VAGs despite the generally poor financial control in the country and in the earlier phases of LIRDP. It is also remarkable how many projects the communities have implemented, especially when compared to earlier phases of the project. The 1999 audit showed that 99.2 percent of the revenue allocated to the forty-three VAGs was fully accounted for (table 12.1). This compared to 40 percent misappropriation by the chief-dominated ADCs. Transparency, participation, and the generation of a proprietary impulse led to increased efficiencies and far more VAG-level projects being completed, far more quickly, than when the LIRDP managed these directly. Between 1996 and 1999, VAGs initiated or completed over 152 projects (table 12.2), and nearly 350 according to the latest report. This compares to about ten projects between 1987 and 1992, including the bus service, agricultural shelters that are seldom used, an unfinished clinic, and at least one classroom block.

The improvements are not an accident. They follow the establishment of effective bottom-up accountability, and represent such a conceptual advance on earlier efforts at CBNRM that they can be termed second generational (see figure 12.5).

As already noted, the election of committees and management and monitoring of finances is entrusted to the whole community (village) through participatory democracy. This contrasts with most CBNRM programs, where benefits (income, grants, training, etc.) are focused on the area, and the primary mechanism of accountability is the implementing agencies—for instance, the audits or institutional monitoring carried out under ADMADE (in Zambia) or LIFE (in Namibia). Although most CBNRM programs are strong on the rhetoric of community participation, this is not institutionalized. Progress is often reversed once the sponsoring agencies' checking mechanisms are phased out or falter.

"Scale Laziness" and the Absence of Downward Accountability

Most community programs succumb to "scale-laziness," with the primary level of intervention being an authority, sometimes democratically elected, that represents a significant number of villages, usually at an area or even district level. CAMPFIRE's "wards," LIFE's "conservancies," and ADMADE's "subauthorities" are examples of this.

A second "laziness" is the tendency to work with (and maintain) easily identified leaders, rather than build and work with institutions through which individuals are cycled by regular elections. The logistical constraints faced by implementing agencies, perhaps real but usually avoidable, encourage CBNRM programs to empower various nonaccountable local bodies (Ribot 2001), usually in one of two forms. As their entry points, agencies often use traditional chiefs (neither elected nor accountable) or committees that are elected at least once but are seldom subjected to close scrutiny and regular reelection (e.g., ADMADE, LIFE).

The probability of a community program being successful is linked strongly to scale, because operating at a village rather than area level allows harnessing the power of participatory democracy and downward accountability. Hulme and Murphree (2001) give examples of two types of devolution. The first they call "aborted devolution." These examples often appear to be linked to systems that rely on upward accountability and representational democracy. When the external champions (individuals or projects) that initiated these programs moved elsewhere, the institutional checks and balances moved with them, and the bureaucratic tendency to centralize took over, to the peril of these programs (see also Sayer and Wells this volume). The second type roots accountability and democracy in the village constituency, and these examples appear to have survived. It is no coincidence that the Mahenya and Masoka areas in Zimbabwe are so often quoted in support of the approach of CAMPFIRE. Both are examples of the second type. Another area, Nyaminyami, is an example of the first type of devolution where organization is centralized at the council level, and is often cited to illustrate the negative aspects of CAMPFIRE.

Getting CBNRM to Work: The Second-Generation Model

By way of shorthand, we characterize CBNRM models that work at an area or council level, and that rely on representational democracy and upward accountability, as *first-generation CBNRM*. It follows that *second-generation CBNRM* works at the village level and encourages participatory democracy and downward accountability. The key difference between these CBNRM models is the way the money flows. The LIRDP model represents second-generation CBNRM and shows that fiscal empowerment of the village is the critical step. This is illustrated in figure 12.5, which we will also use to explain economic concepts.

The real breakthrough in the LIRDP was allowing the upward flow of fully 80 percent of the wildlife income generated in the GMA (cf. Child 1996). This is the cornerstone for overturning the top-down approach and for ensuring that benefits reach communities and are accounted for. The fact that 20,500 people get cash and make decisions about how to use wildlife revenues obviously has a major impact on community empowerment and perception of benefit and proprietorship. Each individual in the community received a cash dividend instead of being subjected to spurious projects selected by chiefs or officials, with no explanation of finances. It is then up to villagers to chose and build their own schools, clinics, water supplies, and other projects. In the event, about 40 percent of income has been used for cash dividends, 40 percent for projects, and 10 percent each for administration and wildlife management.

The Power of Transparency

The LIRDP model provides a framework for community evolution. Change has depended on the combination of an institutional framework, information, and transparency. This also transferred the power for most decisions from the project to the community (see also Brown this volume and Singh and Sharma this volume). But one of the project's greatest challenges was managing the relationship between chiefs and the emerging democratic institutions. The project resisted considerable pressure from the community to intervene directly against predatory chiefs. Instead, it continually provided accurate information on how much money chiefs had taken, the rights of people and their institutions, etc., and then waited for the people to negotiate their own relationships with their chiefs. Changes emerged from within the community after about four years. Chiefs stopped taking money and accepted village-based democratic management, having radically opposed it before. Changes made in this way are both more responsive to local nuance and, also, being locally derived and negotiated, are more robust. This change process is also empowering.

Denationalizing Wildlife and Devolution

Converting Wildlife from a Public to a Private (Community) Asset

The economics of wildlife are also revolutionized, if less obviously, by fiscal devolution: it converts wildlife from a public into a private commodity. In doing so, it replaces command-and-control economics with the energy and sensibility of liberal market democracies. Theoretically, we would predict that the system depicted on the right of figure 12.5 would radically improve the allocation and perception of benefits. Data from the LIRDP's community program confirms this: notably the increase in the number of projects built and the shift in community attitudes (as described earlier).

Converting wildlife revenues into a private benefit significantly improves the "value" of this money measured in terms of projects and dividends. If each individual is able to choose (collectively) how to allocate the money, it will be allocated better (and therefore valued more highly) than if others make this choice. This is true even disregarding the greater efficiency and accountability of the bottom-up system, as illustrated in figure 12.5. Note that the second-generation benefit column (cash, projects, management) is much higher than that for the first generation (projects). Another set of benefits (organizations, empowerment)—entirely lacking in first-generation models—is also added as an extra tier to the second-generation benefit column. The process of organizing communities to receive, allocate, and use wildlife money leads to democratization, institutional and organizational progress, grassroots empowerment, learning, and the development of knowledge and managerial culture. As modern texts on corporate management all emphasize, human and organizational capacity is a significant added benefit and is probably far more valuable than physical assets.

Devolution: Are These Lessons New?

Examples of effective devolution are rare. When we initially compiled the history of LIRDP (Dalal-Clayton and Child 2003), we were not aware of what an important and rare example of downwardly accountable CBNRMs the community program actually represented. Ribot

(2001) argues a powerful conceptual case for downward accountability, and then laments how uncommon this is. He finds (correctly) that most community programs have devolved decision-making powers to midlevel organizations (areas and councils), and argues that this maintains central control and is still nonaccountable to communities—they remain subjects rather than citizens. Even where decentralization has been legislated, requirements for management plans have been used to trump devolutionary intent and to maintain power at a central level (e.g., Namibia; see Taylor et al. 2001). Ribot therefore suggests that the environmental decentralization experiment has not actually begun. A similar conclusion is drawn by Hulme and Murphree (2001). Consequently, the poor have stayed poor, and the elites have neatly fielded the concept, twisting it to their benefit (cf. Holloway 2000 and Shepherd this volume).

While Ribot (2001) emphasizes that the basic institutional infrastructure for decentralization is downwardly accountable local representative authorities, he says little about how to design such mechanisms, presumably because there are few practical examples to draw on. This makes the LIRDP community program invaluable, especially because its confirmation of the value of downward accountability can be supported with quantitative and anecdotal data.

Institutional Economics and Devolution

There is increasing disillusionment with the ability of CBNRM programs to deliver on their initial promise. Murphree (1995) suggested that the technocrats had done an end run around the political system and developed devolutionary conservation. More recently, the polity has reexerted its hegemony. The resulting "aborted devolution" is testimony to this, and to the fact that the principles embodied in CBNRM have never really been tested (Ribot 2001). Those who suggest that CBNRM has failed are actually reflecting the failure to implement CBNRM properly, not the failure of the concept (see also Sayer and Wells, Franks and Blomley, Gartlan, and Brown, all this volume). While we hope that this case study reaffirms that the concept is both powerful and workable, we also warn that the attempts to implement CBNRM properly will be met by powerful opposition, often in subtle ways.

The new system of cash distribution (tyolela) brought significant net benefits to the rural society and to conservation. Yet it was strongly criticized by chiefs and officials. The very people tasked with leading development were actually impeding it. This lends support to the contention of the "new institutional economics" that successful institutions have within them the seeds of greater success, but failing institutions are often predestined to maintain their deficiencies. Failing and insecure administrative cultures tend to perpetuate themselves, especially because leading players often benefit by maintaining an unorganized and nonaccountable system (Chabal and Daloz 1999). This suggests that conventional development aid may perpetuate state failure by allowing the survival of agencies and/or institutions that add no value (or reduce it).

We have a good idea of the end point of CBNRM programs—the emergence of communities with significant rights. Indeed, there is some excellent conceptual thinking in this regard. The real challenge is to get there. Our fallacy is thinking that systems naturally evolve to provide the greatest good for the greatest number. There is ample evidence to the contrary. The challenge is to obtain enough grassroots political support to promote (or at least to defend) major political economic change, before the center exerts its will.

Although LIRDP's new community program has been operational for only five years, it supports an emerging theme that village-level governance offers a powerful opportunity to address many problems. We will come back to this point, for it is important enough to change our whole approach to rural development. Based on our present understanding, a logical end point of this argument is to privatize land resources at this level (see also Kiss this volume). The present program has made only a partial step in this direction. It has given communities rights over wildlife benefits. However, it has not given them the right to manage wildlife, especially to allocate and sell quotas. This is a serious deficiency of the program, though it is difficult to address because of the political strength of the individuals who currently dominate hunting concessions. Communities have also begun to sell and husband other resources, including firewood, timber, grass, fish, and sand, and establishing a price for these resources is a positive force for conservation. But this is, de jure, illegal, though it is certainly legitimate, just, and sensible. The question of land ownership is also unresolved and is one of the reasons why the urban growth in the Mfuwe area is so difficult to plan and organize (see also Tongson and Dino this volume and Gartlan this volume).

Managing Devolution: Tough Love

While good intentions lie behind the concept of decentralized community management, well-intentioned efforts seldom balance devolution with institutional and managerial discipline: devolving powers without the concomitant control systems leads to chaos (Skynner and Cleese 1996). Unfortunately a soft approach (commonly associated with community development) usually does more harm than good (Robinson and Redford this volume). Rules and compliance need to be tough. Otherwise central management, albeit mediocre and ultimately unworkable, will be replaced by equally weak community management, with a high likelihood that the system will be captured by undemocratic, non-accountable, and self-empowering midlevel bodies (Ribot 2001).

Devolution is a powerful concept. It incorporates the paradox of "loosening" the system by giving more authority to lower-tier self-managing teams, but is only effective by "tightening" performance control systems. The nature of control shifts from supervision to reliance on trust, transparency, and peer review. This "loose-tight" concept lies at the core of excellence in business management (Peters and Waterman 1982). Lacking precedence in the conservation literature, the LIRDP program drew from business management experience for its design of community institutions and was tough (cf. Handy 1995) on conformance to a set of procedures ensuring bottom-up accountability: communities were required to make their own decisions but had to follow a sound process in doing so. The making of decisions by certain leaders (especially chiefs) on behalf of the community, and without full participation, was not acceptable (see also Gartlan this volume and Brown this volume).

WHAT IS THE OBJECTIVE OF CBNRM: CONSERVATION, DEVELOPMENT, OR SOMETHING ELSE?

There is considerable debate about whether CBNRM is a conservation program or, rather, uses conservation to fund development (see Robinson and Redford this volume and Franks and Blomley this volume). The objective of the LIRDP is to build institutions and organizational capacity capable of planning and making sound choices in changing circumstances, which is "learning organization."

Careful Planning or Messy Management?

There is increasing rhetoric on the virtues of "adaptive management," but many people concerned with conservation and community empowerment find it difficult not to follow the tradition of detailed, scientific blueprint planning (what the business community terms "Taylorism") (see also Salafsky and Margoluis this volume). Such planning leads to preordained outcomes. Communities may be told they are free to do what they want, but only as long as they produce conservation! The alternative approach is to design "Darwinian" organizations that are capable of adapting to changing circumstances.

The common response to the complexity of CBNRM is to plan interventions that are similarly complex—as clearly evidenced by numerous project plans and reviews. This overplanning presumably reflects risk aversion on the part of the project designers and implementers. But it runs counter to their (usually stated) intention to empower communities by insisting on a range of mechanisms and plans that maintain their technical influence and "ensure" the already prescribed output (e.g., conservation). Wildlife surveys, detailed quota setting, complex land-use-planning requirements, and referral to higher authority for approval of projects spring to mind (see also Singh and Sharma this volume).

We believe that the LIRDP community program is different. It has established what has come to be known as a "learning organization" (Korten 1980; Peters and Waterman 1982; Drucker 1986; Handy 1995; Salafsky and Margoluis this volume). The essence of learning organization is institutional procedures that incorporate transparency, democracy, accountability, and simple performance management. The community must follow proper process, but the outcome decisions are its own.

Assessing and Building Community Capacity: Demand or Supply Driven?

The LIRDP has attempted to drive capacity building in response to demand. The community institutions have been fueled by pushing in wildlife-derived revenues from the bottom, and progress has been monitored. This has enabled limiting factors, leakages, or challenges to be identified and then addressed. This adaptive process has required quality facilitation and monitoring. It accepts that it is impossible to preplan systems as complex as CBNRM. Much of the training capacity was centered on the ten young Zambians who facilitated the field program. Initially this was adequate, but as the program evolved, more specialized inputs were necessary. However, the LIRDP's ability to access such inputs was limited by its budget and by being unable to identify sources of appropriate quality training. This situation stands in contrast to the more common position where capacity building is supply driven—often associated with donor funding where the implementing NGOs need to justify their funding by documenting results (i.e., number of training courses). This often results in an abundance of training courses, often of questionable value, and given in an inappropriate order.

Avoiding Dependency

Communities that have been disempowered for a long time are extremely adept at passing their problems on to development agencies to solve. Such agencies find it hard to resist "filling the silence," and, in doing so, they reinforce the dependency culture that is an underlying reason for underdevelopment. The LIRDP's community facilitators were made aware of the need to continually "pass the monkey," and followed the slogan "we bring only knowledge, and can give you nothing else, not even a pen."

Can CBNRM Work?

The assessment of the community program divides conveniently into two parts, reflecting what we have termed first- and second-generation CBNRM programs (figure 12.5). The project's founding concept of integrating conservation and development was innovative and ahead of its time. Unfortunately, implementation from start-up (1988) to 1995 possessed too many flaws to really test this concept. While we can criticize the first years for a top-down approach, for empowering nonaccountable, nonelected leaders, and for the general absence of accountability, transparency, and democracy, many if not most current programs suffer from the same flaws (Ribot 2001).

It is the reconstituted CBNRM program (post-1996) that is really interesting. It is a rare example of the bottom-up accountability that is now considered essential for successful CBNRM (Ribot 2001) and for avoiding "aborted evolution" (Hulme and Murphree 2001). The project also provides quantitative and anecdotal evidence to confirm the efficacy of the CBNRM concept, when it is implemented with enough adherence to the principles governing devolved community management to actually test them (cf. Dalal-Clayton and Child 2003).

This CBNRM program is also one of the few such initiatives directly linked to the management of a large national park. Records show less poaching by people entering the South Luangwa National Park from Lupande GMA than from other areas. This suggests that it would be in the park's interests to use 5 percent of its annual revenues to support the CBNRM program in perpetuity.

Three conclusions of general importance can be drawn:

First, given the right institutional framework and incentives, even poor, illiterate, and dependent people can contribute significantly to their own development (see also Shepherd this volume). Furthermore, they invariably do a better job than agencies that try to help them by supplanting their responsibilities (e.g., designing and implementing projects for them).

Second, the LIRDP's CBNRM program has been, in many ways, a test of southern Africa's CBNRM principles, and the evidence from the project suggests that, when properly applied, they work. By applying these principles, similar progress can be made elsewhere with a high degree of confidence.

Third, lower-level organizations work best, reinforcing the principles of subsidiarity and scale. The progress of the program reflects the development of sound institutions and organizations that incorporate downward accountability and participatory democracy.

The Challenge of Integrating Conservation and Development

This section compares two systems for integrating rural conservation and development. The early period of LIRDP (1988–1995) epitomizes strong and well-defined central leadership, while the latter period (post-1996) represents the emergence of more messy and decentralized village-based democracy.

Model 1: Strong, Well-Defined District Governance ("Mini-Government" Model)

The LIRDP was originally envisaged as a mechanism to coordinate integrated resource management and eventually to sustain these through a wildlife-based local economy. It set out to manage wildlife and other resources, to involve local communities in decision making, to establish ways in which they could benefit from revenues earned in the area, and to coordinate the activities of other line ministries across a wide array of activities. It has much in common with recent approaches to strengthen local government, especially district councils. These attempts have been disappointing. However, it is usually believed that coordination could work if district councils controlled the operational budgets of line ministries. Yet, despite having reasonable finances, LIRDP largely failed to achieve these coordination goals. We do not have enough information to conclude that these seemingly sensible approaches are invalid, or that the failure simply reflects the difficult circumstances in which the project evolved. However, we would hazard a guess that without bottom-up, democratic accountability, the system is likely to fail.

Reasons for Failure of the Centralized Model

Hindsight provides us with several lessons for why the mini-government model proved ineffective. First, it may have been dominated more by the need for a political strategy for building a support base with a range of ministries and stakeholders than by the technical realities of coordinated economic growth. Thus we have criticized LIRDP's mini-government phase for its excessively wide scope and focus on services, when it should have focused on developing financially sustainable activities to fund these services.

Second, the LIRDP spent too much money on unproductive overheads (over 600 employees) and did not pay for results. The project's effectiveness could have been higher had it taken a supportive role (especially linking payment to performance) rather than taking over the responsibilities of weak ministries or local authorities.

Third, the project had too much money, which encouraged wasteful managerial habits.

Fourth, despite the word "sustainable" peppering plans and reviews, LIRDP was never able to seriously invest in the economy that was supposed, ultimately, to sustain the program. It spent money providing services but was unable to invest money in infrastructure like roads and did little to commercialize the wildlife economy (cf. Dalal-Clayton and Child 2003). Even today, the scale of commercial tourism activity is limited by investments in roads, bridges, and an airport constructed by the 1973 Food and Agriculture Organization of the United Nations (FAO) project. Retrospectively, it is hard to understand why a project that set its original goal around developing a sustainable wildlife economy never defined how it was going to do this. Finally, apart from the chiefs, there was little grassroots participation or buy-in to the project.

How Do We Support District Governance?

It is usually argued that projects are temporary and must work with and strengthen the existing government structures, however imperfect they might appear. Because this is so frustrating, many projects have bypassed inefficient government systems, setting themselves up as mini-bureaucracies, but usually these collapsed once the project ended (Mellors 1988; Dalal-Clayton and Dent 2001; Sayer and Wells this volume; Kiss this volume). In the case of the LIRDP, however, both avenues led to failure. Even government agencies that were strengthened by the project (e.g., agriculture) collapsed once external technical support and funding were no longer available. These desperate circumstances have led donors to emphasize NGOs, but again results are disappointing. Perhaps there is no quick fix, and we have to build a democratic culture and institutions from the village up. Fortunately for the political hegemonies that benefit from disorder and underdevelopment (Chabal and Daloz 1999), this does not fit the donor project cycle.

After the reorientation between 1995 and 1998, LIRDP succeeded in making progress using reduced and more focused objectives (Robinson and Redford this volume; Salafsky and Margoluis this volume). However, the project adopted the multisectoral approach specifically because of the poor performance of line ministries. Ultimately, the LIRDP was too small to have a significant impact on reversing the steady deterioration of government finances and services, and could not substitute these services, having no tax base to sustain these inputs. The reorientation does not overcome the fundamental problem that, without financial, technical, and capacity-building support, line ministries will remain unable to deliver what is expected of them. The area will still suffer—as it did before—from neglect on other fronts. Or so goes conventional wisdom. However, what if we accepted that many line agencies will never be effective and, instead, shifted much of this responsibility (and the requisite funding) to communities?

Model 2: Devolved, Village-Level Governance

After 1996, wildlife revenues were used to initiate and strengthen community organization at the village level. Is there evidence to suggest that village organizations provide a better system for integrating conservation and development?

We have provided evidence that village governance was more effective at using wildlife revenues to develop social infrastructure. Communities that organize themselves can also capture matching funds from other programs. For instance, a community that invests wildlife revenues to build a classroom block can use the European Union's microproject grants (which require a 25 percent matching effort from communities) to ratchet up this investment. With donors increasingly responsive to demand-based delivery (e.g., the World Bank's Social Investment Funds), perhaps the most productive use of wildlife revenues is as a mechanism to promote organizational development. As government services and coordination deteriorate, demand-driven development replaces the conventional supply-driven process. We can postulate that coordination in this manner is relatively effective, mirroring a free-market system. It also rewards success because better-organized communities are better able to obtain funds. This promotes a healthier culture in communities that are more used to accessing support by acting particularly pitiful.

In Zambia, wildlife can become a valuable mechanism for developing grassroots governance structures, provided development planning is accepted as a democratic movement rather than a technical plan (cf. OECD 2001, UNDESA 2002). In most of rural Zambia there are no subdistrict structures at all. All that exists is an elected councilor and a weak council, often with insufficient revenue to pay salaries for months at a time. If, optimistically, the 12 councilors meet once a month, these districts have 144 days of local governance. Compare this to the wildlife governance structures in the LIRDP area, where all 43 VAGs meet quarterly (5,000–7,000 people × 4), VAG committees monthly (43 VAGs × 10 people × 12 months), and ADCs monthly (6 ADCs × 10 people × 12 months), a total of some 40,000 days of local governance.

Institutional Economics—Devolved Systems Work, but Will They Be Allowed To?

The experience of the LIRDP adds to the mounting evidence that the devolution of rights to rural African communities can significantly improve livelihoods and rural governance, including the environment. Our quantitative and anecdotal evidence confirms that where villagers are able to assume the local civil and administrative functions of the state according to the principles of subsidiarity, things can improve (Handy 1995). However, we are wary of all the feel-good talk about community empowerment and how often mismanaged community programs are actually manipulated to disempower rural people (Ribot 2001). This leads us to conclude that only fiscal devolution to a scale that allows face-to-face decision making and accountability and that operates through institutions that are carefully designed to incorporate the principles of democracy, transparency, equity, and accountability to a constituency will be ultimately successful.

Consequently, we are optimistic about the ability of communities to integrate local resource management. However, we are also pessimistic about whether the state hegemony will encourage, or even permit, this to happen. Instead, it tends to truncate the progress of community empowerment and results in the "aborted devolution" referred to by Hulme and Murphree (2001). This reversal is no accident. At a local level, the chiefs initially captured the so-called participatory process, and it would not have evolved into real devolution without careful and tenacious management (cf. North 1990 and Harris et al. 1995). Almost all national programs suffer the same affliction (see also Brandon and O'Herron this volume). Why is such progress consistently stifled? How do we interpret the state's tenacious adherence to institutional structures that result in lower net benefits to society? De Soto (2000) documents officialdom's proclivity to create excessive value-negative rules and reluctance to grant poor people the formal rights to use their property effectively. He hypothesizes that this dampens the effectiveness of capital and entrepreneurial energy so much that it may be the real difference between developed and developing economies. It is no coincidence that the poorest economies have small and conspicuous elites that maintain laws and an inefficient institutional environment suited to their predation. Because wildlife remains a state asset subject to bureaucratic and political control, it is particularly badly affected by this phenomenon.

Unfortunately, when a disproportionate amount of political and economic power is vested in the government-dominated status quo, change is far from assured. Hence, it is no surprise that there is resistance to projects like the LIRDP, even when they clearly lead to social, economic, and environmental progress. This paper provides some insights into how to design, manage, and finance wildlife-based rural development programs. However, while it has no answers to the challenges of political and institutional change, just raising this issue is important: that is, why don't conservation agencies put in place the sensible policies and practices that are available? Conservation is not alone in facing this dilemma. For instance, there is an emerging consensus about how to develop poor nations; the real question is why political systems fail to put these sensible policies into place and, when they do, why don't they work properly (Clague 1997).

It is all well and good to learn about and know how to implement effective CBNRM. But unless we can also get through the door to do so, this will be in vain. Similarly, we are learning that pumping money into development is not the answer. It is the "rules of the game," or institutions, that are critical. They make the difference between the economic growth and stagnation of societies. On the one hand, clear and fair rules encourage most people to direct their assets energetically to fostering progress for themselves, activities that contribute to the general livelihoods of their society. Policies, on their own, are insufficient. They must be underpinned by the institutions of property rights and contract enforcement, as well as by the nature of society and the manner of interactions between government agencies and civil society. The wealth of these societies grows as reliable exchanges allow specialization, and because individual reward is linked to amount of value added. In the alternative, or bad equilibrium, those with the power set and maintain rules that tend to serve their narrow interests, and become wealthy by extracting value from others rather then being productive. This encapsulates the differences that North (1995) draws between productive and pirate economies.

Conclusions

The LIRDP pioneered a people-centered approach to wildlife conservation. It began as a well-intentioned integrated conservation and development project in which wildlife was to play a major role. While the concept was sound, the administrative model, or management, or both, proved ineffective, expensive, and unsustainable. With time, the weaknesses of the centralized model became visible, and it was replaced by more modern systems built around the principles of empowerment and responsibility. The LIRDP, which had picked up bad habits from having too much funding and too little accountability, was slimmed down, employees were given greater authority, and a serious effort was made to develop and outsource tourism to ensure financial sustainability. With the introduction of decentralized performance management and peer review, costs fell dramatically and output also improved. The approach to district development, similarly, represented a shift from a top-down to a bottom-up management philosophy, which also proved advantageous. The LIRDP began with much faith in the ability of competent technocrats to coordinate development with good funding, but evolved through four phases to a much slimmer project that relies more on institutional controls, performance management systems, and wider, if less technical, participation.

In a lesson to donors, NORAD continues to have faith in the project after two decades of support. Its reward is a project that stands at the cutting edge of both park and community management, with perhaps the major benefits being the learning process. Interestingly, one of the potential problems facing NORAD and the project is that the project will become sustainable, making NORAD's financing role redundant. The real threat is not the loss of money but the loss of NORAD's powerful and supportive voice as an advocate of the approach.

Although financially inefficient, the project has been worthwhile. It has conserved a park and elephant population threatened by commercial poaching. It has developed and tested management and commercial systems that will sustain the park and, if the national transformation process is not captured by self-serving elites, is likely to roll out across Zambia.

The project has also advanced the cause and understanding of community development and conservation. Importantly, it validates the principles of CBNRM developed primarily in the CAMPFIRE program. In particular, it tests and corroborates the efficacy and legitimacy of downward accountability. The success of the CBNRM program is no accident. It is based, first, on the development of downwardly accountable village institutions according to sound democratic and management principles. The second pillar is the small team of paraprofessionals (fourteen people) that have facilitated and monitored its implementation, and developed the valuable set of procedures, techniques, and procedural manuals that, over six years, operationalized the CBNRM principles. In the same way that following sound CBNRM principles led to a working success, we predict that following these procedures will allow replication with a high probability of success.

However, we must qualify this prediction. While there is a high probability of implementing a successful CBNRM program if principles such as those in LIRDP's CBNRM policy are applied, true empowerment is almost always resisted. For instance, the tortuous and intrigue-filled decade-long process of transforming the Zambia National Parks and Wildlife Service to the more autonomous Zambian Wildlife Authority (ZAWA) reflects the underlying tension between economic and social progress on the one hand and the entrenched interests of a powerful elite and the status quo on the other. Only time, and the interactions between culture, politics, economics, and donors, will tell if these symptoms of "state failure" (North 1990; Harris, Hunter, and Lewis 1995) will be overcome by the enlightened self-interests of society. The implementation of community programs, and the future of the LIRDP, rests in processes much larger than themselves. State hegemony is threatened by, and often undermines, successful programs.

The experience of the LIRDP has led to the development of methodologies for improved park management and to several key lessons for implementing CBNRM programs. The most important lessons are that fiscal devolution must take place and accountable institutions be developed at the village level. However, the final lesson is that sound techniques and principles are necessary, but not, of themselves, sufficient. Someone has to be prepared to fight the political battles necessary to allow these principles to be applied.

References

Balakrishnan, M. and D. Ndhlovu. 1992. Wildlife utilisation and local people: A case study in Upper Lupande Game Management Area, Zambia. *Environmental Conservation* 19 (2):135–144.

Butler, C. 1996. *The Development of Ecotourism in South Luangwa National Park, Zambia.* Master's thesis, Durrell Institute of Conservation and Ecology, University of Kent.

Butler, C. 1998. Case studies in tourism and its relation to communities: Kakumbi and Nsefu Chiefdoms, Lupande GMA, Zambia. Durrell Institute for Conservation and Development, University of Canterbury. Unpublished manuscript.

Chabal, P. and J. Daloz. 1999. *Africa Works: Disorder as Political Instrument.* Oxford: James Currey.

Child, B. 1996. The practice and principles of community-based wildlife management in Zimbabwe: The CAMPFIRE program. *Biodiversity and Conservation* 5:369–398.

Clague, C., ed. 1997. *Institutions and Economic Development: Growth and Governance in Less-Developed and Post-Socialist Countries.* Baltimore: Johns Hopkins University Press.

De Soto, H. 2000. *The Mystery of Capital: Why Capitalism Triumphs in the West and Fails Everywhere Else*. New York: Basic Books.

Dalal-Clayton, D. B. and B. Child. 2003. *Lessons from Luangwa: The Story of the Luangwa Integrated Resource Development Project, Zambia.* Wildlife and Development Series no. 13. London: International Institute for Environment and Development (IIED).

Dalal-Clayton, D. B. and D. L. Dent. 2001. *Knowledge of the Land: Land Resources Information and Its Use in Rural Development.* Oxford: Oxford University Press.

Dalal-Clayton, D. B. and D. Lewis, eds. 1984. *Proceedings of the Lupande Development Workshop (An Integrated Approach to Land Use Management in the Luangwa Valley).* Lusaka, Zambia: Government Printer.

Drucker, P. F. 1986. *The Practice of Management.* New York: Harper Brothers.

Gibson, C. C. 2000. *Politicians and Poachers: The Political Economy of Wildlife Policy in Africa.* Cambridge: Cambridge University Press.

Holloway, J. 2000. *All Poor Together: The African Tragedy and Beyond.* Johannesburg: Capricorn Books.

Handy, C. 1995. *Beyond Certainty: The Changing Worlds of Organizations.* London: Arrow Business Books.

Harris, J., J. Hunter, and C. M. Lewis. 1995. *The New Institutional Economics and Third World Development.* London: Routledge.

Hulme, D. and M. W. Murphree, eds. 2001. *African Wildlife and Livelihoods: The Promise and Performance of Community Conservation.* Oxford: James Currey.

Korten, D. 1980. Community organization and rural development: A learning process approach. *Public Administration Review* 40 (5):480–511.

Mellors, D. R. 1988. Case Study 27: Serenje, Mpika, and Chinsali Districts Integrated Rural Development Program, Zambia. In C. Conroy and N. Litvinoff, eds., *The Greening of Aid: Sustainable Livelihoods in Practice.* London: Earthscan.

Murphree, M. W. 1995. Optimal principles and pragmatic strategies: Creating an enabling politico-legal environment for community-based natural resource management. In A. Steiner and L. Rihoy, eds., *The Commons Without the Tragedy? Strategies for Community-Based Natural Resources Management in Southern Africa,* 47–52. Harare, Zimbabwe: Regional NRMP for Southern Africa, USAID.

North, D. C. 1990. *Institutions, Institutional Change, and Economic Performance.* Cambridge: Cambridge University Press.

North, D. C. 1995. The new institutional economics and Third World development. In J. Harris, J. Hunter, and C. M. Lewis, eds., *The New Institutional Economics and Third World Development,* 17–26. London: Routledge.

OECD (Organisation for Economic Co-operation and Development). 2001. *The DAC Guidelines: Strategies for Sustainable Development: Guidance for Development Cooperation.* Paris: Development Cooperation Committee, Organisation for Economic Co-operation and Development. www.SourceOECD.org (September 2001).

Olsen, M. 2000. *Power and Prosperity: Outgrowing Communist and Capitalist Dictatorships.* New York: Basic Books.

Peters, J. T. and R. H. Waterman. 1982. *In Search of Excellence.* New York: Warner Books.

Phiri, E. 1999. A case study of consumptive tourism's contribution to conservation and rural development in Lupande Game Management Area, Zambia. Master's thesis, University of Greenwich, England.

Ribot J. C. 2001. Decentralized natural resource management: Nature and democratic decentralisation in sub-Saharan Africa. Summary report prepared for the UNCDF symposium, Decentralization Local Governance in Africa, Cape Town.

Skynner, R. and J. Cleese. 1996. *Life and How to Survive It.* London: Vermilion.

Taylor, G., B. Child, K. Page, B. Winterbottom, and K. Aribeb. 2001. *Evaluation of the Living in a Finite Environment (LIFE) Project.* Windhoek, Namibia: International Resources Group.

UNDESA (UN Department of Economic and Social Affairs). 2002. *Guidance in Preparing a National Sustainable Development Strategy: Managing Sustainable Development in the New Millenium.* Background Paper no. 13 (DESA/DSD/PC2/BP13). Submitted by the Division for Sustainable Development, Department of Economic and Social Affairs, United Nations, to the Commission on Sustainable Development acting as the preparatory committee for the World Summit on Sustainable Development Second preparatory session, 28 January–8 February, New York. www.johnnesburgsummit.org (May 2002).

Wainwright, C. 1996. *Evaluating Community-Based Natural Resource Management: A Case Study of the Luangwa Integrated Resource Development Project (LIRDP), Zambia.* Master's thesis, Durrell Institute of Conservation and Ecology, University of Kent.

Wainwright, C. and W. Wehrmeyer. 1998. Success in integrating conservation and development. A study from Zambia. *World Development* 26 (6):933–944.

World development. Now think small. 2001. *The Economist*, 15 September, 42–43.**Table 12.3** Comparison of community attitudes before and after revenue distribution

13

Ecodevelopment in India

Shekhar Singh and Arpan Sharma

Introduction

Ecodevelopment, as a conservation strategy, is very much like the integrated conservation and development project (ICDP) approach prevalent in many parts of the world. This paper describes ecodevelopment, its rationale, some of the issues that it raises, and traces its genesis and progress. The experience of ecodevelopment in India is discussed in the context of the wider global debate on ICDPs.

The Definition of Ecodevelopment

In India, ecodevelopment is defined as a strategy for protecting ecologically valuable areas (protected areas) from unsustainable or otherwise unacceptable pressures resulting from the needs and activities of people living in and around such areas (Singh 1994a).

It attempts to do this in at least five ways:

1. By identifying, establishing, and developing sustainable alternatives to the biomass resources and incomes and other inputs being obtained from the protected areas (PAs) in a manner, or to an extent, considered unacceptable

2. By increasingly involving the people living in and around such protected areas in the conservation planning and management of the area, thereby not only channeling some of the financial benefits of conservation to them, but giving them a sense of ownership of the protected area
3. By raising the levels of awareness, among the local community, of the value and conservation needs of the protected area, and of patterns of economic growth and development that are locally appropriate and environmentally sustainable
4. By strengthening individual and institutional management capacities at the protected area and individual, institutional, and systemic capacities at the local, state, and national levels
5. By attempting to integrate conservation concerns into national, state, and local plans and activities

Though, by their very nature, ecodevelopment initiatives will differ from area to area (and even from village to village), three basic principles define ecodevelopment:

1. Site-specific, microlevel planning, assessing the adverse impact that protected areas have on the local people and the impact that the local people have on the protected areas, and identifying the options available
2. Sectoral integration, especially of local-level activities and investments
3. People's participation, at all levels, especially in the planning, implementation, monitoring, and evaluation of the project and, through the project, in the management of the protected area and in the planning and implementation of other related activities in the area

Unfortunately, ecodevelopment has often been either blamed for not doing things that it never intended to do or for being something that it is not. Therefore, it is important to clarify not only what ecodevelopment is, but also what it is not.

Ecodevelopment is *not* solely or primarily an effort at rural development, nor is it solely or primarily directed toward the economic development of the rural population for its own sake.

Ecodevelopment is *not* solely or primarily an effort at enhanced policing of the protected area in the sense that it does not seek to protect an area solely or primarily through the enforcement of laws aimed at excluding local people. Rather, it seeks to involve the local people in the process of protecting the park and provides them real options to do so. However, it concurrently strengthens protected area management capacities so that deviant individuals or communities can be deterred.

Ecodevelopment is *not* primarily aimed at minimizing or negating pressures from commercial or development activities and projects. However, it is a reasonable expectation that, as the involvement and stake of local communities in conservation increase, there would be increased capacity to resist those projects and activities that are destructive to the area. The fact that ecodevelopment seeks to address pressures posed only by the local community should not be understood to imply that these are either the most prevalent, the most destructive, or the most illegitimate of the pressures. On the contrary, the importance given to ecodevelopment is a result of the recognition that pressures exerted by the local people are, usually, the most legitimate of all the pressures and, as such, cannot be handled in the conventional, regulatory manner but need a more humane and sympathetic approach so as to ensure that the subsistence needs of local communities are respected and provided for.

Ecodevelopment is *not* a strategy for revolutionizing wildlife management or even for bringing about fundamental changes in the way biodiversity is being conserved. As a strategy, it has been designed and applied with the understanding that it could, within the existing framework, help conserve critical ecosystems and species for a little while longer while minimizing the costs that local communities have to pay for such conservation. It is recognized that over the medium to long term more fundamental changes will have to be made if biodiversity is to stand a chance of surviving. Some of these changes might involve redefining the role of local communities in the control and management of wilderness areas and perhaps redefining what biodiversity conservation involves. Therefore, ecodevelopment should not be seen as necessarily endorsing the prevalent paradigm of conservation but only as an interim measure aimed at minimizing social and environmental costs while a new paradigm is developed, accepted, and applied.

There is a special need in India to develop a paradigm of protected area management that does not presuppose the exclusion of all human use, especially use by tribal people and other local communities. However, before that can be done, various questions of science and strategy, as discussed later, need to be satisfactorily answered. There also has to be a balancing of the needs of the weakest among human beings with those of animals and plants, who are essentially even more disempowered. Though the current debate seems to focus on the need to open up access to protected areas and to shift control and ownership to local communities, perhaps a concurrent effort needs to be made to rationalize the control and access of resources outside the protected area system. If we could ensure a more equitable distribution, among different segments of the Indian population, of the 96 percent of land and land-based resources outside protected areas, perhaps the poorest of the poor would not be forced to commit ecological suicide by overusing the remaining 4 percent.

The Rationale for Ecodevelopment

The objectives, methods, and rationale of wildlife conservation have been a part of the development and social justice debate for the past three decades. However, by the mid-1980s wildlife conservation in India had become exceptionally contentious. The incidence of conflicts and clashes between protected area managers and local communities was on the rise. Also, in many protected areas, the ability to regulate use and extraction to the levels prescribed by law was nonexistent. This was primarily because of the following:

1. There had been a steady increase in human population and a resultant increase in their need for land and natural resources.
2. There had been a concurrent and often a resultant decrease in wilderness areas.
3. Though there had also been much "development," and a consequent growth in economic opportunities and infrastructure, this was not equitable across categories of population, regions, and the urban-rural divide.
4. Similarly, the costs and benefits of conservation were not equitably apportioned, the poor losing the most and gaining the least.
5. However, one effect of "development" was to raise the economic aspirations of people almost uniformly, thereby creating a greater demand for income and resources.
6. The establishment of a democratic process of governance, after India became independent in 1947, made people increasingly aware of their political and economic rights and gave them a voice that could not be easily ignored.
7. Historically, wildlife and forest management was primarily regulatory and was perceived to be oppressive and indifferent to the needs and aspirations of the local people. Under the colonial regime, control and ownership of forests had been taken away from communities and usurped by the government. Not surprisingly, there was a reaction against this.
8. Traditional cultural imperatives for conservation were losing ground, while scientific reasons for conservation were neither widely understood nor universally accepted.

The 1980s was also the period when a very large number of new protected areas were set up, raising the number from a little over 200 at the start of the decade to nearly 500 by the beginning of 1990. Therefore, the creation of all these new protected areas, and the consequent inevitable deprivation for the local communities, further heightened the sense of unrest against this form of conservation.

By the time the process of formulating the eighth five-year plan[1] was initiated by the Indian Planning Commission, in the early 1990s, it was clear that the current system of wildlife protection was not working and that not only were protected areas getting degraded at a very rapid rate, but also there was widespread resentment against them. In fact, many political parties and people's representatives were locally voicing their discontent with the protected area network. Added to that, many powerful lobbies, especially of miners, tourist operators, timber merchants, land developers, hoteliers, industrialists, and contractors constructing dams and other infrastructure projects, were working hard, especially through money power and political patronage, at diluting the protected area network in India and getting access to the land and other resources within protected areas.

The 1980s was also a time of prime ministers and national governments perceived to be sensitive to wildlife conservation, but by 1990 there were new prime ministers and new governments in power who had no such pretensions. Consequently, pressures started building to reverse the process of conservation and to dilute the various laws dealing with wildlife and forest conservation.

On the other hand, a survey of the status of national parks and sanctuaries in India, the first of its kind, had been published in 1989 (Kothari et al. 1989). It revealed that, despite stringent laws and an increasing network, a large proportion of the national parks and sanctuaries in India were not being managed as such and had all sorts of pressures within them. The survey highlighted the urgent need to tackle both pressures from commercial and development interests and those from local communities, and suggested various measures, including the development of an ecodevelopment-type approach.

The Ministry of Environment and Forests of the Government of India and the Planning Commission were, therefore, confronted on the one hand with pressures to lighten the regulatory process associated with wildlife and forest conservation and, on the other hand, with evidence that, even with the current levels of regulation, forests and protected areas were rapidly deteriorating. It was out of such a predicament that ecodevelopment emerged.

Evolution of the Idea

In analyzing the problems of wildlife management in India, it became obvious that the appropriateness of a protection strategy was largely dependent on answers to three types of questions.[2] First, there were the scientific questions, about the level of human use and manipulation that was in consonance with biodiversity conservation and the size, number, and variability of the areas required. Despite a growing disillusionment with the concept of large "pristine" areas with little or no human activity, the predominant conservation philosophy continued to espouse the "protected area" approach, with large protected areas containing viable populations of mammals and of all other species. This was not only the predominant view among wildlifers in the government, but also among many non-governmental conservationists and scientific institutions. Of particular influence was a pioneering study done at the Wildlife Institute of India by Rodgers and Panwar (1987), which identified the gaps in the protected area network and recommended adding new areas, expanding many of the existing areas, and upgrading some areas from sanctuary status to national park status.[3]

The second type were ethical questions. There were, of course, the usual "intergenerational" ethical concerns with the imperative to leave for future generations a working planet. But, added to that, India's stratified society also raised important intragenerational questions. Who benefited from conservation? Who paid the costs? Why? Clearly, any conservation strategy would have to take into consideration these questions and ensure that both costs and benefits were more equitably distributed.

There were also related issues regarding the rights of communities versus the rights of governments, especially regarding natural resources. Because, in the Indian system, national priorities tend to supersede individual or community rights, these were contentious issues, especially where the poor or tribal communities were concerned.

There was also another set of ethical questions, especially relevant to the Indian condition, involving interspecies issues. Compassion for animals is a characteristic of Indian thought (even though it might not always be a part of all Indian action), and wildlife protection had to be carried out in a manner that was sensitive to these sentiments.

Finally, there were questions of strategy, especially about levels of community control and ownership of wilderness areas. The success of joint forest management (JFM) in India had resulted in the expectation that a similar joint protected area management system would work for protected areas. However, prevailing legal and scientific expectations restricted human use of protected areas to a level that would make a JFM type of approach unviable, leaving few options for community control and ownership of protected areas.[4]

The Ecodevelopment Debate

It was in this setting that ecodevelopment started being seriously debated in the early 1990s. Right from the start, the design of an ecodevelopment approach raised many issues and questions.

What role should different institutions play in the design and implementation of the ecodevelopment strategy? While the ecodevelopment scheme and projects were without doubt to be designed and implemented in partnership with the local communities, the role of nongovernmental organizations (NGOs), various government departments, and especially the forest department was much debated. There was a strong view that ecodevelopment should be handled by nonforesters. It was argued that development agencies were more sympathetic to the needs and aspirations of the local communities than forest departments.

However, the Planning Commission rejected this view, mainly because development agencies knew little about the real objective of ecodevelopment: biodiversity conservation. Also, to establish a link between restrictions on protected area use and the alternatives provided, it was considered essential that both be managed by the same agency. As the forest department had the responsibility of enforcing the Wild Life (Protection) Act and thereby restricting access of the local communities, it was the obvious choice. Also, the authority and willingness of protected area managers to impose such restrictions, especially on the abjectly poor, would be greater if the alternatives provided for under ecodevelopment were theirs to offer.

There was also a demand from some quarters that, because NGOs were more sympathetic both to the requirements of biodiversity conservation and to the needs of the local community, were more flexible, and had a better rapport with the local people, they should be the implementers. This was also not accepted by the Planning Commission, partly because of the earlier stated reasons to work through the forest department and partly because NGO capacity to run such programs was not considered adequate.

There was a major debate on whether or not people would stop using protected area resources once alternatives and other inputs were provided and they felt a greater sense of ownership of the protected area. International experiences were studied (Singh 1995) and the problems with ICDPs analyzed. Four conclusions were reached.

First, just offering alternatives to the local communities was unlikely to significantly reduce pressure on the protected area. There needed to be a concurrent strengthening of management capacities, so that there was better enforcement. However, for such enforcement to be effective and just, it must be backed by viable alternatives to livelihood needs. Second, protected areas could not be sustained if they remained isolated islands of conservation without any influence on the landscape around them. There was, therefore, a need to set up appropriate coordination mechanisms (like project coordination committees, district/regional coordination groups), involving representatives of all major government departments in the area, to coordinate government activities toward conservation. The relevant provisions of the Environment (Protection) Act[5] also needed to be invoked to regulate pressures around the protected area.

It was also necessary to ensure a link or trade-off in the minds of the people between restrictions related to the protected area and ecodevelopment inputs (see Brown this volume). For this purpose, there needed to be ecodevelopment committees (EDCs) in each village, to sign, on behalf of the village, a memorandum of understanding (MoU) with the forest department and the protected area managers, laying down the rights and obligations of both parties toward the protected area and its surroundings. These EDCs would also be the principal planning agency for village ecodevelopment.

Finally, fixed-term projects do not promote sustainability (Sayer and Wells this volume). Therefore, if the gains of ecodevelopment are to be consolidated, permanent institutional and financial arrangements have to be made. Consequently, in the long term, ecodevelopment support should become an intrinsic part of regular protected area funding, so that a sustained flow of resources becomes available from the state and central governments. However, as an immediate measure, village-level and protected-area-level trust funds should be set up as revolving funds to support ecodevelopment activities beyond the project period.

Though the initial capital for these funds could come through ecodevelopment projects and programs, innovative methods of replenishing these trust funds should also be developed (Kiss this volume).

Initiating Ecodevelopment Projects

The first formal effort by the Government of India at introducing the ecodevelopment approach was through a centrally sponsored scheme on ecodevelopment introduced in 1991. There was also an effort at introducing ecodevelopment as a part of all the externally aided forestry projects. Some NGOs, most notably WWF–India and the Ranthambhore Foundation, had also initiated ecodevelopment projects. However, this paper deals only with the two ecodevelopment projects taken up with World Bank and Global Environmental Facility (GEF) support.

THE FORESTRY RESEARCH, EXTENSION, AND EDUCATION PROJECT

In 1992, the Government of India decided to include ecodevelopment as a component in the World Bank–funded Forestry Research, Extension, and Education Project (FREEP). Though FREEP was in an advanced stage of preparation, at the request of the government, the World Bank agreed to include ecodevelopment around two protected areas. The Government of India requested the Indian Institute of Public Administration (IIPA), in Delhi, to help design the project.[6]

One interesting feature of the planning process of the FREEP ecodevelopment project was that the World Bank insisted that it be a participatory activity. However, they also required that the project proposal document be complete in all respects and list every activity that was to be taken up in every village or location, along with the detailed costs. This created an interesting dilemma. The IIPA project planning team argued that it was neither fair nor efficient to develop such a detailed proposal at this stage. Essentially their argument was that it was insensitive to go into village after village and use the villagers' time to sit with them and discuss, prioritize, and collectively decide on what they wanted the most, when it was not known when the project would commence and, indeed, whether it would be approved at all. It was insensitive and disrespectful to raise people's expectations and to waste their time, only to tell them that if and when the project was approved and the money came through, they might get what they had so painstakingly identified as their priority need.

Also, local conditions were likely to change in the time it ordinarily took for projects to be considered and approved. Consequently, the priorities determined today might no longer be relevant by the time the project was initiated.

The World Bank, on the other hand, seemed to require details of all activities and expenditures in order to even consider, let alone approve, such a project. Besides, the World Bank argued, if a village-by-village exercise was not done in advance, it would be difficult to justify the proposed project budget, as it would have no empirical basis.

The World Bank ultimately agreed to consider an indicative plan, which would be based on a participatory planning exercise covering only a small sample of representative villages in the project area. The budgets developed for these villages would be extrapolated to determine the overall project budget covering all the project villages. Also, the final project document would provide for an ecodevelopment fund, without a detailed breakdown of expenses, and specify the method to be used in determining the details of expenditure. Essentially, microlevel planning teams, in consultation with village EDCs, would develop detailed budgets during project implementation. Each EDC would be given a predetermined budget constraint within which it would develop its priorities for investments. This not only made prioritization more community driven, but allowed for cost-effectiveness and better overall budget management. It also ensured that there was no sense of discrimination among different EDCs.

The India Ecodevelopment Project

Even as FREEP was being processed, the Government of India decided to propose a larger ecodevelopment project, covering eight protected areas, to the GEF for funding. The government again asked the IIPA to help design this project. As indicative planning was now an accepted process, it was also used for the India Ecodevelopment Project (IEP).

In designing the IEP, the first task was the development of criteria for selecting project sites. The World Bank, which was the GEF implementing agency for this project, engaged a consultant[7] to work with the Indian government and the IIPA team and help develop such criteria. The debate soon settled around one critical issue: should the selected sites be those with poor management capacities and high levels of pressure, or should they be the better managed and less threatened ones? The former sites needed urgent attention and might not survive unless something was done immediately. Also, it was thought that protected area managers might be more enthusiastic about ecodevelopment where traditional methods of conservation were proving inadequate. On the other hand, ecodevelopment was a new initiative and much still had to be learned about it. By starting in very difficult situations, there was a chance that the approach would be discredited without being given a chance to evolve. In the end, a compromise was reached and the eight areas selected were those that had good management capabilities, six of the eight being Project Tiger[8] areas. However, all these areas also had significant pressures that were, collectively, representative of the pressures faced by protected areas across the country. Another consideration was to select not more than one site from any one state so that the ecodevelopment approach could be introduced in as many states as possible.[9]

Once the sites were selected, the process of indicative planning started. Interestingly, at this point the World Bank decided, reportedly at the behest of some Indian NGOs and forest officials, that any financial input to a village must be matched by a financial contribution from the "beneficiary" villagers. Presumably this was "rural development" type of thinking where it was believed that the villagers would not value or own the project unless they also had a financial investment in it.

The IIPA team argued that, whereas rural development projects involved outright investments for village development, in ecodevelopment villagers were being compensated, and not always adequately, for foregoing their use of protected area resources. Therefore, any insistence on financial contributions by villagers would be unfair and weaken their resolve to help conserve the protected area. Besides, the benefits of the rural development projects went wholly or primarily to the village community; in ecodevelopment the benefits were not wholly or even primarily those of the villagers. In fact, the main benefit was biodiversity conservation, which was a benefit to the whole world. Therefore, if beneficiaries were required to contribute financially to the project, then all the beneficiaries, especially the World Bank consultants working on the project, should contribute a part of their earnings!

Finally, as a compromise, it was agreed that village trust funds would be set up and a small percentage of the wages to be paid to villagers for work done under the project would be deposited into this trust fund. A matching amount would be deposited from the project budget, and this fund would be used to sustain village ecodevelopment activities even after project completion. Unfortunately, the final World Bank project document did not correctly or clearly reflect this agreement.

Conceptual Issues

The IEP was perhaps the most widely debated wildlife project ever undertaken in India. It was both supported and bitterly criticized from various standpoints. Broadly speaking, most opposition to ecodevelopment, especially to the IEP, came from two extremes of the ideological spectrum.

THE IDEOLOGICAL DIVIDE

On the one hand, the project was criticized by those who were fundamentally opposed to the system of protected areas as it existed, especially since it appeared to disempower local communities and prohibit or curtail their access to protected area resources. The assumption behind ecodevelopment that local communities were often a cause of protected area degradation was also unacceptable to them. According to these critics, people should not be treated as beneficiaries but as the legitimate owners or rights holders who have a preferred access to all protected area resources. In general, they argued that the strategy to reduce the dependency of local communities on protected areas was based on a mistaken assumption that traditional use of forest and other wilderness resources by the local communities was harmful to wildlife conservation. They claimed that there were no studies to prove this. On the contrary, it was maintained that tribals and other villagers had been living in harmony with forests and wildlife for many generations, and they were not the ones responsible for the loss of forest cover or destruction of wildlife. It was also argued that the local people were the best protectors of biodiversity and that they should be empowered to do so, rather than excluded, as the protected area system aimed to do.

At the other extreme were those who thought that human use of protected areas was disastrous for biodiversity conservation and that any compromise on this "fundamental truth" was unacceptable. Some were explicitly antagonistic to the idea of stakeholder participation or empowerment of local communities and seemed to feel that all human population should forthwith be removed from protected areas and protected area management designed strictly along "scientific" principles. Some of them also thought that any economic development around protected areas was undesirable, as it would encourage market forces around protected areas. They also thought ecodevelopment was diverting, to rural-development-type initiatives, staff-time and money that should rightly be focused on protected area management.

Obviously, there were others who held positions between these extremes. However, all of them had their own answers, though not always coherent or internally consistent ones, to the three questions of science, ethics, and strategy discussed earlier.

Specific Issues

Within this ideological divide, various specific issues were raised, some of which are discussed below.

Participation and Empowerment of Local Communities

The lead in attacking ecodevelopment from the perspective of local communities was taken by the Centre for Science and Environment (CSE), a well-known NGO based in Delhi. In their magazine, *Down to Earth,* they repeatedly published a full-page letter, shown in box 13.1.

Box 13.1. *An Open Letter to the World Bank President*

Dear Mr. James D. Wolfensohn,

"People do not want charity; they want opportunity. They do not want to be lectured to; they want to be listened to. They want partnerships," So you have said.

WE AGREE. But your own staff does not. The result is a Bank sponsored $68 million abomination like the Ecodevelopment Project which aims to protect India's wildlife.

Wildlife—our precious natural heritage—is facing destruction. Our forests are habitats of our people and not wilderness areas. Wildlife management, therefore, demands the active involvement of communities who live in these forests. The Ecodevelopment Project is fundamentally flawed as it is based on doling out charity and does little to make these communities equal stakeholders in the management of our sanctuaries.

"You participate in MY programmes!" That is how your staff defines people's participation.

Past experience shows that such an approach would further alienate people from their lands and turn them against wildlife. It would impoverish the people and the environment.

You have said, "When people are given a chance, the results are truly remarkable."

The Project does not give people a chance in hell!

(Issued in public interest by the Centre for Science and Environment)

The CSE also initiated many other letters and appeals. In one such, signed by the director of the CSE and various prominent persons, including a former cabinet minister, it was stated that "the first problem of forest-based people is not poverty but disempowerment by wildlife laws and programs and the erosion of their environmental right to use their habitat. By alienating the people, the transaction cost of management of parks will inevitably go up. And no amount of dole will help!" The letter contained the demand that the ecodevelopment project be "immediately withdrawn" because it "fails to address the present problems with the conservation policies and sees the people's involvement as only an appeasement strategy, rather than as a recognition of their rights and abilities" (CSE 1996).

A somewhat more moderate interpretation of these types of objections was that though protected areas were legitimate, they could not survive without involving the local communities in their protection. And, such an involvement was not possible unless the rights of local communities over protected areas were recognized. The moderates criticized ecodevelopment because it was still essentially an exclusionary model. It focused, as conventional conservation had done, on excluding people from protected areas rather than on integrating them. It departed from conventional conservation only insofar as this exclusion was not enforced coercively. Thus the net effect of ecodevelopment on biodiversity conservation was unlikely to be very different from earlier exclusionary policies.[10]

It was also argued (Kothari 1998) that one of the reasons for ecodevelopment initiatives remaining rather exclusionary was the inflexible nature of the Wild Life (Protection) Act, 1972, the principal legislation in India that facilitated wildlife conservation. The fact that the act permitted absolutely no resource use from national parks and only very restricted use from sanctuaries, implied that communities living inside such protected areas, or otherwise dependent on such areas for meeting their livelihood needs, had no incentive to protect them because resource extraction from them was prohibited as soon as the areas were gazetted as national parks or sanctuaries.

Kothari (1998) also pointed out three broad areas where ecodevelopment as a concept, and the IEP in particular, failed to assign requisite rights and responsibilities to local communities, thereby undermining the capacity of such communities and, in the process, the success of the project. First and foremost, he pointed out that as there was no provision within the ecodevelopment framework to recognize the rights of local communities over the resources they use, it seemed to condone the colonial takeover of community lands and the denial of tenurial security to local people. Second, local communities were not vested with the power to take part in and be responsible for decisions regarding the areas inhabited by them. Provisions within ecodevelopment for local communities to participate in protected area management planning and the implementation fell far short of any real devolution of power to these communities. Finally, communities were given very little responsibility with regard to handling ecodevelopment funds. The flow of funds of various ecodevelopment projects remained heavily biased in favor of the forest department, and this was reflective of a lack of genuine empowerment of local communities in ecodevelopment projects.[11]

There was also much debate on whether the planning process itself, as envisaged in ecodevelopment schemes and projects, was participatory enough. Many NGOs and activists, and even some forest officers, felt that it was not. Essentially, the IEP and FREEP ecodevelopment components were planned for in two stages. First, there was an indicative plan, which laid down the broad parameters of the project, developed an indicative budget and time frame, and described the methodology to be followed for building up the detailed, microlevel plans, and for implementing and monitoring the project. Once the project had been approved and initiated, the participatory, village-level planning process began.

Perhaps one reason for the dissatisfaction with the planning process was the wide disagreement on how much participation is enough. Also, there was a somewhat unreasonable expectation in the minds of a few that democracy would suddenly appear in societies, overnight, where traditionally the social structure had been very hierarchical and stratified. Critics were not satisfied unless the participatory process they saw in reality conformed to the ideal scenarios they read about in textbooks.

The fact was that, in much of Indian rural society, decision making had been far from democratic (see also Child and Dalal-Clayton this volume). There were distinct caste, gender, and age biases. Another significant barrier was the bureaucracy itself, which was a hierarchical and almost totally nonparticipatory system. To expect that people working in such systems would suddenly become totally democratic when they started dealing with the village communities was unrealistic. Ecodevelopment envisaged training and orientation for the protected area staff. It also envisaged selecting protected area managers who were more inclined to work in a participatory manner. However, it would be a long time before the expectations of many of the NGOs, especially the more radical ones, could be met on this count.

To try and minimize this problem, the project envisaged that at least one NGO, and where required more than one, would be involved in each protected area to facilitate the participatory process. However, even where consultations were managed by NGO representatives, the age-old and well-known divisions of caste, class, gender, and age still made real participation difficult. Besides, NGO representatives had their own biases, which also fed into the process.

In short, to make the process genuinely democratic and participatory was perhaps the greatest challenge of ecodevelopment. Clearly, there were no easy answers. All that could be claimed was that the ecodevelopment project had taken some big steps toward a participatory model of decision making, though there was still quite some distance to go.

Many, including S. Deb Roy, a former director of Wildlife Preservation, Government of India, and a senior wildlifer and forest officer, attacked the project from the opposite standpoint. In a letter to the Ministry of Environment and Forests (MEF) he stated: "I don't see any necessity of consultation with the so called 'stake holders' as the prescription in the management plan should and must follow only one course, that of ecological considerations and nothing else, as far as the core areas of the Tiger Reserves are concerned, which enjoy the status of National Park.... Though it is true that the protected areas are (directly or indirectly) adversely affected by biotic influences from near and far, yet no purpose will be served by consulting the people of the impact zone. On the other hand, scientific views are likely to be compromised in the process, which will dilute wildlife management interests" (Deb Roy 1994).

The response from the IIPA team went something like this: While people's participation and devolution of power are desirable ends in themselves, the process of invoking such devolution and participation has to be a gradual one. This is primarily because the Indian society continues to be stratified and hierarchical. It is also not prone, traditionally, to participatory decision making, particularly in terms of involving disadvantaged groups like the "lower castes," tribals, and women. A sudden devolution of power could lead to the strengthening of the hegemony of dominant groups in a village, such as members of the so-called upper castes and those who are financially well-off. Also, the forest department, like most bureaucracies, is itself hierarchical and has historically been nonparticipatory. For such structures to become truly democratic and participatory, a fair amount of time is needed. Such a process cannot and should not be rushed, if it is to be sustained and genuine. What ecodevelopment does is to initiate this process. It attempts to achieve higher levels of participation and greater levels of empowerment than have ever been achieved in wildlife management in India. However, it will be a long time, if ever, before "perfect" participation and total empowerment is achieved.

Besides, the debate on what human use should be allowed in protected areas and, indeed, whether there should be protected areas at all, is an important one that still has a long way to go before it runs out of steam. Admittedly, the ecodevelopment project has been designed within the context of the prevailing law and policy in India. When that law and policy changes, certainly all sorts of new possibilities will open up for ecodevelopment.

To those who felt that the funds being used for ecodevelopment should instead have gone toward strengthening protected area management, the response was that the objective was better management, and if they could show a more efficient way of doing this then, certainly, the funds should be used to promote that way. However, enforcement and regulation by themselves had proved to be ineffective in the past, and they were unlikely to succeed in the future.

The ecodevelopment project does not attempt to change social norms; it only tries to get as much space as possible for animals, plants, and human beings within the existing norms. Perhaps the important thing is to ensure that it neither inhibits the debate on social justice nor compromises the position of those who rightly believe that animals and plants also have rights (Singh 1999).

INDICATIVE PLANNING

As earlier mentioned, the Ministry of Environment and Forests proposed to plan for the IEP and FREEP in two phases: first, a somewhat quickly formulated indicative plan, on the basis of which the project would be approved; then a set of more detailed and participatory protected area and microlevel plans.

By persuading the World Bank to depart from its earlier practice of preplanning for every *paisa* or cent, the Government of India had succeeded in introducing the sort of flexibility into World Bank projects that had not been seen before. This also opened up the way for other projects and projects in other countries to demand and get similar flexibility.

The fact that all plans had to be developed in consultation with the local communities did not mean that there were no constraints on the local communities. The project plan prescribed certain guidelines that had to be followed in determining what types of activities could be supported by the project. The guidelines prescribed for income-generation activities insisted that all such activities must

- demonstrably reduce pressure on the protected area;
- be economically viable and sustainable;
- not be socially and morally oppressive;
- not be illegal.

Interesting examples of activities that violated one or more of these conditions emerged during the initial planning phase. For example, from one village there was a demand that ecodevelopment funds be used to provide street lighting on the main street. This proposal was objected to because it was not clear how providing streetlights would reduce pressures on the protected area. However, the village elders argued that many young villagers sneaked out at night to poach animals in the protected area. If the streets were lighted, they would be more easily spotted and prevented!

Similarly, in a high-altitude village, the villagers agreed to stop extracting resources from the protected area if the project helped them in cultivating and marketing *charas* (cannabis)!

Displacement of People Living Within Protected Areas

The Wild Life (Protection) Act, 1972, makes it incumbent on the government to remove all human populations living within national parks. It allows, by the amendment of 1991, some limited human habitation to continue within sanctuaries. Most of the protected areas selected under the IEP and FREEP were national parks, and this resulted in the apprehension that these externally funded projects would result in the displacement of hundreds of families, mostly tribal and nontribal poor, from protected areas. Though the World Bank had already announced that only voluntary relocation would be allowed under the project or in the project areas and would be determined on a family-by-family basis, because of a general distrust in certain quarters of government and World Bank pronouncements, accusations and counter-accusations continued.

The conflict became especially heated in the case of three protected areas: Simlipal National Park in Orissa, Nagarahole National Park in Karnataka, and Gir National Park in Gujarat. In Simlipal, which had been selected as one of the eight sites under the IEP, the protected area authorities decided to finish all the displacement prior to the start of the project, for they felt that once the conditionality of only voluntary displacement became applicable, it would be impossible to shift out many of the people living within. However, the World Bank took the view that this was a violation of the project conditions, even though the project had not been formally initiated. Consequently, Simlipal was dropped as a site under IEP. A battle was also brewing in Nagarahole. In an SOS e-mail (June 1995) sent across the world and also to the World Bank and other concerned agencies, Walter Fernandes of the Indian Social Institute raised the alarm: "Dear Friends, You are probably aware of the situation in Nagarahole, in Karnataka. Based on the report of Shekhar Singh, the World Bank is funding a tiger reserve there and they seem to be determined to displace the tribals. The local tribals have worked out an alternative to it in which they are demanding joint sanctuary area management. But the World Bank does not seem to be prepared to listen to them. The sanction is expected to be given in early July. Once it is given it is extremely difficult to change it. So it is very important to create public opinion against it immediately. The plan worked out by the local tribals does not need World Bank funding."[12]

Even "voluntary" relocation was not acceptable to some, for they argued that the forest department resorts to a process of slow strangulation of the populations living inside, in order to force them to volunteer (e.g., Cheria n.d.). Similar objections were raised by many other organizations, both for Nagarahole[13] and for Gir.

On the other hand, there were people protesting against the project because it was making the shifting out of people from protected areas too difficult. In a letter to the GEF, K. Ullas Karanth, an associate research zoologist with the Wildlife Conservation Society, New York, working in the state of Karnataka, India, expressed his concern and echoed the concern of many others because "the urgent issue of reducing human population densities ... inside the targeted parks through well planned and executed voluntary resettlement schemes is avoided by the GEF document. It is very likely that many ongoing resettlement schemes such as the one in Nagarahole Park, Karnataka will be shelved to comply with the GEF concerns. Under the GEF guidelines, resettlement cannot be based on group, or majority decision of the people who want to go out, but has to be on 'a case by case basis.' In reality this means that even if one individual does not want to go, no resettlement scheme can begin to operate" (Karanth 1996).

In actual fact, the major problem was that the relevant Indian laws did not provide for voluntary relocation, and the Wild Life (Protection) Act made it mandatory to remove all human populations from national parks (see also Gartlan this volume). The only loophole was that the act did not specify a time frame within which relocation had to be completed.[14]

The IIPA team expressed the view that the only practical way was to resettle those who were willing, and to do it so well that others would also soon become willing. Even if some elected not to shift, the pressures on the protected area would be significantly reduced because many had left.

In reality, this approach was not as difficult as it might sound, for in each of the protected areas selected there were invariably at least a few families who wanted to shift out. In order to induce the remainder to voluntarily move, these few, initially rehabilitated families would have to be so well provided for that their experience would tempt the rest.

Involvement of local NGOs, as monitors and contact agencies, was seen as the way of ensuring that people living inside the protected area were not forced to "volunteer" by the protected area managers making their life inside difficult. NGO involvement would also ensure that people were not officially shown to be willing to shift out even when they were not.

To prevent people from being forced out because of deprivations, it was also proposed to make those who opted to stay inside the protected area eligible for some of the benefits of ecodevelopment. Obviously these benefits would have to be in consonance with the requirements of a wildlife protected area. Also, the young people living inside a protected area could be helped to develop skills such that they would have much greater opportunities for employment and incomes outside the protected area. This would encourage at least the younger generation to seek a life outside, thereby gradually but surely solving the problem.

Apart from the high financial costs of such an approach, which were certainly justified, the main problem was the reaction of the host communities. In order to compensate the displaced people for all they had left behind once they relocated, they had often to be provided with a level of lifestyle that was higher than that of the host community or of people living outside the protected area. This could create social tension and encourage members of the host community to encroach into the protected area and then demand to be relocated.

This is not to suggest that all those living within protected areas were encroachers. Many of them, or their ancestors, had been brought and settled there by the government in order to assist in "working" the forests. Some of them, especially the tribals, probably lived there long before the forests were taken over by the government and certainly long before the protected area was established.

To minimize the host community problem, the project sought to ensure that ecodevelopment benefits flowed to the host communities also, so that even though they might not get as much as the relocated families, at least the gap between the two was lessened (Singh 1999).

Economic Development and Conservation

Another attack on ecodevelopment came from those who believed that one could not have both economic development and conservation. They believed that ecodevelopment promoted a market economy around protected areas, thereby encouraging consumerism, which was among the greatest threats to conservation. Some also demanded that communities living around protected areas be allowed only traditional, low-consumption lifestyles, like their forefathers, so that they were less of a threat.

One champion of such a viewpoint was Bittu Sahgal, a well-known environmentalist, member of the Project Tiger Steering Committee of the Government of India and editor of the popular magazine

Sanctuary Asia. He wrote repeatedly to the Ministry of Environment and Forests, criticizing the project. "The underlying premise of the IIPA seems to have been that it is both possible and desirable to integrate economic growth with the preservation of natural resources" (Sahgal 1994:para. 4). A similar point was made by him in comments, sent on 22 February 1994, on the draft GEF-ecodevelopment projects drawn up by IIPA.

S. Deb Roy, former director of Wildlife Preservation, Government of India, in a letter to the World Bank made a similar point when he said that "conservation and market economy forces are, in effect, invited through this plan. This will surely raise the level of consumption of renewable resources. This is exactly opposite of the underlying aim of this plan" (Deb Roy 1996).

Responding to the point about conservation and development, the IIPA team had the following to say:

> It is true that the integration of economic growth with conservation of the environment is a premise of the document. In my mind, it is a premise of ecodevelopment itself.... The alternative that seems to be suggested, appears to be very dangerous. If we were to work with the assumption that these two cannot be integrated then we are presenting, to the local communities and to the nation, an either/or choice: either economic growth or environmental conservation. Surely an ideology that offers only one of the two cannot be conducive to conservation. (Singh 1994b)

About "market forces," the response was the following:

> in actual fact the market economy and the consequent forces of consumerism have penetrated almost all parts of India, without the help of ecodevelopment. In these circumstances, all that ecodevelopment can attempt to do is to help provide the people living around protected areas with a legitimate way of earning their living, so that they can satisfy their market needs without adversely impacting on the protected area.
>
> However, even more significantly, the local communities living in and around protected areas must have the right to decide what type of a lifestyle they want to live. It is not for ecodevelopment planners, NGOs and officials, most of whom are themselves willing members of the consumerist society, to foreclose options for others. (qtd. in Singh 1999)

External Aid and the Debt Trap

There were also objections from those who opposed taking money from external sources, especially the World Bank, primarily because they felt that such lenders attached unfair conditions to loans and, in meeting these conditions, countries often compromised their own interests.

These groups and individuals also raised the specter of the debt trap and expressed the worry that when these loans were paid back, the wildlife sector would have little or no money left for its regular activities.

The objection to external funding, per se, was a larger ideological issue that could not be resolved in the context of any one specific project or sector. Insofar as the Government of India thought fit to accept external resources, the real questions were, should they be accepted for the wildlife sector and, specifically, for ecodevelopment?

Whatever the experiences of other countries, or other sectors and projects, the design of the ecodevelopment project was almost completely Indian. The task manager from the World Bank was especially sensitive to this point, and there was hardly any occasion when she asserted the views of the bank over those of the Indian planners. Besides, the view taken by the Indian planners was that Indians were not especially gullible or corruptible. Therefore, it was wrong to think that, even if the donors wanted to impose their own agenda, this would have been acceptable to the Indian government. Besides, in the overall size of the Indian budget, the inputs coming for these projects were too small to give any special leverage to the World Bank and the GEF, even if they wanted such leverage.

It was also factually wrong to think that the repayment of the loan component would be at the cost of future fund availability in the wildlife sector. Loans were not recovered from specific-sector allocations, but from the consolidated funds of the Government of India. Also, given the fact that the loan component came from the country-committed funds of the World Bank, if these funds were not channeled to the wildlife sector, they would most likely have gone to make dams or other infrastructure projects, which would have further depleted biodiversity.

The apprehension about the debt trap was also misplaced. In various responses to NGOs in India, who had expressed similar misgivings, the World Bank task manager, Jessica Mott, explained the situation as follows: "Total project costs (for the IEP) are estimated at US$67 million, of which US$20 million would come from a GEF grant, and US$28 million from an International Development Association (IDA) credit. IDA credits have soft terms which give them the equivalent of an 80 percent grant content to cover the US$48 million incremental portion of the costs" (Mott 1996a). In a subsequent letter to the Centre of Science and Environment, she went on to clarify: "In other words, the project financing involves the equivalent of US$42.4 million in foreign grants, and US$5.6 million as loan (which is less than 10% of the total project costs)" (Mott 1996b).

THE MAGNET SYNDROME

Considering that ecodevelopment strategies resulted invariably in investments around the protected area, some felt that such investments would encourage the immigration of poor people from elsewhere. The resultant increase in population around protected areas would heighten rather than lower pressures on the protected area (see Gartlan this volume).

As a planning exercise, experiences from other parts of the country and from other countries were reviewed (Singh 1995). It was recognized that immigration could be a major problem where surpluses are created because of large investments in infrastructure projects. Such projects created a demand for labor that could not be met locally, thereby encouraging immigration.

However, a study done as a part of the planning exercise for the IEP established that most often the areas around protected areas were much less economically developed than the rest of the region. Historically, forested areas had gotten less than their share of development inputs, and this itself, in many cases, was the reason why some wilderness survived there. Consequently, the investments that came through the ecodevelopment project would not even bring the protected area surroundings up to par with the larger region, let alone make them into magnets.

Another way of preventing the magnet syndrome from operating was to keep investments under ecodevelopment as low as possible and certainly of the sort that did not suddenly create a large number of jobs or wealth.

TRADE-OFFS VERSUS "ADDITIONALITIES"

The basic philosophy of ecodevelopment was that local communities, who negatively affected protected areas because of livelihood imperatives, should be helped to develop alternate, environmentally and socially sustainable sources of incomes and biomass, of their own choosing, *so that they could phase out their dependence on the protected area.* However, in order for this to happen, ecodevelopment inputs had to be seen as alternate, and not as additional, to protected area resources. The fact that most communities dependent on protected areas were desperately poor, and would remain so even after ecodevelopment, made their wanting to consider all inputs as additional both likely and understandable. Unfortunately, it also meant that the protected area would continue to be degraded.

One way in which this was to be prevented was, as discussed earlier, by entering into a MoU with villages. If the village went back on the MoU, it would not only lose the inputs but would be subject to action under the law, and detection and prosecution would be much more likely, since the protected area management had, in the meantime, been strengthened and the number of villages violating the law had been significantly reduced. Obviously this threat could work only where a small proportion of the villages violated their agreement. However, in the long run, the protected area could be saved only if the local communities had a stake in conserving it. For this purpose, it was not only important to minimize the deprivations they faced because of the protected area, and to involve them in its management, but also to ensure that they were the first and primary beneficiaries of the revenue forthcoming from the protected area, primarily through tourism. This would give them a further stake in the protected area and its maintenance.

The Sustainability of Financial Support

Very early in the design phase it became obvious that, given the different conditions in each of the selected protected areas and the varying constraints and advantages, it would be impossible to ensure that all the sites successfully completed ecodevelopment at the end of the five-year project period. However, the life cycle of the project was finite. Therefore, it was suggested by the IIPA team that, apart from the village-level trust funds described earlier, there should be a national-level trust fund where project funds could be deposited (see also Kiss this volume). This would, on the one hand, help ensure that expenditures were made according to real needs and frugally, not wastefully, as is often the case when there is a threat of unspent funds lapsing. A trust fund would allow flexible funding, enabling continued support even after project completion. Also, if there were savings, these could be used for other areas.

The amount of funds committed for the project had also become an issue because the initial budget prepared by the IIPA team was about a quarter of the budget that was finally agreed to between the World Bank and the Government of India. There was, therefore, the apprehension that either much of the funds would be wasted, or they would remain unutilized. This apprehension seems to have been well founded, for the latest assessment of the IEP reveals that the disbursement under the project remains below 25 percent after more than 60 percent of the time has passed. The midterm review goes on to say that it expected the Government of India to request "cancellation of approximately $12 million (25 percent of the original grant/credit) now" (IEP MTR 2000:2). It is unlikely that most of the remaining 50 percent would also be spent in the last part of the project.

Unfortunately, efforts to set up an ecodevelopment trust fund were frustrated by the Ministry of Environment and Forests itself, where the old guard had changed and the new setup did not appear to be either as supportive or as knowledgeable about ecodevelopment as their predecessors.

MISCELLANEOUS ISSUES

Some of the other issues that were raised from time to time included the following:

- *Scale of the project*. Should the project cover all villages that have an impact on the protected area, or only a few? If all villages were not covered, it was likely that the protected area resources freed by the villages covered by ecodevelopment would be "expropriated" by those not so covered, with no residual benefits to the protected area. Also, there could be resentment from both those covered and those not covered, for the former could resent this selective regulation of access to protected area resources, while the latter could resent being denied ecodevelopment benefits.
- *Integration between different government departments and sectors*. In India, as perhaps elsewhere, different government departments do not always work well together. An Indian activist has described the Indian government as an organization that has vertical loyalty but horizontal animosity! Yet, for the success of ecodevelopment, interdepartmental coordination was critical. Consequently, apart from the earlier described coordination mechanisms, it was also proposed that the project implementation team should have, on deputation, professionals from other relevant departments. These professionals could help solve many of the coordination problems with their erstwhile departmental colleagues.
- *Integration of resources*. For ecodevelopment to succeed, development inputs of various departments had to be applied in a focused and integrated manner so that they could do the most good. The age-old practice of thinly spreading available resources, and thereby achieving nothing anywhere, had to be avoided. Of course, this was a difficult proposition. As one politician remarked, while discussing the ecodevelopment approach, "politically, if one has enough resources to dig one well and provide a pump to pull out water, it makes sense to dig the well in one village and to provide the pump to another one. This ensures that both villages vote for you, partly in the hope that if you got re-elected you would provide each with the missing component and, partly, out of gratitude for what you had provided."

In Retrospect

FREEP ended in 2001 and IEP in 2002. Though it is too soon to judge the long-term and sustained impacts of the projects, some preliminary assessments can be made.

Based on the midterm reviews and various other reports on the progress of the two ecodevelopment projects, the following can be said:

- The budgets sanctioned for the project were far too large, and it is very unlikely that, by the end of the project, even 75 percent of the budget will be spent.
- One critical precondition of the project, that there be adequate staff posted for the implementation of the project and that all the staff be oriented and trained, was not fulfilled.
- Baseline data were not collected, and monitoring and evaluation systems were mostly not in position.
- There was poor supervision from the Government of India, primarily because of inadequate staff and facilities.
- The strengthening of protected area management was focused on more than the village-level ecodevelopment activities.
- Microlevel planning at the village level was mostly behind schedule.
- Though NGOs were identified and in some cases appointed to assist with the microlevel planning and village-level implementation, they mostly performed poorly. This might have been due partly to the selection of inappropriate NGOs and partly to unclear or inadequate supervision.
- Most of the research and professional inputs to be provided by institutions and consultants were not commissioned in time or not commissioned at all.
- In some protected areas there were problems regarding the disbursement of funds, which inhibited project implementation.
- In some other protected areas there were protests by local NGOs and community groups, mainly around the issue of rehabilitation, and this also inhibited the progress of the projects.
- The ecodevelopment concept and project details were not effectively disseminated to field staff and community institutions, despite provisions for translating the project document into local languages.

However, on the plus side what emerged was that in those protected areas, and parts of protected areas, where microlevel planning was done and activities started, there appeared to be significant achievements. Though the absence of baseline data and a systematic monitoring of the biological, socioeconomic, and institutional parameters did not allow for definitive comparisons in terms of before and after the project, there were many indications that the pressures on these parks had significantly lessened and that the economic situation of the villagers had improved or, at worst, not deteriorated as a result of ecodevelopment.

Most important, in those areas where ecodevelopment actually got started, there was a significant and positive change in the interaction between the protected area staff and the local communities, and in the perceptions of the local communities toward the protected area.

Discussion

What lessons can be learned from the IEP and FREEP? The Indian experience suggests that ecodevelopment as an approach has great possibilities, for wherever it was implemented with even a little diligence it seems to have had dramatic results, at least in the short term. However, it is possible that certain specific conditions in India have made this so. Perhaps the most important of these is that, mostly, rural communities are themselves inclined to conserve forests and wildlife. Many people of these communities are either vegetarians or do not eat wild animals, and most of them have cultural and religious links with forests and other wilderness areas and have strong traditions of protection and even worship of nature. Essentially, all they need is the real option to conserve, a situation where they can both meet their basic needs and also protect their forests and rivers. Ecodevelopment, where it is properly planned and implemented, provides them this real option.

Second, the forest department in India, as a rule, is committed to safeguard the protected areas. In fact, the most common criticism against it is that it is more concerned about wildlife than it is about people! Therefore, the strategy of both strengthening protected area management and providing alternative resources to the local people seems to be the right approach. Without ecodevelopment, the ability and willingness of protected area managers to protect the area is not adequate to offset the resolve, born out of desperation and real need, of the communities to get access to protected area resources. In those few cases where protected area managers, without access to ecodevelopment inputs, try to stop the access of local people, the usual result is violent clashes, which are not only by themselves undesirable but also have political ramifications that are rarely in the interest of conservation. Ecodevelopment narrows the gap between the desperation of the local people to access the resources and the capacity of the protected area managers to regulate and control such access.

And, finally, India has a vibrant democracy with strong and articulate groups espousing all possible perspectives and viewpoints. Though this might slow down the process of decision making and occasionally disrupt implementation, it ensures that every aspect of ecodevelopment is closely monitored and errors highlighted.

The major weakness has been the relative apathy and disinterest of the higher echelons of bureaucracy, especially those who do not have to face the day-to-day conflicts involved in protected area management. There seems to have been relatively poor support from the centralized bureaucracies at the central level and, especially in the matter of flow of funds and coordination, from the various departments in the state governments. There has also been too little appreciation of the value of research and monitoring.

At the field level, the major fault appears to be the relatively weak effort at developing the individual, institutional, and systemic capacities needed to implement ecodevelopment (cf. McShane and Newby this volume). In those instances where adequate and appropriate staff was provided right from the start, as in the case of Kalakad Mundunthurai Tiger Reserve, the results have been very good. However, even there, as elsewhere, the orientation and training of field staff and the building up of suitable institutions could have been better. This is especially true of the involvement of NGOs. Despite there being a large number of very committed NGOs in India, it appears that most of the NGOs engaged to assist in the ecodevelopment projects did not perform up to expectations. There are many reasons for this, and one could be the fact that as the two ecodevelopment projects became very controversial, many NGOs did not want to be associated with them, as they themselves did not want to become controversial. However, there are other, more fundamental reasons.

What is the way forward? Clearly the Government of India must make up its mind whether it wants to go forward with ecodevelopment and, if it so decides, then it must develop its resolve and capacity to support the program vigorously. Innovative institutional and procedural mechanisms must be developed to resolve some of the more important constraints to the success of ecodevelopment that have emerged through the lessons learned so far. It must also be recognized that ecodevelopment is a new approach in many ways and especially in the way local communities and protection are to be looked at. In order to make it work, the implementers, mainly forest officers and NGOs, have to be reoriented and trained. In fact, even the planning process must not be initiated until officers and NGOs with the correct perspectives have been identified and given the skills and knowledge that they need to plan for and implement such a program.

A national debate must be initiated, involving all relevant sectors and departments of the government, both at the central and state levels, as well as NGOs and other concerned and interested persons, including local community groups, to ponder the experience of ecodevelopment and to discuss and decide what is the best way to conserve biodiversity. Alternative strategies must also be critically examined and a final consensus, though not necessarily unanimity, be arrived at.

Acknowledgment

The authors are particularly grateful to Jessica Mott of the World Bank, formerly the task manager of the India Ecodevelopment Project, for her detailed and very useful comments on an earlier draft.

Endnotes

1. The development process in India is steered by five-year plans based somewhat on the Soviet model of planning.

2. For a more detailed discussion of these questions, see Singh et al. 2000, ch. 3.

3. Under the Wild Life (Protection) Act, sanctuaries have a lower level of protection, and various human-use activities, including grazing, are allowed there, while in national parks no human-use activity is allowed—somewhat as with the strict nature reserves of the IUCN categorization.

4. For a more detailed discussion of this point, see Singh et al. 2000, ch. 5.

5. Under the Environment (Protection) Act of 1986, activities in any specified area can be restricted or banned subject to the clearance of an authority appointed for the purpose.

6. Of course, the design of the ecodevelopment projects was significantly influenced by past experiences within the country. Perhaps one of the earliest attempts to use an ecodevelopment approach for conserving a protected area was in the Kanha National Park, in the state of Madhya Pradesh, where, in the late 1960s, H. S. Panwar, the then director of the Kanha National Park, had moved villagers living in the heart of the PA to the periphery or, in some cases, to outside the PA, and also made an effort to ensure that, as a consequence, they did not suffer economically.

7. Dr. W. A. (Alan) Rodgers was hired for the purpose. Dr. Rodgers had extensive experience in India and had spent over ten years at the Wildlife Institute of India as an FAO consultant. He had been the lead author of the very influential report *Planning for a Protected Area Network in India,* and had contributed immensely to developing the capacity of Indian PA managers. He was, therefore, the happiest of choices.

8. The Government of India initiated Project Tiger in 1972, and tiger reserves under this project were provided additional financial support by the central government. Consequently, they usually had better management capacities than other PAs.

9. Of the eight areas selected, only one was selected because of political considerations extraneous to the project. Though this area was not unsuitable for ecodevelopment, it might ordinarily not have featured in the list of eight.

10. See, for example, Kothari 1998.

11. Though this was a valid criticism of the project document, in practice all ecodevelopment investment funds started flowing through EDC accounts in all but one (Buxa) of the IEP protected areas and one (Kalakad) of the two FREEP protected areas within a year of project initiation.

12. The author of the report, and one of the authors of the present paper, subsequently wrote to Walter Fernandes asking him to specify where and how the "report" suggested displacement of tribals from Nagarahole. In a reply sent on 1 October 1996, Walter Fernandes stated:

> Dear Shekhar, It is a long time since I received your letter saying that I had stated in my email that based on your report the World Bank has worked out an ecodevelopment project which will displace people. You wanted to know where in your report you had stated that people should be displaced.
>
> As you have quoted correctly from my message, I have only said that the project of the World Bank which is based on your report, will displace people. Nowhere have I stated that your report suggests or in any way encourages displacement of the tribals.
>
> With best wishes I remain sincerely yours, Walter Fernandes.

13. The World Bank Inspection Panel looked into the charges of inappropriate rehabilitation practices in Nagarahole National Park. Though the inspection panel's report recommended further investigation, the World Bank management did not agree with this. Serious doubts were raised, in various quarters, about the methodology and recommendations of the inspection panel.

14. A subsequent order of the Supreme Court of India made it incumbent on all state governments to clear human habitation from national parks within a period of one year. However, this order was never fully implemented.

References

Cheria, A. n.d. Ecodevelopment in Nagarahole, India—A critique. 26 VGP Shanthi Nagar, Pallikarnai, Madras, India. Mimeograph.

CSE. 1996. Circular letter from the Centre for Science and Environment. 3 August.

Deb Roy, S. 1994. Letter to the director, Project Tiger, Ministry of Environment and Forests, Government of India. 12 September.

Deb Roy, S. 1996. Letter to the World Bank. 17 June.

IEP MTR. 2000. India-Ecodevelopment Project (Credit 2916-In/GEF TFO28479-IN), Mid-term Review. 1 June.

Karanth, K. Ullas. 1996. Letter to Dr. Hemantha Mishra, GEF, Washington, D.C., 20 July.

Kothari, A. 1998. Ecodevelopment and joint management of protected areas: Legal and policy implications. Paper presented at the workshop on Ecodocumentation, Wildlife Institute of India, November, Dehradun. Mimeograph.

Kothari, A., P. Pande, S. Singh, and D. Variava. 1989. *Management of National Parks and Sanctuaries in India: A Status Report.* New Delhi: Indian Institute of Public Administration.

Mott, J. 1996a. E-mail to various NGOs and the Government of India. 10 July.

Mott, J. 1996b. Letter to the Centre for Science and Environment. 23 July.

Rodgers, W. A. and H. S. Panwar. 1987. *Planning for a Protected Area Network in India*. Dehradun: Wildlife Institute of India.

Sahgal, B. 1994. Letter to Ministry of Environment and Forests, Government of India. February.

Singh, S. 1994a. Tiger conservation through ecodevelopment—A conceptual framework. Paper presented at the first meeting of Tiger Range States to set up a Global Tiger Forum, March, New Delhi.

Singh, S. 1994b. Letter to the secretary, Ministry of Environment and Forests, Government of India, July.

Singh, S. 1995. Integrated conservation development projects (ecodevelopment) for biodiversity conservation: The Asia Pacific experience. Washington, D.C.: Economic Development Institute, The World Bank. Mimeograph.

Singh, S. 1999. Ecodevelopment: Many questions, some answers. Paper presented at the UNESCO Meeting for South-South Cooperation, 19–23 May, at Xalappa, Mexico.

Singh, S., V. Sankaran, H. Mander, and S. Worah. 2000. *Strengthening Conservation Cultures: Local Communities and Biodiversity Conservation.* Paris: UNESCO.

14

Conservation Landscapes: Whose Landscapes? Whose Trade-Offs?

Stewart Maginnis, William Jackson, and Nigel Dudley

Introduction

A common approach to dealing with a complex problem is to break it down into a suite of small, easy-to-understand elements; this approach is called the problem-isolation paradigm. In theory, each element of a problem can be examined, tested, and solved as necessary so that when the constituent parts are reassembled—be it in a computer program or assembly line—the whole unit functions seamlessly. While this approach has been central to the philosophy of industrial economies, being used to help companies optimize productivity and nations collaborate successfully on highly complex projects such as the international space station, its application is really only useful where the relationships between the component parts are discrete, limited, and well understood.

Over the past few centuries land use has been increasingly shaped by an approach akin to that of the problem-isolation paradigm. Food and fiber production were dealt with via intensification of agricultural activities on the best land; rivers were diverted, wetlands drained, and forests cleared. Timber production followed either the intensive agricultural production model via industrial plantations, or intensive management of natural resources through the manipulation and simplification of those natural stands that were particularly rich in timber species.

Wildlife was corralled into "protected areas," which were chosen for a number of reasons: sometimes on the basis of biological importance, on other occasions because the land had previously been protected as a hunting reserve, or sometimes simply because the land was of no interest to anyone with power or influence. A crude, but not entirely inaccurate, parody of this model of land-use compartmentalization is that wildlife is removed from agricultural areas and (poor) people removed from protected areas.

One of the problems of determining land use according to the problem-isolation paradigm is that the biophysical and social relationships that shape both "natural" and "modified" ecosystems cannot be neatly divided into discrete components and are subject to endless permutations, many of which remain poorly understood (see also Brown this volume). Conventional top-down land-use planning, which supposedly addresses these layers of complexities in a systematic manner, usually ends up reinforcing the social and biological compartmentalization of land use through strict zoning.

Charland (1996:7–8) illustrates the problem of compartmentalized land-use planning by criticizing the simplistic assumption that biodiversity impacts from the destruction of wetland areas can be offset by the preservation or restoration of small areas of water. He observes that such an approach results in "a small patch of water, reeds and grass in a sea of concrete where a few ducks may splash down for the occasional drink" and is brought about because "policymakers and developers are so adept at isolating a few characteristics of wetlands from the whole picture." This paper will attempt to address what exactly constitutes "the whole," who determines it, and what all this means for integrated conservation and development projects (ICDPs).

The Spatial Context of Conservation

Modern conservation also has its roots in the problem-isolation paradigm. The creation of Yellowstone National Park illustrates not only how the problem-isolation paradigm was employed, but also why it has proved such an inherently unsatisfactory approach for conservation. When the national park was established in 1872, it amounted to a square block of some 810,000 hectares (2 million acres), an area of sufficient extent and configuration that it was assumed that "wilderness" could be preserved forever. Yet it quickly became apparent that, despite its relatively large size, the national park was not large enough to preserve the wildlife it was created to protect. The Greater Yellowstone Ecosystem was defined by the range of the grizzly bear, which now extends to between 7.3 million hectares (18 million acres) and 10.5 million hectares (26 million acres) (see also Glick and Freese this volume). Paradoxically, the surrounding private lands, which are essential to the integrity of the core "wilderness" area, are now experiencing some of the fastest population growth in the United States (Stohlgren 1996, qtd. in Miller and Hobbs 2002:333), and conservation remains a highly controversial issue in the region.

The fact that a substantial proportion of species and healthy ecological processes cannot be maintained indefinitely in isolated areas, even if these are as large as 1 million hectares, has resulted in a "scaling up" of conservation thinking and ambitions. It is now recognized that most protected areas remain viable only if they exist within a mosaic that contains at least some other land uses that are broadly compatible with biodiversity conservation. The fact that the Convention on Biological Diversity directs governments to adopt an "ecosystem approach" to conservation is a tacit rejection of a "problem-isolation" approach to conservation.

Just as biodiversity cannot be contained within the confines of a protected area, neither people (nor economic development) can easily be kept out of areas required for the conservation of biodiversity (areas that may, but usually do not, coincide with the boundaries of national parks). From the rural development perspective, the problem-isolation paradigm has also proved a particularly unsatisfactory model to shape land use, not least because the identification of the "problem" has tended to be top-down and centralized. Government bureaucrats choose the location for a plantation to resolve a timber shortage problem, a hydroelectric scheme to resolve an energy problem, and an irrigation scheme to solve a foreign exchange problem, with little if any consideration for the needs of the communities that will be most affected. Paradoxically, the problem-isolation paradigm fails rural development for the opposite reason that it fails conservation: namely, the level of resolution that government planners or private investors prefer is often too coarse-grained to take into account the needs, concerns, and aspirations of local communities.

If the above critique of a problem-isolation approach to land use is accepted, then it raises a real challenge for conservationists, especially those such as IUCN–The World Conservation Union, whose philosophy revolves around the integration of social issues into conservation action. What are the attributes of a land-use-planning system that can simultaneously deal with the large-scale requirements of conservation and the medium-scale requirements of economic development and still remain cognizant and supportive of local people's needs, which must often be dealt with at much smaller human scales? We propose that, in order to identify workable approaches capable of reconciling both top-down and bottom-up needs, conservationists first must gain a fuller understanding of the spatial context, or "landscapes," within which key stakeholders define their aspirations and make decisions.

Power and Patronage as the Determinants of Land Use

If the problem-isolation paradigm is problematic as a land-use-planning *framework,* it has in many cases also been fatally flawed as a *process*. Throughout the world, there has been an almost unvarying tendency for a small elite group of powerful people to control decisions relating to the competing demands on land. In its crudest manifestation, this has simply meant that the strongest individuals or groups have grabbed land, resources, and use rights—a situation that still exists to varying degrees both in countries with a poorly developed democratic framework and a mature democratic one (see also Child and Dalal-Clayton, Gartlan, Glick and Freese, Singh and Sharma, Tongson and Dino, all this volume).

More conventionally, in most societies recent land-use and landscape patterns have been shaped and modified by a small cadre of specialists acting at the behest of political and economic elites. While varying degrees of consultation have permitted some level of popular participation, the priorities of central or regional governments, or those best placed to actively lobby them, tend to win out over local concerns. Although democracy generally increases the number of voices that can influence decision making, active participation is still often very limited and the real decisions remain in the hands of a few.

An uncomfortable fact for most conservationists is that the establishment of many protected areas has followed a similar pattern: dispossession of the many to suit the desires of a few. Several of the national parks in India, for example, were established on the whim of Lady Curzon, while as late as 1940, in the United States, 4,000 people were forced to leave their homes in order to establish the Great Smokey Mountains National Park. Even the establishment of Yellowstone National Park, for many the ultimate symbol of how natural areas should be protected, resulted in the killing of the 300 Shoshone residents in conflict with the U.S. Army after it was decreed that people should not live within the boundaries of the park (Haines 1977). The pattern of top-down planning has been repeated throughout much of the world, and the sight of displaced people eking out a living on the edge of national parks is still unfortunately a common sight in some countries.

Annexation of hunting reserves by nobility with sanctions for poaching

↓

Colonially imposed national parks and wildlife reserves with sanctions for poaching and/or encroachment

↓

Postindependence presidential decree—often in response to lobbying from the international community and/or urban elites

↓

Establishment of "softer" protected areas—typically with centrally determined limitations on land-use activities (e.g., UNESCO MAB reserves and IUCN Category V protected areas).

Figure 14.1.
Evolution of approaches to protected area establishment.

At the present time, while decisions are in most cases made in a subtler manner, major protected areas are still often established autocratically, under pressure from the international NGO community or domestic urban elites, neither of which generally understands much about the impacts of such land annexation on local communities. For example, the Government of Botswana is currently trying to "persuade" the last remaining fifty San people to abandon their traditional way of life, leave the 52,000-square-kilometer Kalahari Game Reserve, and move into resettlement camps. While no force has been used, special hunting permits have been revoked, and water tanks placed to help the Kalahari San have been removed.

In response to criticism of what has been perceived as "green imperialism," people-centered approaches have emerged, including the UNESCO Man and the Biosphere Reserve program and the eclectic collection of protected landscapes listed under IUCN's Category V protected areas, a categorization that specifically recognizes the importance of "living landscapes" and seascapes in conservation strategies. Nevertheless, while the managers of these areas often make a genuine effort to reflect people's needs, the establishment of many such parks can still be seen, as illustrated in figure 14.1, as part of an evolutionary continuum of conservation decision making primarily by a politically powerful minority ignoring local residents.

The preceding discussion is not intended to question the value of biodiversity conservation or protected areas, or to imply that local needs and values must always take precedence over wider interests or global public goods. However, given that protected areas have in some cases resulted in local people losing access to resources, we suggest that conservationists need to consider whether existing conservation approaches can be adapted and better applied to produce solutions that are beneficial for both biodiversity conservation and human well-being. Indeed, attempting to preserve biodiversity without taking account of other human priorities is increasingly difficult as human populations expand. A growing number of conservation professionals challenge the "zero sum" thinking that pits humans against nature, pointing out that by necessity conservation has to be a social and political process. This assumes that cultural landscapes can often retain importance for biodiversity. For example, Blumler (1998) in a review of biodiversity and Near East landscapes, points out that anthropogenic change does not necessarily cause irreversible degradation, and that encouraging flexible and varying land use might be a more appropriate conservation response than trying to eliminate any form of human disturbance.

These insights have resulted in three key responses. First, conservation organizations have been switching an increasing amount of their efforts to conservation in areas that are being managed primarily for other purposes: The involvement of the World Wildlife Fund (WWF) in certification of good forest management and IUCN's interest in the role of organic agriculture in biodiversity conservation are two examples among many. Second, there has been an increasing recognition of the importance of reflecting people's needs and desires within conservation projects—even if the extent to which this has been achieved remains the subject of debate. And lastly, there is increasing recognition of the importance of spatial context in land-use decision making, including in decisions relating to biodiversity conservation. Our conclusions so far have broadly mirrored those reached by Zimmerer and Young (1998), who argue that effective conservation programs will be based on a pragmatic approach, utilizing a larger suite of conservation and development tools and focusing on broader landscapes that allow for and incorporate human settlement.

Integrated Conservation and Development Projects—The Missing Link or Missing the Point?

So far we have critiqued the application of the problem-isolation paradigm to land use in general, and conservation in particular, both as a

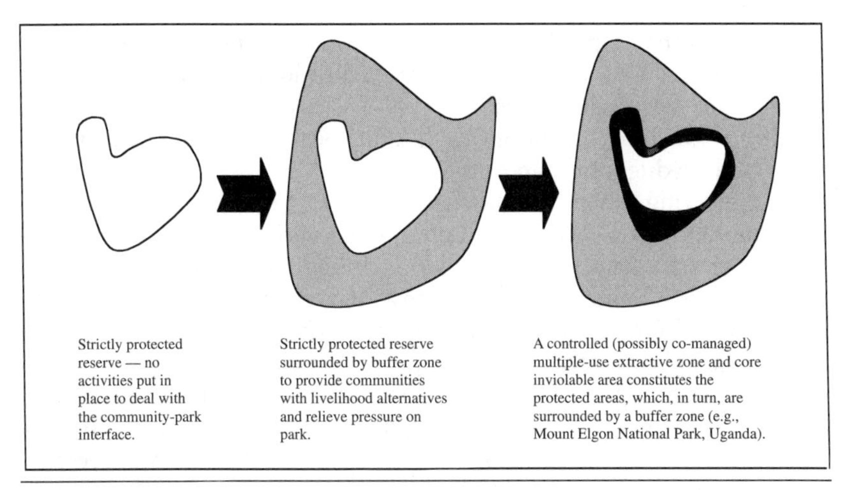

Figure 14.2.
The evolution of the ICDP approach.

framework (because it attempts to compartmentalize that which cannot easily be subdivided into discrete elements) and as a process (because the definition of the problem and the identification of the solution have tended to be the preserve of an elite few with their own vested interests). ICDPs have emerged as one way of putting these wider issues into a coherent package linked to development concerns. Although originally conceived to some extent as a way of buffering protected areas by persuading local communities to practice more sustainable forms of resource management in surrounding areas, ICDPs quickly developed a wider set of aims. They have provided a significant conceptual advance in conservation thinking and marked a genuine response to the problems created by the top-down planning, establishment, and management of protected areas. In particular, they have helped conservationists to recognize that conservation is as much a social and political process as a biological one. In recent years, ICDPs around protected areas have evolved from being a means of buffering the core protected area from the impact of people, to a means of helping people live with the impact of the park, as illustrated in figure 14.2.

Despite being generally regarded as a positive step forward, ICDPs have also attracted critics. A proportion have failed in achieving their aims, which may have attracted particular attention because of the exaggerated claims that had been made by some proponents. As Robinson and Redford (this volume) point out, we remain unsure exactly what ICDPs are or whose agenda they are pushing. Are they primarily a conservation vehicle, or a development vehicle, or promoting both in equal measure? Many development agencies regard ICDPs more as a soft conservation tool than a mainstream development approach, while conservationists such as Terborgh (1999) assert that ICDPs are actually counterproductive, on the grounds that local development increases rather than reduces human pressure.

In answering the question posed at the start of this section, we contend that while ICDPs are a valuable conservation tool in a landscape-focused pragmatic approach, it is misguided to promote them as the only approach. Some proponents may inadvertently have done ICDPs a disservice by overplaying the extent to which they can be vehicles for "win-win" conservation-development outcomes (i.e., good conservation in an area that also allows good economic and social development). While this may be possible in a few cases, results are generally likely to involve a large measure of compromise (see Brown this volume). Perhaps because ICDPs were initially promoted as win-win solutions, any stepping back from this rather utopian ideal is sometimes regarded as a failure. ICDPs are often portrayed as the vehicle to address all development and conservation challenges within a particular landscape, whereas in reality they address a specific set of development issues that pertain to conservation of a particular area—usually a protected area.

Casting ICDPs in a framework in which they are designed to address a specific set of challenges would help to reduce the unrealistic expectation that they are the purveyor of win-win outcomes. Reality dictates that one cannot have one's cake and eat it too—hard choices have to be made, especially when it comes to land use. Given this, we believe that the concept of multiple use makes sense only when firmly embedded within the appropriate spatial context—the larger the area, the more likely that one can accommodate a broader range of specific activities or multiple uses. Alternatively, the smaller the area, the more likely that one specific activity, or land use, will dominate. This means that at the site level *trade-offs* have to be made (cf. Brown this volume).

For example, an area of 500 hectares of secondary forest within a buffer zone of a national park could be left undisturbed in perpetuity—in this case the *trade-off* is foregoing future timber production for the sake of an eventual return of near natural levels of biodiversity. Alternatively, the forest could be cleared and replanted with fast-growing exotic species, which may return high yields of timber but further reduce local levels of biodiversity. A third option might be that trees in the secondary forest are selectively harvested and the forest sympathetically managed under a continuous-cover silvicultural system; the result would likely be more limited timber production, but biodiversity would be maintained or even increased. Judging the acceptability of such trade-offs is possible only if we also know what is happening in the rest of the landscape.

We believe that trade-offs are in themselves no bad thing—in fact, conservation depends on trade-offs, for if society was not prepared to trade one value against another it would be impossible to secure the particular form of specialized land use called a protected area. So we propose that rather than pursuing an elusive win-win outcome, ICDPs should intentionally promote trade-offs that will help maintain and improve livelihoods of local communities without undermining the integrity of a broader conservation goal. To be explicit, this might mean sacrificing the biodiversity of some areas to support development, while sacrificing some of the economic potential of other areas to support conservation. For example, in Mount Elgon National Park, the Ugandan Wildlife Authority has negotiated co-management and access rights to certain areas of the park with the local communities. Undoubtedly this will have some impact on biodiversity within the co-managed areas, yet this trade-off helps to secure the integrity of the core area (see also Tongson and Dino this volume). From the communities' point of view, trading off access to the remote mountain area has allowed them to gather forest products legally from the co-managed zone. Therefore, while Terborgh (1999) correctly asserts that in many cases local development around protected areas increases human pressure, we believe that he is incorrect when he says that local development is automatically counterproductive to conservation aims. Linking local development to conservation is likely to be the most pragmatic way to ensure the long-term conservation objectives of many protected areas. The likely alternative, local development not linked with the aims of protected areas, is likely to be worse.

While recognizing the value of trade-offs, there is one major caveat not yet mentioned: trade-offs are acceptable only if they are balanced at an appropriate scale. Attempting to balance trade-offs at a relatively small scale—for example, to expect one small area of forest to fulfill all possible functions ranging from conservation of the most sensitive species to timber production—precludes the benefits of site specialization and often results in suboptimal conservation and human development results. Conversely, allowing specialization of land uses over large areas can produce uniform, sterile landscapes. For example, a 2,000-hectare forest plantation certified to the Forest Stewardship Council (FSC) standards of good management is likely to be an asset to a landscape; however, this does not mean that the landscape will be improved further by establishing block upon block of similarly certified plantations on adjacent sites. In most cases land-use trade-offs should be balanced at the landscape level. It is this broader process that constitutes Zimmerer and Young's (1998) landscape-level pragmatism.

Understanding Landscapes

At the outset of this paper, the application of the problem-isolation paradigm to land-use planning was critiqued because its compartmentalized approach failed to allow for the importance of maintaining the integrity of "the whole." In the discussion on ICDPs, we concluded that a compartmentalized approach to land-use planning should not be rejected in principle; rather, in practice it has often failed to acknowledge the trade-offs inherent in land-use decisions and failed to provide a mechanism to ensure that those trade-offs are subsequently balanced at an appropriate scale. We introduced the concept of the "landscape" as the place where the integrity of "the whole" can be maintained. However, what is a landscape? And can it be defined precisely enough to provide a workable basis for land-use decisions?

There are a number of different perceptions of landscape. Geographers use the term quite precisely to describe a particular area based around enduring features, while ecologists have developed an approach to ecosystem analysis based on "landscape ecology." Yet common usage depends more on cultural attributes and thus reflects the history of the land. In 1836, the American essayist Ralph Waldo Emerson wrote the following in his book *Nature:* "The charming landscape which I saw this morning, is made up of some 20–30 farms. Miller owns this field, Locke that and Manning the wood beyond. But none of them own the landscape. There is a property in the horizon, which no man has, but he whose eye can integrate all the parts. This is the best part of these men's farms, yet to this their land deeds give them no title."

He encapsulated a view of landscape that refers to a whole greater than the individual parts. Along with geographical and ecological conceptions, then, the wider definition reflects a combination of land uses, each of which has its own specialized function, yet whose unique combination delivers an additional set of functions, both utilitarian and spiritual, including what economists call public goods.

The suite of ecological, geographical, and cultural perspectives is helpful in understanding the complexity of the landscape concept, but it fails to describe what a landscape actually is, where its boundaries lie, and who sets those boundaries. In practice, the definition of a landscape lies largely in the eye of the beholder. This is not to deny that large groups of stakeholders can agree on what broadly constitutes "their landscape," a definition often based, at least in part, on the geographers' "enduring features." Nevertheless, as Grieder and Gurkovich (1994) point out, landscapes reflect the self-definitions of the people within a particular cultural context—and that applies as much to a Western conservationist as it does to a Kalahari San hunter. All landscapes are social constructs, and therefore to try and set rigid common boundaries is impossible. The real value of the landscape concept is that it allows one group of stakeholders to perceive how another group of stakeholders regards the space within which they live and work (their particular landscape), and this provides a context for understanding the landscape functions that they are most concerned about. In other words, it provides a framework for helping groups of stakeholders agree on how to balance the trade-offs inherent in land use.

Why use such a diffuse term? We favor the use of the term *landscape* precisely because it comes with a certain amount of its own scientific and cultural momentum. When we talk about landscapes, we are implicitly suggesting that the area concerned is likely to be both physically and socially heterogeneous and assume that it has an overall quality more complex than a simple summing of its various parts. For these reasons, for the purposes of conservation planning that attempts to take account of development issues, we define a landscape as "a contiguous area, intermediate in size between an 'ecoregion' and a 'site,' with a specific set of ecological, cultural, and socioeconomic characteristics distinct from its neighbors."

Within ICDPs, two types of landscape are likely to exist: those determined by conservationists and those determined by local people. A *conservation landscape* can be defined by using ecological factors, such as the area required to maintain viable populations of key species and/or to maintain healthy ecological processes. A conservation landscape is likely to be a relatively large area, particularly if large herbivores or carnivores are found there. However, landscapes defined by conservationists often overlap with several *livelihood/cultural landscapes* that conform to the realities of other key stakeholders, whose engagement and consensus is needed for practical conservation implementation. Such landscapes can be very limited in size, being composed of farms and woodlands within a valley bottom, or very extensive, such as the annual range used by nomadic pastoralists, or even defined on the basis of spirituality, such as the Uluru region in central Australia. Livelihood/cultural landscapes are often, but not invariably, smaller in area than conservation landscapes.

At first glance, the deconstruction of landscapes based on different stakeholders' needs, priorities, or personal beliefs might appear an interesting intellectual exercise of little practical application. However, our own experience suggests that developing a good understanding of multiple, overlapping landscapes offers real opportunities to avoid some of the more conventional pitfalls of top-down conservation. Crucially, a *landscape approach* provides a broader context within which ICDPs can be designed and implemented, allowing practitioners not only to identify the "best bet" options for balancing trade-offs, but also to scale back expectations as to what the ICDP can realistically be expected to deliver on the ground.

Making Conservation Work Within a Multiple-Landscape Context

From a conservation perspective, conservation landscapes exist within continuous and interrelated ecoregions: much larger units of land or water that contain distinct assemblages of natural communities and share similar environmental conditions. Over the last few years, large conservation NGOs have been increasingly working on this larger "ecoregional" scale, often developing priorities and conservation visions for an entire ecoregion (Redford et al. 2003). However, although the vision for an ecoregion and its component conservation landscapes represents the conservationists' best option for conservation, it is often less meaningful to other stakeholders, and it is possible that some elements of a conservation vision could be incompatible with broader social and economic aspects of sustainable development (see also Franks and Blomley this volume). For example, the need to maintain and enhance the integrity of a key migratory pathway may conflict with the need to develop better infrastructure to help isolated communities get agricultural produce to markets. How can decisions be made about which trade-offs to make? Are balanced trade-offs even possible in such circumstances, and if they are not, who chooses the winners and losers? These questions are or should be at the heart of ICDP planning and landscape approaches to conservation (see also Sayer and Wells, Franks and Blomley, and Brown, all this volume).

Over time, practitioners have started to qualify the "win-win" optimism of the early ICDP proponents and look at realistic options for integrating conservation and development. For example, in the Alternative to Slash and Burn Program of the International Center for Research into Agroforestry (ICRAF), Tomich et al. (2001) outline a process of trade-off assessment among six principal stakeholder groups (indigenous peoples, small farmers, large estates, midsize absentee landowners, public institutions, and the international community) with respect to various land-use systems in Sumatra, Indonesia. One lesson is worth stressing: trade-offs occur independent of whether a formal planning process exists or not. For example, small farmers will make their trade-offs with respect to forest clearance based on returns to land and labor. If there is no action to help direct these trade-offs—say, by changing the incentive basis or fairly compensating farmers for the provision of other goods and services—then forest loss will continue. This reinforces our conclusion that trade-offs, if they are to be balanced in a meaningful way, cannot be separated from the spatial context within which they take place. Figure 14.3 is a preliminary matrix that illustrates the extent to which goods and services can be traded off against each other at different scales without undermining either biodiversity or other aspects of sustainable development.

In advocating an approach that is based on balancing trade-offs rather than seeking out win-wins, we are not suggesting that trade-offs are a panacea or that win-win opportunities do not exist. There are opportunities for taking economically feasible actions that will benefit both people's livelihoods and nature conservation. However, the successful combination of both is much less common than frequently suggested, and trade-offs can be balanced within a landscape context only under certain situations. In some cases land-use conflicts cannot be resolved at the landscape level, either because key groups hold extremely polarized opinions or, more likely, because the key factors that influence land-use practices are outside the control of local players. For example, commodity investment decisions taken in European boardrooms that impact on landscapes in East Kalimantan, Indonesia, cannot be addressed at a local level.

Where intractable situations exist at landscape and ecoregional scales, their resolution can come about only by virtue of action at a national, regional, or global scale—for example, shareholders in an investment bank demanding higher social and environmental impact assessments on all future investments involving primary commodities. Figure 14.4 illustrates the options for dealing with land-use conflict.

We believe that a common problem with ICDPs is that the project executants' desire to combine maximum gains in both conservation and development encourages them to adopt inappropriate strategies. Win-win options to maximize conservation and development outcomes have been pursued in situations where they simply do not exist, with the result that stakeholders become disillusioned or even disenfranchised and the opportunity for negotiation of trade-offs is lost. Even worse, conservationists have sometimes ignored the needs and interests of local communities by negotiating with regional or national authorities, in effect creating an intractable situation, which severely reduces options to negotiate trade-offs in the future.

Unit of Scale	*Acceptability of net losses of environmental functions (goods, services, future options, ecological processes) at a particular unit of scale*	*Trends in land-use specialization*
Regional/ national (10^5 km^2)	Ethically unacceptable—even if possible and economically rational. Examples include Lawrence Summer's option of shipping industrial countries' toxic waste to developing countries; phosphate mining in Nauru. Would inevitably lead to greater social inequity and species extinction.	Multiple use
Provincial/ eco-regional (10^4 km^2)	Strongly discourage in virtually all circumstances. Examples would include conversion of most peninsular Malaysian lowland forest to estate crops and the dramatic division between plantations and natural forests in New Zealand. Directing one provincial administration to pursue industrial development and the other to pursue conservation will raise practical governance issues that will be virtually impossible to resolve democratically. Strong likelihood of social inequity and species loss.	⇕
District/landscape (10^3 km^2)	Discouraged in the majority of circumstances. If trade-offs are to be made at this scale, strong justification is needed along with robust mechanisms to balance loss functions elsewhere. Governance issues likely to be a problem, and ensuring social equity is properly addressed will be a major challenge.	
Stand/site (10^1 km^2)	Acceptable in the majority of cases as long as good management practices are pursued and there is reasonable likelihood of social equity issues being adequately addressed.	Dominant use

Figure 14.3.
Acceptability of trade-offs at different spatial scales.

Within any particular conservation landscape, it will probably be necessary to negotiate trade-offs with all major stakeholder groups, possibly sometimes on a case-by-case basis. Particular care must be taken to understand the various cultural landscapes and how stakeholders use these landscapes before negotiating trade-offs within a con-

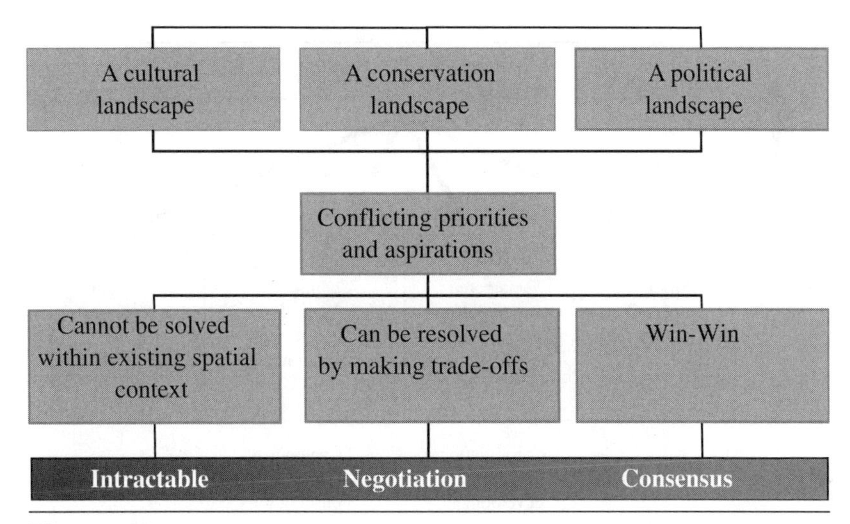

Figure 14.4.
Strategies for dealing with landscape-level conflict resolution.

servation landscape. As Brown (this volume) points out, homogeneous stakeholder groups do not exist in ICDPs.

Placing ICDPs Within the Landscape

If our hypothesis about multiple use being scale dependent is correct, then by implication so is sustainable development. We believe that there is already a tacit understanding among most practitioners that the three pillars of sustainable development—social equity, ecological integrity, and economic viability—cannot be addressed to the same degree at every single site. Sustainability entails getting the right balance at the right scale. ICDPs can provide a useful tool for delivering conservation through development, but the use of ICDPs needs to be balanced with other tools such as protected areas, community-based natural resource management (development through conservation), and economic development. While enthusiasts for any one of these different tools have, on occasion, tended to overstate the role of particular vehicles in sustainable development, we suggest that sustainable development, and therefore conservation, needs a more extensive "toolbox" of land uses, management strategies, and approaches. Figure 14.5 is a first attempt at representing how these approaches fit together.

ICDP and community-based natural resource management (CBNRM) approaches are often seen as the same; however, we argue

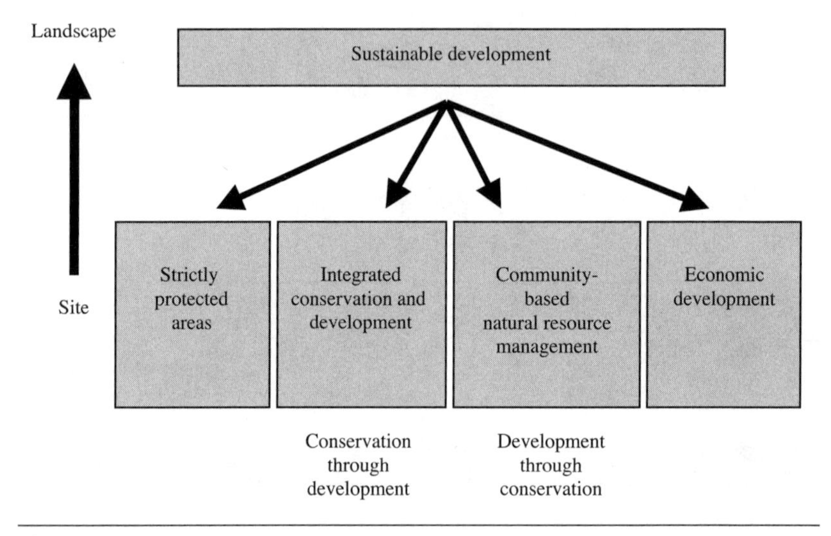

Figure 14.5.
Simplified scale of interventions within the landscape.

that in practice they often differ because ICDPs emphasize *conservation with development*, while CBNRM emphasizes *development with conservation* (see also Robinson and Redford, Franks and Blomley, Child and Dalal-Clayton, all this volume). While there is no doubt a considerable degree of overlap between the two approaches, it is important to recognize that each can have different priorities.

A Practical Example: Conservation at a Landscape Level in Sichuan, China

The ideas presented above are being tested through the application of a landscape approach to planning protected areas, sustainable forest management, and forest restoration in Sichuan, China.

The Minshan area of Sichuan covers 33,000 square kilometers of high mountains and temperate forests lying within the forests of the Upper Yangtze ecoregion. Around a million Chinese, Tibetan, Qiang, and Baima people live in the area, which is also home to about half of the world's giant pandas and to rare species such as the golden monkey and clouded leopard. Rural poverty, soil erosion, overlogging, and declining wildlife all pose major problems, while a rapid increase in tourism and road and dam building projects is bringing fundamental changes to people and nature. In the last few years, international attention has focused on threats to the giant panda, but in reality this is only one visible manifestation of wider underlying problems. The government of China has responded to environmental degradation by setting up a growing network of protected areas and through new policies, including a Grain for Green (converting steep slope cultivated land to forests or pasture) program that encourages and funds reforestation, and a logging ban introduced after severe flooding. Conservation efforts have not come without costs; for example, job losses in the logging industry have hit many of the poorest members of society.

The WWF is working with the Forestry Department of Sichuan, the Chinese Academy of Forestry, and other key stakeholders to develop a shared vision of landscape-scale forest management strategies that address human needs, while leaving enough space for wildlife and maintaining long-term ecological benefits from forests. The work is challenging and is only possible because of support from the seven county forestry departments in Minshan that are participating in the effort. Funding has been made available from donors in Europe and North America. As a first step, the project is collaborating with the government to assess the impact of initiatives of the Great Western Development Program, including forest policies such as the logging ban, the Grain for Green program, dams, mines, roads, and tourism.

Researchers from the Chinese Academy of Forestry are collaborating with local stakeholders to prioritize sites. Different stakeholders have been asked what is important to them about forests—ranging from the price of timber to spiritual values of individual trees—and information is being assembled to both track and understand land-use changes. The techniques involved range from satellite imagery to biodiversity surveys, participatory mapping, and oral histories. This is helping to develop an assessment of the forest quality of Minshan from the perspective of key stakeholders, to draw up broad conservation targets, and to produce a map of forests that may offer the best conservation potential based on social and ecological criteria. These potentially high conservation value forests will be assessed with the assistance of stakeholders to identify those priority areas where conservation interventions can help make a positive difference. Further work will identify the most ecologically and socially valuable forests at the level of stakeholders' cultural landscapes. A series of studies, in collaboration with local communities and other stakeholders, will look at these forests' potential in terms of nontimber forest products, forest certification, agriculture, and tourism. Staff trained in participatory techniques will work with communities to develop a series of different scenarios for how the landscape might develop in the future—for example, a collection of protected areas, restored forests, community-managed forests, and commercial enterprises interspersed with farms, settlements, and tourism developments. The various scenarios will form the basis of negotiations between local government, communities, conservationists, and other affected groups to choose the best landscape options for the future—probably resulting in a suite of formal and informal agreements about future land use. These negotiations will provide the basis for conservation activities across the whole Minshan landscape.

It would be naïve to think that such an approach can fully reflect all stakeholders' interests, and practitioners need to be careful not to raise false hopes in their approach. The approach differs from conventional conservation efforts in terms of its scale (which is intermittent between the site and the ecoregion), in the fact that it explicitly aims to address human needs within the landscape, and in that it starts from the assumption that conservation interests will, inevitably, have to be traded against legitimate social needs.

Conclusions

ICDPs can be a useful tool for conservation and development when they are used in association with other tools and provided that land-use trade-offs are negotiated within the context of a broader conservation landscape. Win-win solutions that maximize both conservation and development gains can occur in certain specific cases, but most conservation-development solutions will be a compromise made through negotiating trade-offs with key stakeholders, particularly local communities who are most affected by land-use decisions. Improving the process of identifying options and negotiating land-use trade-offs should be a major focus for future ICDPs and more generally for all efforts at broad-scale conservation. If we are to move away from the current top-down approaches, the question of who decides land uses and the related question of who has the opportunity to influence decisions are both critical and need to receive further attention. We recognize that identifying stakeholders and involving them in effective dialogue at these larger scales presents a considerable challenge for conservationists and development agencies. However, we are confident that a concerted effort over the next few years will yield new approaches and new tools for more effective integration of conservation and development.

References

Blumler, M. A. 1998. Biogeography of land-use impacts in the Near East. In K. S. Zimmerer and K. R. Young, eds., *Nature's Geography—New Lessons for Conservation in Developing Countries,* 215–236. Madison: University of Wisconsin Press.

Charland, J. W. 1996. The "problem-isolation paradigm" in natural resource management. *Journal of Forestry* 94 (5):6–9.

Emerson, R. W. 1836. *Nature.* Boston: James Munroe and Company.

Grieder, T. and L. Gurkovich. 1994. Landscapes: The social construction of nature and the environment. *Rural Sociology* 59 (1):1–24.

Haines, A. 1977. *The Yellowstone Story.* Yellowstone National Park, Wyo.: Yellowstone Library and Museum Association.

Miller, J.R. and R. J. Hobbs. 2002. Conservation where people live and work. *Conservation Biology* 16 (2):330–337.

Redford, K. H., P. Coppolillo, E. W. Sanderson, G. A. B. Fonseca, C. Groves, G. Mace, S. Maginnis, R. Mittermier, R. Noss, D. Olson, J. G. Robinson, A. Vedder, and M. Wright. 2003. Mapping the conservation landscape. *Conservation Biology* 17:116–132.

Stohlgren, T. 1996. *The Rocky Mountains.* Washington, D.C.: National Biological Service.

Terborgh, J. 1999. *Requiem for Nature.* Washington, D.C.: Island Press/Shearwater Books.

Tomich, T. P., M. van Noordwijk, S. Budidarsono, A. Gillison, T. Kusamanto, D. Murdiyarso, F. Stolle, and A. M. Fagi. 2001. Agricultural intensification, deforestation, and the environment: Assessing tradeoffs in Sumatra, Indonesia, In D. R. Lee and C. B. Barrett, eds., *Tradeoffs or Synergies? Agricultural Intensification, Economic Development, and the Environment,* 221–224. Oxford: CABI.

Zimmerer, K.S. and K. R. Young, eds. 1998. *Nature's Geography—New Lessons for Conservation in Developing Countries.* Madison: University of Wisconsin Press.

15

Poverty and Forests: Sustaining Livelihoods in Integrated Conservation and Development

Gill Shepherd

Introduction

This paper focuses on how integrated conservation and development programs and projects (ICDPs) could do more to address poverty issues, and the potential benefits to them of doing so. Because of the abundance of data on which to draw, I concentrate on forests, ICDPs, and poverty.

There are probably two main reasons why conservationists should care about local livelihoods. First, there is the moral argument—that conservation efforts have often harmed people, and that those being harmed are not those who do the most damage. Conservation measures often impose grave livelihood costs on those least able to bear them, and interventions often have unintended consequences. Second, there is self-interest. Short of the use of force, there is often no practical alternative to engaging with local people: the choice is between people illegally living in a protected area and accessing its biodiversity, or doing so in cooperation with its managers. Forest managers also have an increasing interest in the importance of livelihoods and a growing belief that only through meeting people's legitimate interests will their cooperation be obtained.

In order to meet conservation goals, I argue, ICDP project managers, and the conservation organizations that fund them, need to conduct a much more rigorous analysis of the social, economic, and political situation in which they find themselves. They need to focus more analytically on the immediate local social situation of the ICDP; need to set that local situation into larger spatial contexts (see Maginnis, Jackson, and Dudley this volume); and also need to focus far more explicitly on the relationship of the ICDP to institutions and policies at the national level. Conservation organizations have always operated very comfortably at the international level, but at this level, too, they now need to broaden their horizons and think harder about how their own concerns fit into those of other major players.

All of these levels provide opportunities and incentives for a greater focus on poverty.

Fitting ICDPs to Local Realities

Although I argue that ICDPs should be more alert to interactions at several different levels, the local level remains the one at which the most intensive project–poverty interactions occur, and where the most friction can exist.

Poor people and forests are inextricably linked in the poorest parts of the world—the very areas where there is often most enthusiasm for creating new protected areas, or ICDPs. There are three main reasons for this.

First, poverty in developing countries is most acutely a rural phenomenon—the rural poor being poorer than the urban poor for many reasons (Lipton 1977; Bird et al. 2002).

Second, as table 15.1 shows, 60 to 70 percent of people live in rural areas in the poorest regions—sub-Saharan Africa and parts of Asia.[1] Other characteristics of the populations of these two regions that make it very difficult to escape from poverty (as defined in the UN Development Program's Human Development Report 2001) include low life expectancy and a high proportion of very young societal members, both of which factors are also exacerbated in these regions' rural areas.

Third, while not all poor rural people live near forests, the majority of them do. Wealthier rural people are more likely to own the most desirable land—land of better quality, nearer to roads and markets, and more capable of supporting commercial agricultural crops. They are more likely to have legal title to this valuable land, which will make them more eligible for bank loans and more able to bear risk. Forests have usually already long been cleared from such areas. By contrast, both poorer people and forests tend to be found in more remote places (see Dove 1996, and Tongson and Dino this volume).

Table 15.1 *Rural and urban populations in various world regions, with societal age profile and life expectancy*

Area	*% rural in 1995*	*% urban in 1995*	*% under age 15 in 2000*	*Life expectancy, 1995–2000*
Africa	65	35	43	53.8
Asia	65	35	30	66.2
Central America	34	66	35	71.7
South America	23	77	30	69.0
Oceania	30	70	26	73.9
North America	24	76	21	76.9
Europe	26	74	18	72.6

Source: WRI et al. 1998, ch. 5.

Poorer rural people are much more likely to be food insecure. Most are farmers whose livelihoods are, in Lipton's (1977) famous phrase, "complex, diverse, risk prone," and who must supplement the food and cash that their farms can supply. They tend to live in more marginal and inaccessible areas, further from roads and markets. Though agriculture (sometimes with a livestock component) is still the most important economic activity in most of these areas, off-farm resources form a key supplement to agriculture and an essential part of sustainable livelihoods. Some of these resources may be supplied by wage laboring elsewhere, but many come directly from nearby forest resources. Finally, there are those who live in forests, mixing some agriculture with hunting, gathering, and trading.

Even among the forest-dependent poor, it is useful to distinguish two strands. On the one hand we find the very poor, with too few assets to take advantage of many livelihood opportunities. These may draw on forests mainly for subsistence and very low level commercial activities, and most value the livelihood-buffering aspects of forest. These are often referred to as the *chronic poor*. On the other, there are the slightly less poor, who can take more advantage of market and income opportunities from the sale of forest products when they need to. These are sometimes referred to as the *transient poor*.[2] The two kinds of dependence may be found in the very same household: the wife falling into the first category and the husband into the second. Individuals and households may also swing between the two levels seasonally, or over a lifetime. The point is that both forest functions are vital for the poor.

Agreeing on the Concept of Poverty

What do we mean by poverty? Poverty has most commonly been assessed against income or consumption criteria. Such indicators can give a picture of the extent of poverty at the national level and can be aggregated internationally.

But there is nowadays a broad consensus about the multidimensional and dynamic character of poverty. In consequence, more detailed analysis and planning on the ground means more qualitative measures and approaches are required, the challenge being to achieve a trade-off between measurability and capturing local complexity. "Basic needs" approaches were one way forward, and more recently "basic capabilities" approaches have taken these one step further, to include social inclusion—the ability to have a role in society and to influence decision making. The basic-capabilities perspective has in turn generated an approach to analyzing poverty now commonly termed *sustainable livelihoods.*

A livelihood comprises the capabilities, assets (both material and social), and activities required for a means of living and is sustainable when it can cope with and recover from stresses and shocks and maintain or enhance its capabilities and assets both now and in the future, while not undermining the natural resource base (Carney 1998). The concept is useful in part because it starts from an analysis of existing assets.[3] These local livelihoods are embedded in the larger institutional and regulatory environment, which provides opportunities for or, more often, impediments to greater sustainability.

However, sustainability can be a fragile concept. We know that the poor are found where they are more vulnerable, more exposed to risk, and more powerless. So their focus is on reducing vulnerability, on risk buffering, and on securing access to key local resources and to some income for key needs—in short, on improving their resilience to shocks (World Bank 2001).

Who Depends on Forests?

The ways in which people use forests as a complementary resource vary according to their own dominant livelihood base. This section examines these.[4] Table 15.2 sets out some of the categories of people with some dependence on forest (including urban dwellers, since the rural poor supply them, as well as themselves, with forest products).

Those Who Live in Forests

Though most forest hunter-gatherers also practice agriculture, a few (like the forest dwellers of the Congo Basin) gain access to the agri-

Table 15.2 *Nature of dependence on forest*

Main economic activity	*Main livelihood source*	*Forest foods*	*Wood fuel*
1. Those who live in forests			
Farmers practicing rotational fallowing (shifting cultivators)	Forest products + forest-based agriculture	***	**
Herders in tropical dry forests (mainly Africa)	Forest-fed livestock and agriculture	***	*
Hunters and gatherers	Forest products	***	*
2. Agriculturalists drawing on the forest for key inputs			
Wealthier farmers	Mainly commercial agriculture	** Seasonal and specialist	**
Poor farmers	Agriculture, migrant laboring	*** Emergency, seasonal, specialist	***
Landless families	Agricultural or other wage employment	*** Emergency, seasonal, specialist	***
3. Those basing livelihoods on commercial forest product activities			
Artisan traders and small enterprises	Cash incomes	**?	***
Employees in forest industries	Cash, some agriculture	**?	***
4. Urban dwellers			
Wealthier townspeople	Cash incomes	Some items	No
Poor townspeople	Cash incomes	Some items	***

Source: Based on Byron and Arnold 1997.

Forest medicines	*Housing materials + furniture*	*Forest soil fertility*	*Religious + cultural values*	*Fodder*
***	*** Forest gathered and homemade	***	***	*** If animals kept
*** Inc. vet.	*** Forest gathered and homemade	***	***	***
***	*** Forest gathered and homemade	***	***	—
**	** Some components purchased	**	*	0
***	*** Forest gathered and homemade	***	**	***
***	*** Forest gathered and homemade	—	**?	***
**?	Some components purchased	*?	—	—
**?	Often forest gathered and homemade	**?	—	—
Some items	Purchased forest components only	—	—	—
Some items	Mainly purchased forest components	—	—	—

cultural products they need by exchanging forest products and labor for grains and root crops. The two rather larger categories of forest-dwelling people are those who practice sustainable fallowing systems and rotational agriculture in tropical moist or tropical dry forests, and the herders of the Sahelian and east African dry forests whose camels, cattle, sheep, and goats browse trees rather than graze for most of the year. All three of these categories have rough-and-ready forest management methods, which sustain the resource so long as there are no more powerful outside competitors for it. As table 15.2 shows, the forest supplies a tremendously wide range of total needs for this category. The quasi-religious significance that forests have for people in this category is seen in the fact that key life cycle and religious ceremonies are linked to special forest places and forest foods.[5]

Agriculturalists Drawing on the Forest for Key Inputs

Those who practice agriculture outside the forest as their main activity may still be very dependent on the forest for nutrient inputs via animal manure/bedding (parts of India, Nepal, and semi-arid/sub-humid Africa), for the feeding and locating of animals off-farm, for housing materials, and for emergency and seasonal complements to the farm diet. They often need more fuelwood from the forest than the first category of people, and more house-building materials if their houses are more substantial and destined to last longer. Wealthier farmers may purchase some house materials, however, such as corrugated iron roofs, concrete for the floor, doors, and windows. The most important rituals, marking life crises and the relationship of humans to natural resources, use the symbolism of domestic animals or farm crops rather than the forest, as a rule.

Many people in both of the above categories make some cash from forest products directly or indirectly, either selling products gathered directly from forests or selling crops or livestock for which forest nutrients were an essential indirect input.

Those Basing Livelihoods on Commercial Forest Product Activities

Traders and artisans whose businesses depend on forest products have a rather different relationship to forest. They are not in charge of farming enterprises, which exist to some extent in symbiosis with forest (though in some countries their wives might be), and they are likely to buy rather than collect many of the forest products they do use. Similarly, forest enterprise employees, while they are a poorer category that needs to maintain some farming activities as well as their wage employment, have a more indirect relationship to forest.

Urban Dwellers

Wealthier townspeople are likely to purchase some forest foods and medicines and some forest-based housing components (not necessarily from the nearest forest). Poorer urban people are also dependent on all these things if they can afford them, but their chief dependence on forest is most probably for fuelwood.

How Can the Forest Help Poor People Build Livelihood Assets?

Poor people have the chance to build on all five kinds of livelihood asset if national-level institutions and processes grant a framework within which they are legally able to do so.

Natural Capital

Most traditional forest management systems gave those who lived in and near forests assured rights over the resource through the customary tenure systems that were in operation. While many state tenure systems formally annulled these rights many years ago, forests still exist as de facto natural capital in many areas, and in some countries legislation has been rethought to return forest in whole or in part to local owners (see Tongson and Dino this volume, Gartlan this volume).

In Nepal, for instance, the 1993 Forest Act gives rural groups the legal right to form a community forest-user group, to have it officially recognized as an autonomous corporate body with perpetual succession and the legal power to develop, manage, and utilize the forest in perpetuity. In Tanzania, villages have village land titles that contain both private farm plots and communal land, often including forest. "Joint forest management" arrangements in some Indian states give similar rural groups legal part shares in forests as assets (see also Singh and Sharma this volume). Some forests have been legally passed to indigenous peoples in South America.

Physical Capital

Building supplies from the forest represent an important source of physical capital, and include timber, poles, bamboo, rattan, thatching grass, stones, and other materials needed for fencing, housing, and outbuildings. Agricultural inputs are collected from the forest, acquired indirectly when they are converted into compost and animal manure, or incorporated through fallowing systems. Farm households often also depend on forests for ecosystem services such as maintaining water flows and will manage upland forests and watersheds quite intensively in order to protect agriculture lower down (see also Brandon and O'Herron this volume). Improving the quality of the forest increases the availability to poor people of many forms of physical capital.

Financial Capital

The forest provides a range of goods that can be translated into financial capital for very large numbers of households. These may take several forms, from fuelwood and charcoal for the poorest, to biomass for particular trades such as blacksmithing or furniture making. Many non-timber forest products are especially important for the poor, providing a range of raw materials sold as forest foods, as nonfood traded items (such as wrapping leaves in West Africa and leaf plates in India), or for handicraft industries (see Tongson and Dino this volume). Traded medicinal products are far more important than is often recognized, from the ayurvedic medicinal plants of South Asia to the forest medicines that flow out to urban markets from the forests of the Amazon, the Congo Basin, and from Africa's tropical dry forests.

The finance generated may be of special value if it tides a household over seasonal or unforeseen shortfalls, provides lump sums that pay off a debt, acts as working capital for some new enterprise, or enables farmers to sell crops at a more advantageous time of year. It is commonly noted that forest incomes are more important for their timeliness than for their magnitude.

Such forest products often make up a higher percentage of the annual income of the poorest, of women within households, and of certain age groups. For instance, among the Akan in southern Ghana, while the profits from any on-farm activities go to the (male) household head, both wives and teenage children also wish to generate income that they control themselves, to safeguard their future. Unmarried sons may look for off-farm employment, but daughters and wives gather and sell forest products and make cooked food for sale. Wives often choose to make remittances to their natal families, for instance, as security in case of divorce (Milton 1998). In Cameroon and Benin, women increase their collection and sale of nontimber forest products right before children's school fees are due, at times of year when ill health is more common, and during the preharvest period when food is scarce. It is an especially important activity for elderly women (Schreckenberg 1996; Schreckenberg et al. 2002).

Human Capital

The most basic forms of human capital are health and food security. Forests provide a range of consumer goods that contribute to these ends in the shape of traditional medicines, food, and the fuelwood to cook it. As in the case of physical capital, improving the guaranteed access of the poor to the natural capital of the forest will increase their human capital in the form of food and medicines. There is often no clear line between forest foods and forest medicines, and both are part of acquired knowledge about the foods and food supplements, which help with specific ailments. Woodfuels, too, influence what is cooked and how often, and are related to infant nutrition.

Many forest users rely on the forest for wild foods. Forest areas often provide the protein, vitamins, wild vegetables and fruits, relish, spices, and condiments without which the starchy diet of grown or traded farm crops is very unpalatable (Falconer 1989, 1990, 1994). Even wealthy farmers, who can buy what they need, often retain affection for the flavor of particular forest foods. This wild harvest includes foods that the poor cannot afford to buy from the market, their qualitative contribution to diet being probably much greater than their quantity. Such foods may be particularly important for women and girls, who traditionally suffer from intrahousehold discrimination in food distribution, and for those with special nutritional needs such as children and pregnant and lactating women.

A second vital function of wild food is that it often has to cover dietary shortfalls during seasonal food shortages before the next crop is ready to harvest, when agricultural labor demands peak and there is no time for cooking. Some poorer farmers in the hills of Nepal may have to make do with these foods for three months or more each year (Shepherd and Gill 1999). In North Kasungu District in central Malawi, mothers and daughters in all households gather and store some wild food as a variation to their diet all year round. Smaller, poorer households collect wild food far more frequently than wealthier households, while the rich plant more of their own crops and eat wild foods less, farm size being inversely correlated with wild food gathering (Abbot 1997).

Finally, in emergencies, when normal crops fail or bear poorly for some reason, those who cannot buy food fall back on forest foods whatever the time of year.

Social Capital

Where individuals come together to manage forest in groups (through their own independent decisions or through the political arrangements that some countries have made for them), mechanisms including the extended family and community networks are strengthened through action, and there is an accumulation of social capital for group members. In Nepal, it has been noted that forest user groups may have more authority—as a result of their shared asset—than village development committees (Shepherd and Gill 1999). Some of these groups have even devised special dispensations for the elderly or the very poor, which do not apply to all. Similar arrangements are very widely reported from Africa (Shepherd 1992).

Conclusions: How Can ICDPs Make Themselves More Pro-Poor at the Local Level?

Understanding Poverty Better

Those who plan and manage ICDPs need to better understand the poor with whom they work. They need to know how they put their livelihoods together, how they bridge lean and difficult periods, and what their time frames are. They need to know the difference between the poor and the very poor, and to be aware that the poor have a greater proportion of their total livelihood invested in forest benefits than the less poor. Similarly, they need to understand that poverty alleviation and the meeting of subsistence, survival, and reduced vulnerability needs are just as important as poverty reduction and increased participation in market activities, and that perhaps they need to restructure the balance of any poverty-related initiatives to reflect this.

Finally, they need to be aware of the assets the poor already have, so that they are not thoughtlessly removed. As we have seen, at least four forms of capital can be acquired and built up if those who live near forests have assured access to resources. Furthermore, if institutional arrangements make it possible for social capital to be built as well (through shared decision making and management of the resource), the poor can build all five through their relationship to forests.

However, by the same token, all five forms of capital may well be lost if ICDPs take resources from local people. The loss of natural capital can lead to a loss of physical capital, to risk-increasing financial losses, to poorer health and a poorer diet, to less sustainable agriculture, and to a withering of social capital. All these forms of capital, built up over many years, can be dispersed rapidly. ICDPs may be left with impoverished and resentful neighbors, far less able than previously to escape from rural poverty traps, and even more likely to sink from transient to chronic poverty.

Prioritizing Poor Local Stakeholders

Often a very wide range of stakeholders is listed for a particular ICDP, from local people to the international community. It is essential to prioritize local stakeholders, who do not have alternative livelihoods and who are least able to bear new costs, over those for whom the ICDP is cost-free and represents only benefits.

Building on Existing Knowledge, Management Skills, and Priorities

Ask about, learn more about, and build on where possible rather than destroying functioning common property forest management regimes. Opportunities for the poor to benefit from and contribute to biodiversity may be better in their own forests than in protected areas. Tighter tenure creates tighter control; tenure denied creates an open-access resource (see also Tongson and Dino, Gartlan, Child and Dalal-Clayton, Singh and Sharma, all this volume).

Learn more about, and, where possible, build on rather than destroying, local environmental knowledge, values, and practices. Protect this unique knowledge both informally and by seeking out better intellectual property protection methods and fair and equitable benefits that might arise from the knowledge of biodiversity.

At the same time, recognition of local management methods and knowledge allows local people the freedom to manage forest for their own goals. Outside agencies can do much to ensure that the needs of the poor in particular (who value a wider range of biodiversity than the less poor in most places) are factored into management regimes.

While such approaches demand compromises and negotiation (some local choices will not be the best from the conservation point of view), managers need to compare actual outcomes with the realistic alternatives that would have eventuated rather than with a fantasized but unrealizable perfect protected area (see Brown this volume).

Tailoring Management to the Poor

To address poverty effectively, choose poverty-oriented strategies, ways of managing, product bias, and time frames that fit with the needs of the poor. Many of these will be positive for biodiversity.

Fine-Tuning Protection and Sustainable Use

In areas of high biodiversity and high livelihood dependence, use more fine-grained patch-by-patch planning. Ensure as much multiple use as possible and that areas for complete protection are kept relatively small. Apply landscape principles of this kind even within protected areas (see also Robinson and Redford this volume).

Land Tenure

While ICDP projects genuinely seek in many cases to improve the lives of the populations living in the immediate vicinity of a protected area, fundamental land tenure issues are rarely addressed, as Steve Gartlan points out in this volume. Indeed, he argues that conservation projects cannot succeed where there is ambiguity and antagonism over land tenure, and that probably nothing can deliver conservation better than addressing basic underlying injustices and championing better local tenure arrangements (see also Tongson and Dino this volume).

The "Do No Harm" Principle

It is becoming increasingly common to hear the commitment: "No forest-dependent people should be worse off as a result of forest conservation activities." Profound knowledge of the sorts suggested here is probably needed before one can know whether harm is being done or not. But if compensation is being considered, then it must be much more realistic than has often been seen. Ideally, like should be replaced with like, and new forest (with land rights) offered for old. Financial compensation would need to reflect a lifetime's loss of livelihood capital—and such sums have rarely been worked out, let alone paid to displaced people (cf. Shepherd 1993a; Gartlan this volume). The reality is probably that it is so difficult to offer adequate compensation that finding an accommodation is the more realistic alternative.

Beyond the ICDP: The National Level

In order to address poverty more effectively at the local level, while also conserving biodiversity effectively, ICDPs and the agencies that fund them must realize that they not only need to "step outside the park" but be prepared to consider several layers of action at once. The previous section has suggested a journey into the surrounding landscape, but this section and the one that follows make it clear that the local level is impacted so profoundly by the national and the international level that different commitments must also be made there if a transition to a more pro-poor approach is to occur.

It is only a slight caricature to say that the environment and conservation world "thought globally" rather than "acted locally" in the twenty years preceding the Rio Earth Summit, while the forestry world dealt primarily with the national and then the local. Table 15.3 illustrates this by setting out some milestones along the forestry and environment pathways that met at Rio in 1992.

The table shows that forestry has at least a ten-year head start on environment and conservation in its experience of working with local people. Forestry project staff also began with little trust in local people, and with a too exclusively site-based focus. Over time their trust grew, and they began to see that in order to create more opportunities for forests and local people, they needed to get involved in national and indeed international policy. The conservation world needs to capitalize on the forestry world's greater experience here, and to work closely with that world at these levels, in order to enhance its own goals and to achieve a greater focus on poverty.

Most forest is of course owned by governments and controlled by state legislation and regulations controlling use. So access to forests by the poor has often been subordinated to industrial and conservation interests. In state forests, there have been restrictions in favor of industrial use at the expense of local users, and heavy regulation/taxation to curb use or generate revenue for the state. Use of forests may be restricted in a number of ways, or revenue sharing with the forest department be imposed. Protected areas have similarly attempted to impose stringent restrictions on what local people can use and do in the area.

But over the last decade or two, in the forest if not the environment and wildlife sectors, there has been a process of devolution of the control of forests. The process began in drier or lower-value forests in the Sahel (Kerkhof 2000) and East Africa (Wily and Mbaya 2001; Barrow et al. 2002), and in Nepal and India (Hobley 1996), but is now being extended much more widely to tropical moist forests. This process has often been linked overtly to poverty concerns.

Through the efforts of donors, NGOs, and community-based organizations, and in the name of better governance and greater focus on poverty, there have been attempts to constrain the power of the state and strengthen the ability of civil society to make its voice heard (see also Child and Dalal-Clayton this volume).

Working at the National Level to Strengthen Local-Level Land Rights

Often, conservation organizations managing protected areas have been content to accept national laws that fail to recognize customary land rights at the local level. The state has often been most reluctant to acknowledge these rights, since they have implications for national sovereignty, the right to grant concessions at will, etc. In the past, conservation organizations may also have found such laws helpful, since they give legal support for removing local people from state forestland and clearing them for conservation.

The reasons for revisiting this view are twofold. First, as we have seen, it has been impossible for conservation organizations to make such laws stick. Second, as a result of international pressure, forest policy and law is currently being redrafted in many countries to devolve more ownership and management rights to local people.

Table 15.3 *Forestry and environmental milestones, 1972–1992*

Date	*Forestry milestones*	*Environmental milestones*
1972–82	1973–74 Oil price shocks—a likely developing-world woodfuel crisis? Solution seen as large-scale tree planting by and with local people.	1972 UN Human Environment Conference, Stockholm. Environmental affairs first advanced as an international political issue. UNEP created.
	1978 World Forestry Congress, Jakarta. "Forests and People."	1979 Myers's *The Sinking Ark* published, linking mass species extinctions to forest loss.
	1978 First World Bank forestry policy paper. Large increases in forestry sector lending, and a shift from northern model industrial-scale plantations to "social" forestry.	1980 *World Conservation Strategy* published. Concept of "sustainable development" launched (IUCN, UNEP, WWF).
	1978 FAO Forestry for Local Community Development Programme launched. Early stress on villager tree planting. Collaborative forest management to come later.	1982 Bruntland's *World Commission on Environment and Development* identified issues for action and pressed for a higher profile for problems.
		1982 World Parks Congress in Bali, Indonesia. First explicit linking of conservation and development.
1972–82	*Summary: woodfuel crisis precipitates first intensive interactions between foresters and farmers and rural people.*	*Summary: Importance of the environment and of sustainable development articulated and promoted internationally.*
1985–92	1985 Tropical Forestry Action Plan launched at World Forestry Congress, Mexico, to promote sustainability, increase forest sector aid. National NFAPS initiated at country level.	1987 *Bruntland Report (Bruntland + 5)*. Promoted the environment as an essential part of sustainable growth.

Date	*Forestry milestones*	*Environmental milestones*
	1985 ITTA signed, ITTO created. Commodity agreement on both timber trade and resource, signed by main producers and consumers.	1987 National Environmental Action Plans launched by World Bank to identify key environment problems in all sectors. Broader than NFAPs in concept, more driven by external interests, but usually linked to conditionality.
	1985 BOSTID National Research Council CPR Conference, U.S.A. Launched experiences of Common Property Resource Management to wide audience.	1992 World Congress on Parks and Protected Areas, Caracas. The important role of the people who live in and near protected areas highlighted. The concept of ICDPs launched.
	1988 *No Timber Without Trees.* A report from ITTO stated that less than 1% of tropical timber was from sustainable sources.	
	1990 *ITTO Guidelines for Natural Forest Management* produced.	
1985–92	*Summary: Intense national- and subnational-level donor focus on forests. Evidence that local people may manage natural resources intensively themselves. ITTO begins to define and promote sustainable forest management.*	*Summary: environmental planning follows forestry to the national level. The complex relationship between protected areas and people identified, but little practical experience at this point.*
1992	Earth Summit Conference, Rio de Janeiro. Conventions on Biological Diversity and Climate Change agreed. Agenda 21's chapter 11, "Combatting Deforestation," recognizes the role played by forests, and drafted *Forest Principles* develop into IPF, IFF, and finally UNFF.	

Supporting Policy and Legislation That Enables Local Decision Making and Management

Generally, forestry managers have progressed further than the managers of ICDPs with solutions for the problems that get in the way of sustainable natural resource management by local people. For some time they have been tackling obstructive legal, policy, and institutional frameworks (e.g., Talbott and Khadka 1994; Ribot 1995; Thomson and Schoonmaker-Freudenberger 1997; Kaimowitz, Graham, and Pacheco 1999), unhelpful tenure arrangements (e.g., Raintree 1987; Shepherd 1993b; Wily and Mbaya 2001), and the conditions for successful common property resource management (e.g., Berkes 1989; Ostrom 1990; Bromley 1992; Arnold 1998). They have become aware that they will often need to address the competing interests of different government agencies in this process. In many cases, protected area and wildlife policies are more retrogressive than forest policies: for instance, most of the latter have been revised in the last decade, while few of the former have.

Many countries have decentralized forest management to more local levels, along with other government functions. A variety of deals with local people are in place—from legally based handover and sharing arrangements in Nepal and India, to complete devolution in Bolivia. Some countries have hybrid arrangements, having decentralized the management of forest but kept wildlife, protected area management, and tourism as national functions (e.g., Indonesia, Tanzania), greatly complicating coherent integration of different forest land uses.

These shifts in international thinking have important implications for the management of ICDPs, and for the role of local people within and around them. ICDPs will have to consider their relationship to local people in the light of other models now in existence, if they are not to be left behind (see also Child and Dalal-Clayton this volume).

One of the rare examples of successful negotiation has been the East Usambara Catchment Forest Project in Amani, Tanzania. The project negotiated with the Tanzanian government (with great difficulty, it has to be said) a legal status for the area as a nature reserve, which avoids some of the confrontations that often attend protected area projects. It opens up the possibility for "patch" management—the creation of a mosaic of subareas within the Amani reserve—in which varying levels of protection, sustainable use, and management are agreed on with local people. The project has created a way of combining good conservation with good local public relations in an increasingly densely settled area (Shepherd et al. 1999).[6]

Engaging with Available Tools for Forest Land-Use Planning

It is important that, at the national level (and still more importantly at the regional and local level), different kinds of land use are balanced through proper land-use planning. If too many logging concessions and/or plantation areas are granted near a national park/heritage site, the pressure on the park from adjacent people may be unfairly great.

Trade-offs between different kinds of forestland use become more complex at higher levels. Conservation areas, commercial production areas, and local people's use and ownership areas all need to find their place. Some countries have embarked on processes intended to address this. They include the National Forest Program planning process and work on criteria and indicators for sustainable forest management. The importance of such processes is that all forest uses have to be considered, not only those that directly concern protected areas. The protected area/buffer zone paradigm is overtaken by a more inclusive concept that takes an interest in all the forestland use into which conservation concerns are embedded. In this way, ideally, some of the bigger decisions about the future of the nation's forests can be made in the context of conservation, development, and sustainable use (Prabhu, Colfer, and Shepherd 1998; Mayers et al. 2001; Maginnis, Jackson, and Dudley this volume).

Harmonizing Diverse International Commitments at the National Level

Though international issues will be addressed in the next section, conservation organizations, and the protected areas and ICDPs they are responsible for in-country, need to take increasing note of the international commitments particular countries have made and their harmonization with national processes. The relevant stakeholder landscape at the national level is much more complex than it was a decade ago.

There are two key clusters. First, many poorer countries have embarked on "poverty reduction strategy processes" (PRSPs) in pursuit of the Millennium Development Goals. These have been initiated by the World Bank and supported by a variety of donors. There are opportunities for integrating forest and biodiversity issues into these, and indeed some fear that donors will take less and less interest in forests and natural resources if they are not so integrated. This is an area where an increased focus on forests and poverty is essential.

Second, there is a need, explicitly recognized at the recent Convention on Biological Diversity (CBD) COP-6, for the greater harmonization of the priorities of the CBD and UN Forum on Forests (UNFF). In particular, there is now a welcome commitment to bringing national biodiversity action plans and national forest programs into a more synergistic and beneficial relationship.

Both these policy clusters will tend to adjust the balance back in the direction of national and local environmental values and practices, and away from previously overprivileged global environmental values. Again, there will be indirect benefits for the poor.

Beyond the ICDP: The International Level

ICDPs themselves encounter the impact of international decisions from two angles. They must respond to the agreements reached through UN conventions and processes, and signed up to by the nation-states within which ICDPs are found. And they must also respond to the dialogue that goes on, just as intensively, among international conservation organizations. They are a good deal more familiar with the second debate than the first. These two forums are discussed in the next section.

INTERNATIONAL PROCESSES AND ICDPS

First, as we have seen in the previous section, ICDPs must respond to the implications of international conventions and donor policies in a country context. The most important of these at the moment from many points of view are the Millennium Development Goals: the commitment to poverty reduction by donors to meet the International Development Targets for halving poverty by 2015 (see box 15.1).

At the Millennium Summit in September 2000, the states of the United Nations reaffirmed their commitment to working toward a world in which sustainable development and the elimination of poverty would

Box 15.1. *The Millennium Development Goals and Targets*

Goal 1: Eradicate extreme poverty and hunger.

Target 1: Halve, between 1990 and 2015, the proportion of people whose income is less than $1 a day.

Target 2: Halve, between 1990 and 2015, the proportion of people who suffer from hunger.

(continued)

Goal 2: Achieve universal primary education.

Target 3: Ensure that, by 2015, children everywhere, boys and girls alike, will be able to complete a full course of primary schooling.

Goal 3: Promote gender equality and empower women.

Target 4: Eliminate gender disparity in primary and secondary education, preferably by 2005, and in all levels of education no later than 2015.

Goal 4: Reduce child mortality.

Target 5: Reduce by two-thirds, between 1990 and 2015, the under-five mortality rate.

Goal 5: Improve maternal health.

Target 6: Reduce by three-quarters, between 1990 and 2015, the maternal mortality ratio.

Goal 6: Combat HIV/AIDS, malaria, and other diseases.

Target 7: Have halted by 2015 and begun to reverse the spread of HIV/AIDS.

Target 8: Have halted by 2015 and begun to reverse the incidence of malaria and other major diseases.

Goal 7: Ensure environmental sustainability.

Target 9: Integrate the principles of sustainable development into country policies and programs, and reverse the loss of environmental resources.

Target 10: Halve, by 2015, the proportion of people without sustainable access to safe drinking water.

Target 11: Have achieved, by 2020, a significant improvement in the lives of at least 100 million slum dwellers.

Goal 8: Develop a global partnership for development.

Target 12: Develop further an open, rule-based, predictable, nondiscriminatory trading and financial system.

The Millennium Goals Web site is at www.developmentgoals.org.

be the highest priorities. The Millennium Development Goals grew out of the agreements and resolutions of world conferences organized by the United Nations in the past decade. The goals have been commonly accepted as a framework for measuring development progress and are being monitored by the United Nations, the Organisation for Economic Co-operation and Development (OECD), and the World Bank. The goals focus the efforts of the world community on achieving significant, measurable improvements in people's lives. They establish yardsticks for measuring results, not just for developing countries but for the rich countries that help to fund development and for the multilateral institutions that help countries implement the goals. The first seven goals are directed at reducing poverty in all its forms. The last—global partnership for development—is about the means to achieve the first seven.

For the poorest countries, many of the goals seem far out of reach. Of all regions, Africa has the largest share of people living on under one dollar a day, high infant mortality, and the lowest primary school completion rates. South Asia relies more on agriculture than any other region, has the highest child malnutrition rates, and the highest youth illiteracy rates. Both areas are experiencing HIV/AIDS and tuberculosis epidemics. Though targets will not be fully met by 2015, countries in these regions need to make progress toward them, and above all to ensure that poor people are not left further and further behind.

The CBD has had a decade to develop its conservation and development thinking, but its implementation capacity remains weak, and it has to link more closely with the best of UNFF experience on the ground. Its awareness of other international and national policy processes has grown, but the CBD has little experience of effectively mainstreaming its biodiversity concerns into their agendas. Insofar as conservation organizations draw their inspiration for better ICDPs from CBD processes (and Global Environment Facility funding), they need to round out their own understanding with other sectoral and international experience.

International Conservation Organizations and ICDPs

Second—and this is where this section will focus more closely—conservation organizations themselves form their own international environment in which they talk to and argue with one another. Because they spend so much of their time in this company, there is too little exchange of ideas with those engaged in the forestry and poverty worlds. Yet, challenges from the perceptions of those other worlds are now urgently needed if creative, innovative, lateral thinking and experimentation are to take place. There are several key areas for action, if a poverty focus is to be taken seriously.

Researching New Approaches

More work is needed on partnerships, on access, and on property rights.

Rethinking the Conservation Publicity Machine

Conservation organizations have had an important role in creating public awareness and commitment to biodiversity in the West. Now they must adapt responsibly to generating messages about biodiversity and livelihoods, and to explaining the positive synergies that can exist between the two. More rigorous and empirical evidence must be gathered to make this possible. There has been far too much fund-raising from what are in effect anti-people messages.

More Attention to Country Context in Choosing ICDP Sites and Approaches

When conservation organizations select areas in need of protection, identify "hot spots" and "ecoregions," or perform a gap analysis, they are guided purely by biodiversity criteria—which is the right first step. However, they often fail to perform the next logical step—or perform it poorly—which is to take into account the economic and political nature of the countries they are selecting for action (see also Sayer and Wells this volume, McShane and Newby this volume). The global context within which they work has tended to make them treat all countries as socioeconomically identical.

In fact, different strategies are urgently needed for different kinds of countries. Countries on the OECD or World Bank's "least developed country" (LDC) list—often the countries with substantial remaining areas of important forest—cannot respond to the presence of protected areas in the same way that "middle income" or "newly industrializing countries" (MICs and NICs) can. The latter countries, with most of their population living in towns, can host protected areas far more easily than can countries where 70 percent of the population lives rurally. In the case of MICs and NICs, protected areas are gradually becoming sites for recreation, while in the case of LDCs they are mainly inhabited areas with vital livelihood functions. Latin American models for protected areas, for instance, will rarely be appropriate for Africa or for many of the poorer parts of Asia.

Much more imaginative donor strategies are thus needed to design appropriate ICDPs with the needs of the poor in mind in LDCs with dense rural populations and few livelihood alternatives. By the same token, LDCs often have weak and cash-short government institutions, including forest and wildlife departments, from whom effective long-term action cannot be expected. There is also a higher likelihood of corruption in the granting of concessions and other turnkey activities. The degree of political stability may be low. For all those reasons, too, poor local people in LDCs live in a much riskier and vulnerable socioeconomic environment.

This section, then, is a plea for fewer global "one size fits all" conservation solutions, and for more attention to be paid to country specificities and to the particular needs of poor people in poor countries as well as to biodiversity categories.

Choosing the Appropriate Protected Area Category

An IUCN report (IUCN 1999) makes it clear (though it does not set out to do this) how little relationship there is between the IUCN category of protection and the effectiveness of protected area management. There is little evidence that categories 1–3 protect more effectively. There are too many extrinsic variables for a pattern to emerge—no doubt including some of those set out in the previous section as well as a protected area's relative remoteness.

From a poverty standpoint, if it is impossible to show that greater protected area status (IUCN categories 1–3) creates greater intactness, then IUCN Protected Area Category 5 (which allows people to live in and use the resource) should be selected more often. Table 15.4 shows this lack of relationship (Shepherd and O'Connor 2001).

If this table is summarized to its essentials (see table 15.5), it becomes even clearer that intactness seems to bear no relation at all to the quality of management, the degree of local involvement, or the protected area category selected. Other factors are dictating the observed patterns. More research is needed to understand these other factors, but in the meanwhile there is no justification here for excluding people from protected areas.

Timelines

With practice, it will be seen that appropriate protected area and ICDP solutions fit along a kind of time and space continuum. The poorest countries need the most complex attention in the planning of protected areas and ICDPs, with the greatest role for poor local people and the greatest accommodation to their needs. The issues get simpler the richer the country is: complete protection threatens fewer people, and recreation needs are easier to accommodate than livelihood needs (see also Glick and Freese this volume).

Table 15.4 *IUCN protected area categories and resource status*

Category	*Name*	*Country*	*Status of wildlife*	*Management status*	*Degree of threat*	*Local involvement*
II	Kayan Mentarang	Indonesia	Intact	None	Facing serious threat	Some
II	Pico da Neblina	Brazil	Fairly intact	None	Facing serious threat	None
II	Morro do Diabo	Brazil	Fairly intact	Minimal	Some degradation	Some
II	Cutervo	Peru	Fairly intact	Minimal	Facing serious threat	None
II	Bahuaja Sonene	Peru	Fairly intact	Minimal	Facing serious threat	Some
III	Tumbes	Peru	Fairly intact	Serious gaps	Facing serious threat	Some
IV	Wonga-Wongue	Gabon	Fairly intact	Minimal	Some degradation	None
VI	Pacaya Samiria	Peru	Fairly intact	Serious gaps	Facing serious threat	Some
VI	Garu	PNG	Fairly intact	Minimal	Some degradation	Co-managed
Ia	Gurupi	Brazil	Degraded	Minimal	Considerable degradation	None
Ia	Dong-zaigang	China	Degraded	Serious gaps	Considerable degradation	None
II	Jacupiranga	Brazil	Degraded	Minimal	Considerable degradation	None
II	Mata do Rio Vermelho	Brazil	Degraded	None	Some degradation	None
II	Ujung Kulon	Indonesia	Degraded	Serious gaps	Some degradation	Limited
II	El Chico	Mexico	Degraded	Minimal	Some degradation	None
II	El Tepozteco	Mexico	Degraded	None	Degradation throughout	Some
II	Tingo Maria	Peru	Degraded	Minimal	Considerable degradation	None
III	Machu Picchu	Peru	Degraded	Minimal	Some degradation	Some
IV	Moukalaba	Gabon	Degraded	Minimal	Some degradation	None
IV	Bien Lac-Nui Ong	Vietnam	Degraded	Partial	Some degradation	Some

Source: Shepherd and O'Connor 2001.

Table 15.5 *Analytic summary of table 15.4*

	1a		*II*		*III*		*IV*		*VI*	
Status of wildlife	—		Fairly intact	45%	Fairly intact	50%	Fairly intact	33%	Fairly intact	100%
	Degraded	100%	Degraded	55%	Degraded	50%	Degraded	67%	—	
Management status	Minimal	50%	Minimal	91%	Minimal	50%	Minimal	67%	Minimal	50%
	Serious gaps	50%	Serious gaps	9%	Serious gaps	50%	Partial	33%	Serious gaps	50%
Local involvement	None	100%	None	55%	—		None	67%	—	
	—		Some	45%	Some	100%	Some	33%	Some	100%

Some once-poor countries have escaped from poverty and can now accommodate protected areas more easily. One can hypothesize that, over time, similar evolutions will occur elsewhere. Nevertheless, in the case of Africa in particular, there is no sign of this transition in the short term.

For all these reasons, international conservation organizations need to nuance their initiatives far more than they have done so far, to fit with the reality of country situations. They will not succeed in their mission if they do not take this step from the global and universal to the specific and local.

Improving the Capacity of ICDPs to Sustain Livelihoods

This paper has suggested a series of changes needed at various levels in order to create an environment in which ICDPs can address poverty more effectively. It has focused in particular on the local level where conservation areas and poor people cluster together in the remoter regions of the world. In these areas, natural resources are a fundamental component of livelihoods, which cannot be withdrawn without gravely damaging those livelihoods.

But it has also been suggested that work at the local level must be complemented both by action at the national level in support of local decision making and management, and by action at the international level to bring conservation and poverty concerns into a closer, more complementary relationship.

Committing to Greater Engagement with Local Livelihoods

Some of these changes will require a far more thorough analysis of local livelihoods than has been conducted by ICDPs in the past. This will involve mastering new skills, but it is changed attitudes that are needed above all. In the past, conservation organizations have been, perhaps understandably, reluctant to become deeply involved in the socioeconomic aspects of the places where they work. Their key employees and managers are usually trained in the biological sciences, and they want to get on with conservation. Involvement with local people has been reluctant and limited for the most part, and with biodiversity concerns at the forefront.

But experience to date with ICDPs, well illustrated in this volume and elsewhere (e.g., Brandon, Redford, and Sanderson 1998), demonstrates conclusively that this has not worked, and will no longer do.

The answer in many of the poorer parts of the world is that conservation organizations and local people should make common cause against the much larger forces that could easily sweep away the forested landscapes that both want to retain.

In order for conservation bodies to rise to the challenge, they will again need to look to the last fifteen years of forestry experience for guidance, hiring from a wider range of disciplines and acquiring new expertise. There is still, for instance, a too limited familiarity among conservationists and protected area managers with local people's independent forest management, common property resource management, or joint forest management (e.g., Ostrom 1990; Shepherd 1992; Saxena 1997; Arnold 1998; Hill and Shields 1998). This is a pity, since these literatures make it clear when local management regimes can and cannot be successfully built on. Thus the circumstances in which people do or do not want to manage forests, and how they manage, is not yet sufficiently familiar ground. Yet collaborative protected area management is much more difficult to implement than collaborative forest management, giving local people less and changing their lifestyle more.

ICDPs also need to master the experience to date on land-use intensification (e.g., Hazell 1995; Lee and Barrett 2000), participatory poverty analysis techniques (Norton et al. 2001), and a host of other topics. It will be necessary to go well beyond biodiversity concerns, to engage with the agricultural and economic matrix into which local livelihoods and the forest itself are embedded.

Committing to More Collaborative Engagement at the National Level

Changes at the national level will require a closer, more constant, and more collaborative engagement with government and other poverty and natural resource donors and their institutions and policies. After all, larger-scale approaches are likely to be part of the conservation world's future armory, and these must imply a more sophisticated understanding of the political and economic landscape—the art of the possible—as well as better understanding of the physical landscape. Through such linkages, the situation of local people, and the national policy, legal, and regulatory changes that might improve their situation, will become clearer.

Committing to Broader Engagement at the International Level

At the international level, conservation organizations first need to move outside their normal circles and engage more with other players—especially those with strong forestry and poverty reduction implementation experience. Above all, out of this will grow an ability to ally global analyses of biodiversity priorities to a disaggregated, site-specific approach that recognizes and adapts to the constraints inevitable in dealing with poor people and poor countries.

Committing to Better Analysis, Monitoring, and Evaluation

There is a further area where improvements would make an even greater difference, and that is in the area of analysis, monitoring, and evaluation. Conservation organizations have been surprisingly weak in this area and, in contrast with bilateral agencies, have displayed very poor rigor (Shepherd and O'Connor 2001). Why this is so is not entirely clear, but it seems to be tied to perceived fund-raising needs. There has been the temptation to give the public positive messages about agency capacity, coupled with dire warnings about the state of the world's biodiversity, to trigger giving. In the process, the facts that agencies themselves need to know have not been sufficiently carefully gathered, or analyzed.

Much ICDP-received wisdom has thus remained stuck in the realm of the normative rather than the actual, with not even trial and error playing a sufficient role. The disadvantaging of the poor through the selection of excessively exclusive IUCN Protected Area Categories is perhaps an example of the perils of poor data gathering. This is clearly an area in need of improvement. Poor monitoring of both biodiversity successes and of livelihood impacts has meant that ICDPs have fallen down on both of their missions, yet have not been able to understand why or learn from earlier efforts. Continual monitoring and evaluation, and the adaptive management that should follow from it, must now be an urgent priority (Salafsky and Margoluis this volume).

Conclusion

ICDPs and the conservation bodies that manage them have been out of the mainstream of changing thought about forest management and rural livelihoods and now risk getting stuck with an old-fashioned paradigm. Though it will always be difficult to persuade people to abandon some of their old assumptions, it is now urgent that they consider doing so.

Confronted with the continuing disappearance of the world's forests and wild places, it is easy to panic at the global level. It is harder, but now essential, to step down to the specific and the local, and patiently, patch by patch, build up a safely conserved world from there.

Endnotes

1. Low-income countries are heavily clustered in sub-Saharan Africa, in South Asia, in parts of East and Central Asia, and in Vietnam, Cambodia, and Laos in Southeast Asia.

2. The livelihoods of the transient poor fluctuate around the national poverty line, while the chronic poor always exist well below it. They are more likely to have been born poor and to bequeath poverty to their children.

3. *Natural capital* is defined as natural resources such as land, forests, water, and pastures. *Physical capital* means *(a)* privately owned assets that can be used to increase labor and land productivity, such as farm animals, tools, machinery; and *(b)* publicly owned infrastructure (e.g., roads, electricity supply, schools, hospitals). *Financial capital* means cash income, savings, and readily convertible liquid capital. *Human capital* means health, nutritional levels, educational standards, and skills. Finally, by *social capital* we mean the social relationships on which people can draw to expand livelihood options. These include kinship, friendship, patron-client relations, exchange arrangements, membership in groups and organizations that provide information, loans, training, etc.

4. This section is based on Byron and Arnold 1997.

5. Whereas those who depend primarily on farming accord a similar religious role to the grain staple, to cattle, etc.

6. Project Web site: www.metsa.fi/eng/tat/usambara/.

References

Abbot, P. 1997. The supply and demand dynamics of miombo: A household perspective. Ph.D diss., Aberdeen University.

Arnold, J. E. M. 1998. *Managing Forests as Common Property*. FAO Forestry Paper no. 136. Rome: Food and Agriculture Organization of the United Nations (FAO); London: Overseas Development Institute (ODI).

Barrow, E., J. Clarke, I. Grundy, R. Kamugisha-Jones, and Y. Tessema. 2002. *Analysis of Stakeholder Power and Responsibilities in Community Involvement in Forest Management in Eastern and Southern Africa*. Nairobi: IUCN Regional Office for Eastern Africa (EARO).

Berkes, F., ed. 1989. *Common Property Resources: Ecology and Community-Based Sustainable Development*. London: Belhaven Press.

Bird, K., D. Hulme, K. Moore, and A. Shepherd. 2002. *Chronic Poverty and Remote Rural Areas*. CPRC Working Paper. Manchester, U.K.: Institute of Development Policy and Management, University of Manchester.

Brandon, K., K. Redford, and S. Sanderson, eds. 1998. *Parks in Peril: People, Politics, and Protected Areas.* Washington, D.C.: Island Press.

Bromley, D. W., ed. 1992. *Making the Commons Work: Theory, Practice, and Policy.* San Francisco: Institute for Contemporary Studies.

Byron, N. and J. E. M. Arnold. 1997. *What Futures for the People of the Tropical Forests?* CIFOR Working Paper no. 19. Bogor, Indonesia: Center for International Forestry Research (CIFOR).

Carney, D., ed. 1998. *Sustainable Rural Livelihoods: What Contribution Can We Make?* Papers from the Department for International Development's Natural Resources Advisers' Conference. London: Department for International Development (DFID).

Dove, M. 1996. "So far from power, so near to the forest": A structural analysis of gain and blame in tropical forest development. In C. Padoch and N. L. Peluso, eds., *Borneo in Transition: People, Forests, Conservation, and Development*. Kuala Lumpur: Oxford University Press.

Falconer, J. 1989. *Forestry and Nutrition: A Reference Manual*. Rome: Forestry Department, Food and Agriculture Organization of the United Nations (FAO).

Falconer, J. 1990. *The Major Significance of "Minor" Forest Products: The Local Use and Value of Forests in the West African Humid Forest Zone*. Community Forestry Note 6. Rome: Food and Agriculture Organization of the United Nations (FAO).

Falconer, J. 1994. *Non-Timber Forest Products in Southern Ghana: Main Report*. Chatham, U.K.: Natural Resources Institute for the Republic of Ghana Forestry Department and Overseas Development Administration (ODA).

Hazell, P. 1995. *Managing Agricultural Intensification*. 2020 Vision Brief 11. Washington, D.C.: International Food Policy Research Institute (IFPRI).

Hill, I. and D. Shields. 1998. *Incentives for Joint Forest Management in India: Analytical Methods and Case Studies*. World Bank Technical Paper 394. Washington, D.C: The World Bank.

Hobley, M. 1996. *Participatory Forestry: The Process of Change in India and Nepal*. Rural Development Forestry Study Guide 3. London: Overseas Development Institute (ODI).

IUCN. 1999. *Threats to Forest Protected Areas: A Survey of 10 Countries Carried Out in Association with the World Commission on Protected Areas*. Gland, Switzerland: IUCN–The World Conservation Union.

Kaimowitz, D., T. Graham, and P. Pacheco. 1999. The effects of structural adjustment on deforestation and forest degradation in lowland Bolivia. *World Development* 27 (3):505–520.

Kerkhof, P. 2000. *Local Forest Management in the Sahel: Towards a New Social Contract*. London: SOS Sahel International.

Lee, D. R. and C. B. Barrett. 2000. *Tradeoffs or Synergies? Agricultural Intensification, Economic Development, and the Environment*. Wallingford, Oxon, U.K.: CABI.

Lipton, M. 1977. *Why Poor People Stay Poor: A Study of Urban Bias in World Development*. London: Temple Smith.

Mayers, J., J. Ngalande, P. Bird, and B. Sibale. 2001. *Forestry Tactics: Lessons from Malawi's National Forestry Programme*. London: International Institute for Environment and Development (IIED).

Milton, R. K. 1998. Forest dependence and participatory forest management: A qualitative analysis of resource use in Southern Ghana. Ph.D. diss., School of Development Studies, University of East Anglia.

Norton, A., B. Bird, K. Brock, M. Kakande, and C. Turk. 2001. *A Rough Guide to PPAs: Participatory Poverty Assessment—An Introduction to Theory and Practice*. London: Overseas Development Institute (ODI).

Ostrom, E. 1990. *Governing the Commons: The Evolution of Institutions for Collective Action*. Cambridge: Cambridge University Press.

Prabhu, R., C. Colfer, and G. Shepherd. 1998. *Criteria and Indicators for Sustainable Forest Management: New Findings from CIFOR's Forest Management Unit Level Research*. Rural Development Forestry Network Paper 23a. London: Overseas Development Institute (ODI).

Raintree, J. B., ed. 1987. *Land, Trees, and Tenure*. Nairobi: International Council for Research in Agroforestry (ICRAF); Madison, Wis.: Land Tenure Center (LTC).

Ribot, J. 1995. *Local Forest Control in Burkina Faso, Mali, Niger, Senegal and the Gambia: A Review and Critique of New Participatory Policies*. Washington, D.C.: The World Bank/Review of Policies in the Traditional Energy Sector (RPTES).

Saxena, N. C. 1997. *The Saga of Participatory Forest Management in India*. Bangor, Indonesia: Center for International Forestry Research (CIFOR).

Schreckenberg, K. 1996. Forest fields and markets: A study of indigenous tree products in the woody savannas of the Bassila Region, Benin. Ph.D. diss., University of London.

Schreckenberg, K., A. Degrande, C. Mbosso, Z. Eoli Baboule, C. Boyd, L. Enyong, J. Kanmegne, and C. Ngong. 2002. The social and economic importance of *Dacryoides edulis* in S. Cameroon. *Journal of Forests, Trees, and Livelihoods* 12 (2):15–40.

Shepherd, G. 1992. *Managing Africa's Tropical Dry Forests: A Review of Indigenous Methods*. London: Overseas Development Institute (ODI).

Shepherd, G. 1993a. *The Rural Development Component of the Korup National Park Project, Cameroon*. London: Overseas Development Administration (ODA).

Shepherd, G. 1993b. Local and national level forest management strategies—Competing priorities at the forest boundary: The case of Madagascar and Cameroon. *Commonwealth Forestry Review* 72 (4):316–320.

Shepherd, G. and G. Gill. 1999. *Community Forestry and Rural Livelihoods in Nepal: Issues and Options for the Future. The Challenges to Community Forestry from a Sustainable Livelihoods Framework*. London: Overseas Development Institute (ODI).

Shepherd, G. and H. O'Connor. 2001. *People and Conservation: A Review of WWF Policies and Processes*. London: Overseas Development Institute (ODI).

Shepherd, G., P. Paalanen, A. Launiala, M. Richards, and J. Davies. 1999. Forest sector development co-operation in the sub-Saharan African and Central American regions: An analysis of the Finnish forestry approach. In *Thematic Evaluations on Environment and Development in Finnish Development Co-operation*. Evaluation Publication Number 1999:4. Helsinki: Ministry for Foreign Affairs of Finland, DIDC.

Talbott, K. and S. Khadka. 1994. *Handing It Over: An Analysis of the Legal and Policy Framework of Community Forestry in Nepal*. Washington, D.C.: World Resources Institute.

Thomson, J. and K. Schoonmaker-Freudenberger. 1997. *Crafting Institutional Arrangements for Community Forestry*. Community Forestry Field Manual 7. Rome: Food and Agriculture Organization of the United Nations (FAO).

UNDP (UN Development Program). 2001. *Human Development Report 2001.* New York: Oxford University Press.

Wily, L. A. and S. Mbaya. 2001. *Land, People, and Forests in Eastern and Southern Africa at the Beginning of the 21st Century.* Nairobi: IUCN Regional Office for Eastern Africa (EARO).

World Bank. 2001. *World Development Report 2000/2001: Attacking Poverty.* New York: Oxford University Press.

WRI (World Resources Institute), UNEP (UN Environment Program), UNDP (UN Development Program), and the World Bank. 1998. *World Resources 1998–99: A Guide to the Global Environment.* New York: Oxford University Press.

16
Using Adaptive Management to Improve ICDPs

Nick Salafsky and Richard Margoluis

> If the only tool you have is a hammer, every problem looks like a nail.
>
> —PROVERB

Introduction

One of the main questions of this book is:

"Do integrated conservation and development projects (ICDPs) work?"

At first glance, this seems to be a straightforward question about a conservation tool that should have a straightforward answer. But let's consider an analogous question about an actual tool:

"Do hammers work?"

If you think about trying to answer this question, at least three problems emerge:

1. *What is a "hammer"?* The first problem is with the word *hammer.* If you consider it, there is no single tool that is a hammer. Instead, there are small claw hammers, large sledgehammers, and power jackhammers. *Hammer* is just a word for a category of tools. So

clearly, to answer the question, you need to have a much better definition of what specific tools you are talking about within this broad category.

2. *What does "work" mean?* The second problem is with the word *work.* Obviously, whether or not a given type of hammer will work depends on the goal of the task at hand. If you are trying to pound in or pull out small nails, then a small claw hammer might make sense. If you are talking about breaking concrete, a sledge or jackhammer might be okay. But if you are trying to sew clothing, cook food, or fly across the ocean, a hammer will not work.
3. *In what context is the hammer being used?* The third problem is with the lack of specificity as to the conditions under which the hammer is being used. Even if you define the problem as breaking concrete, you still need to know what other factors might affect your assessment as to whether this is the right tool for the job. Is a manual sledgehammer better than a power jackhammer for breaking concrete? The answer depends on your budget, how much concrete you have to deal with, how fast you need it broken up, and your knowledge about and skill in using the different kinds of hammers.

By this point it should be obvious why the original question about ICDPs is difficult to answer in its current form. There are at least three fundamental questions embedded in this original question:

1. What specific strategies and tools do ICDPs use?
2. What is the goal of an ICDP?
3. Under what conditions does each strategy or tool work?

We believe the answers to these fundamental questions can best be found by taking an adaptive management approach. Thus, before we examine each of these questions, we first provide an introduction to adaptive management in conservation.

An Introduction to Adaptive Management in Conservation

In this section, we first consider the conditions that warrant an adaptive management approach. We then propose a definition for adaptive management and outline the steps in the adaptive management process at a project level. Finally, we discuss applying these concepts to groups of projects in a learning portfolio.

Conditions That Warrant an Adaptive Management Approach

Conservation takes place in complex systems. Systems theory holds that there are two main sources of complexity. *Detail complexity* refers to the large number of variables in the system. For example, conservation practitioners must deal with ecological, geophysical, social, demographic, economic, political, and institutional factors. *Dynamic complexity* refers to the unpredictable ways in which these factors interact with one another. In the systems in which conservationists operate, change is not only constant, but also often nonlinear and not predictable, and takes place over many different time scales. Since conservation involves combining both natural ecosystems and human societies, we are dealing with systems that are extremely complex both in detail and in dynamic.

As conservationists, we are also dealing with competitors for access to natural resources who are constantly changing their tactics and strategy. Furthermore, with rare exceptions, conservation projects are most often managed by governmental agencies or non-governmental organizations that have far fewer financial and human capital resources than their competitors (see Brandon and O'Herron, Tongson and Dino, Gartlan, Singh and Sharma, Child and Dalal-Clayton, all this volume). We thus have to be smarter and more flexible to get and stay ahead of our competition.

Finally, the current rates of biodiversity loss demand that we take immediate action despite the risks inherent in our lack of certainty about how best to proceed. We may not have the best available information. We may not know exactly what the outcomes of our proposed actions will be. But we cannot wait to take action until we do know for sure. We need to act now.

Over the past few decades, different fields dealing with complex systems have developed convergent approaches for deciding how to take action in the face of risk and uncertainty. Examples of these approaches include adaptive management of ecosystems (Lee 1993; Gunderson, Holling, and Light 1995), reflective practice (Schön 1984), social learning (Argyris and Schön 1978), and the theory of learning organizations (Senge 1994). We use the term *adaptive management* to refer to this type of approach within conservation.

In a book written with Kent Redford, *Adaptive Management: A Tool for Conservation Practitioners* (Salafsky, Margoluis, and Redford 2001), we used these different sources to develop a definition and preliminary framework for adaptive management. We then tested this framework with conservation projects in Canada, Zambia, and Papua New Guinea.

Table 16.1 *Framework for adaptive management of conservation projects*

Definition of adaptive management
Adaptive management incorporates research into conservation action. Specifically, it is the continuous integration of design, management, and monitoring to systematically test assumptions in order to adapt and learn.
Conditions that warrant an adaptive management approach
Condition 1: Conservation projects take place in complex systems Condition 2: The world is a constantly and unpredictably changing place Condition 3: Our "competitors" are changing and adapting Condition 4: Immediate action is required Condition 5: There is no such thing as complete information Condition 6: We can learn and improve
Steps in the process of adaptive management
Start: Establish a clear and common purpose Step A: Design an explicit model of your system Step B: Develop a management plan that maximizes results and learning Step C: Develop a monitoring plan to test your assumptions Step D: Implement your management and monitoring plans Step E: Analyze data and communicate results Iterate: Use results to adapt and learn
Principles for the practice of adaptive management
Principle 1: Do adaptive management yourself Principle 2: Promote institutional curiosity and innovation Principle 3: Value failures Principle 4: Expect surprise and capitalize on crisis Principle 5: Encourage personal growth Principle 6: Create learning organizations and partnerships Principle 7: Contribute to global learning Principle 8: Practice the art of adaptive management

Source: Salafsky et al. 2002.

The final framework that we developed is presented in table 16.1 and summarized in the remainder of this section.

Definition of Adaptive Management

Adaptive management has recently begun to gain popularity in the mainstream conservation community. But what is it?

Some people ask, "Isn't adaptive management simply good management? Doesn't it merely involve trying something and then if it doesn't work, using your common sense to adapt and try something else—trial and error?" We believe that adaptive management is good management, but that not all good management is adaptive management. We also believe that adaptive management requires common sense, but that it is not a license to just try whatever you want. Instead, adaptive management requires an explicitly experimental—or "scientific"—approach to managing conservation projects.

On the other hand, although early proponents of adaptive management (e.g., Holling 1978; Gunderson, Holling, and Light 1995) used sophisticated ecological modeling techniques, adaptive management does not necessarily require such sophisticated methods. Instead, we propose the following definition:

> Adaptive management incorporates research into action. Specifically, it is the continuous integration of design, management, and monitoring to systematically *test assumptions* in order to *adapt* and *learn.*

This definition can be expanded. *Testing assumptions* is about systematically trying different actions to achieve a desired outcome. It is not, however, a random trial-and-error process. Instead, it involves first thinking about the situation at your project site and developing a specific set of assumptions about what is occurring and what actions you might be able to take to affect these events. You then implement these actions and monitor the actual results to see how they compare to the ones predicted by your assumptions. The key here is to develop an understanding not only of which actions work and which do not, but also why (see also Robinson and Redford this volume, McShane and Newby this volume).

Adaptation is about taking action to improve your project based on the results of your monitoring. If your project actions did not achieve the expected results, it is because either your assumptions were wrong, your actions were poorly executed, the conditions at the project site have changed, your monitoring was faulty—or some combination of these problems. Adaptation involves changing your assumptions and your interventions to respond to the new information obtained through monitoring efforts.

Learning is about systematically documenting the process that your team has gone through and the results you have achieved. This documentation will help your team avoid making the same mistakes in the future. Furthermore, it will enable other people in the broader conservation community to benefit from your experiences. Other practitioners are eager to learn from your successes and failures so that they can design and manage better projects and avoid some of the hazards and perils you encountered. By sharing the information that you have learned from your project, you will help conservation efforts around the world.

Perhaps the key feature of adaptive management is its incorporation of research into action. If one were to define a spectrum with pure research at one end and pure practice at the other, adaptive management is in the center. Pure researchers seek to understand how the world works and are successful if knowledge increases, regardless of what happens to the system they are studying. Pure practitioners seek to change the world but do not have the time or inclination to invest in trying to understand the system in which they are working. Adaptive managers attempt to reconcile these viewpoints—they are people who want to change the world and achieve a defined goal, but who are also willing to invest in systematically learning about whether their actions work or do not work, and why.

Project-Level Adaptive Management

A project can be defined as a group of people interested in taking action to achieve defined goals and objectives. Projects thus range from actions taken by villagers on a small island to restore their traditional resource management systems to a large multilateral-funded initiative. Figure 16.1 presents the steps involved in doing adaptive management at a project level (Margoluis and Salafsky 1998).

The starting point of the cycle involves determining who will participate in your project and your overall mission (see also Brown this volume). Once these are clear, step A involves assessing the conditions and determining the major threats to biodiversity at your project site. Using a *conceptual model* (cf. Margoluis and Salafsky 1998), your project team defines the conditions and relationships between key factors at your project site. Step B involves using this model to develop a project *management plan* that outlines the results that your team would like to accomplish and the specific actions that your team will undertake to reach them. Step C involves developing a *monitoring plan* for assessing your progress in implementing the project. Step D involves implementing your actions and monitoring plan. Step E involves analyzing the data collected during your monitoring efforts and communicating the information that you obtain to the appropriate audiences. Finally, you use the results of this analysis to change your project and learn how to do projects better in the future. Based on feedback information, you may want to modify your conceptual model, management plans, or monitoring plans.

The key to this process is that project design, management, and monitoring cannot be separated (see Sayer and Wells this volume). Instead,

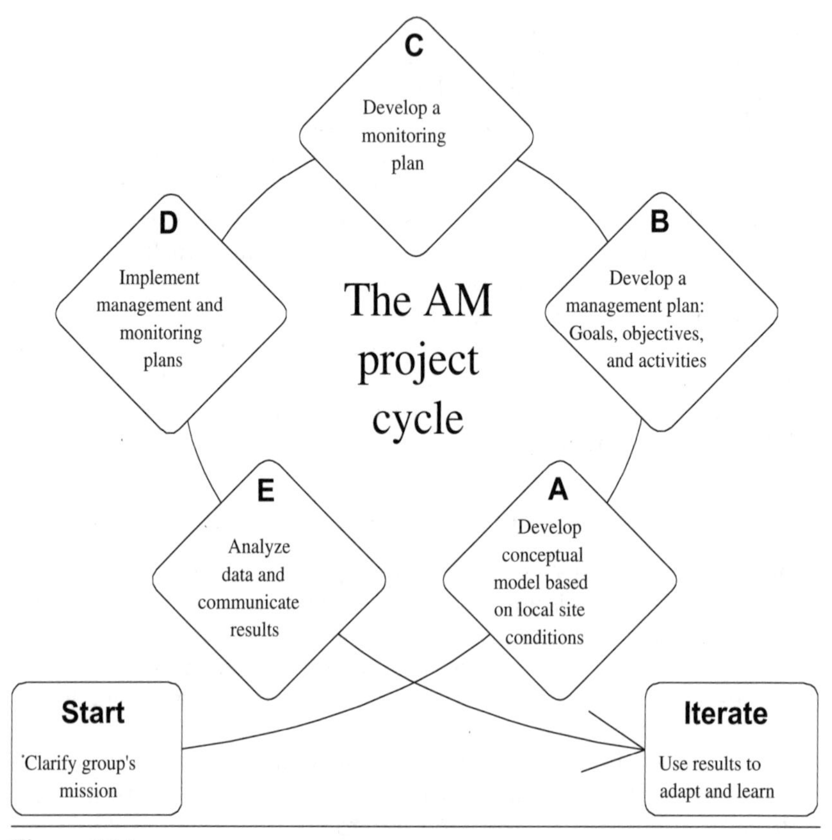

Figure 16.1.
The adaptive management project cycle. *Source:* Margoluis and Salafsky 1998.

monitoring must be integrated into the overall project cycle because it is essential for generating sound information on which management decisions are made. By systematically moving through the cycle, practitioners can test different assumptions about their actions to adapt and learn, thereby increasing their effectiveness and efficiency over time.

Portfolio-Level Adaptive Management

Project team members can learn about a conservation strategy they are using based on their own experiences. But there is probably another team a few kilometers down the road, two more teams in the same province, and a dozen or so teams around the world using and learning about a similar strategy. A *learning portfolio* seeks to bring together these different project teams to go through the adaptive management cycle together, as shown in figure 16.2 (Salafsky and Margoluis 1999; Foundations of Success 2001). By developing a common learning framework, collecting an agreed-on set of data, sharing stories and experiences, and promoting peer mentoring, members of the learning portfolio can much more efficiently improve and learn.

Any network of projects needs to have at least one common theme that holds the projects together—a network can be oriented toward a specific region, theme, donor, or conservation tool (see McShane and Newby this volume, Franks and Blomley this volume). We believe, however, that learning is most effective when the projects in a network have a similar goal and are organized around a common conservation strategy or tool to achieve this goal. Projects in a learning portfolio come together in a collaborative fashion in order to learn about the conditions under which the strategy or tool they are using is most effective. Project managers and other stakeholders in the learning portfolio work together to develop a *learning framework.* This document clearly articulates the major underlying assumptions about the common strategy

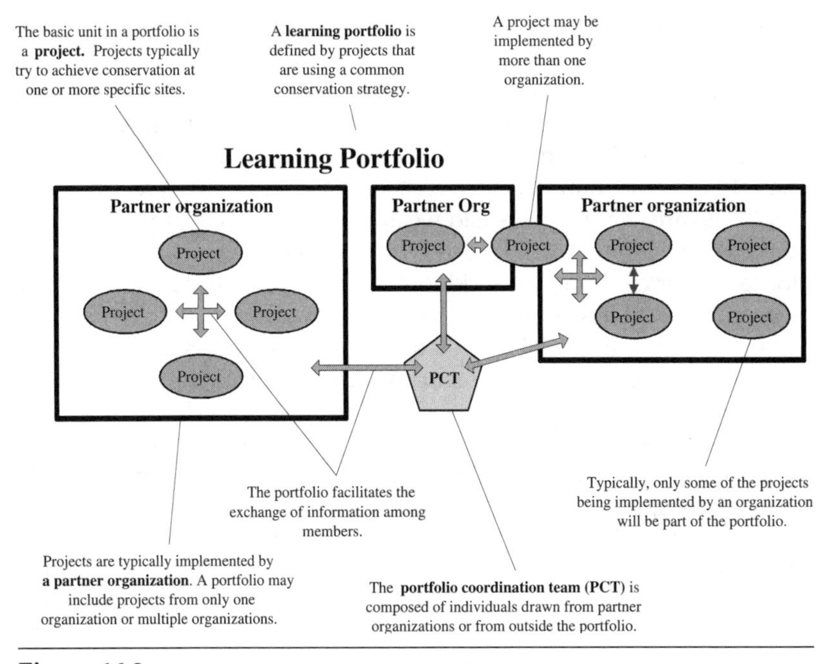

Figure 16.2.
Learning portfolio structure. *Source:* FOS 2001. See www.FOSonline.org for a more detailed discussion of learning portfolios and how to develop and implement them.

or tool that the portfolio wishes to test and learn about. The learning framework provides the basis for planning, implementation, monitoring, and analysis. It is the central tool in the adaptive management process conducted by the portfolio.

Using Adaptive Management to Improve ICDPs

At this point, we would like to return to the three fundamental questions posed at the start of this paper regarding the effectiveness of ICDPs:

1. What specific strategies and tools do ICDPs use?
2. What is the goal of an ICDP?
3. Under what conditions does each strategy or tool work?

In this section, we explore how adaptive management at project and portfolio levels might help answer these questions.

WHAT SPECIFIC STRATEGIES AND TOOLS DO ICDPS USE?

One of the first steps in either project- or portfolio-level adaptive management is to define the intervention we are using—what an ICDP approach to conservation is. This definition is important so that we can specify our assumptions as to how the approach works to reach our stated goals. ICDPs often seemed to be viewed as a specific conservation intervention. For example, a project team might say, "We are not using a protected area strategy; instead we are trying an ICDP approach." We believe, however, that ICDPs can be more accurately characterized as a loose cluster of strategies and tools brought together to achieve both conservation and development goals.

As shown in the row labels in table 16.2, conservation actions can be broadly grouped into four categories: direct protection and management, law and policy, education and awareness, and changing incentives (Salafsky et al. 2002). The bold font entries in each cell in the table are broad *approaches* to doing conservation. The italic font entries in each cell are more specific *strategies* under each broad approach. And finally, beneath each of these strategies (but not shown in this table) are the actual *tools* that organizations use to promote conservation. Although they are at a level of detail beyond the scope of this table, these tools are ultimately what conservation practitioners use. Conservation projects do not set up enterprises or even linked enterprises in a generic sense. Instead, they set up specific businesses such as community homestays or forest guiding services.

Specific strategies and conservation tools under each approach can be defined based on the scale on which the tool is being used and the actor using the tool, among other factors. For example, under the strategy "reserves and parks," we could use tools that include large parks administered by a national government, medium-sized reserves managed by a private organization, and small community-run sacred groves. Under the strategy "media campaigns," we could use tools that include global television advertising campaigns on CNN or efforts to write letters to the editor of a small-town newspaper.

Almost all of the strategies shown in this table have probably been used at one time or another in an ICDP somewhere in the world. However, as a starting point, we have marked the approaches in the table that we think are most commonly associated with ICDPs. ICDPs almost always involve the simultaneous implementation of multiple strategies and tools (see McShane and Newby, Kiss, Glick and Freese, Tongson and Dino, Singh and Sharma, Child and Dalal-Clayton, all this volume). For example, community-based enterprise approaches to conservation are usually implemented in tandem with other interventions such as environmental education and enforcement.

What Is the Goal of an ICDP?

Another early step in the adaptive management process is to define the goal you are trying to reach. This is a problem that is especially germane to ICDPs, which by definition are trying to reach both conservation and development goals. The direction that a project takes will depend on where the project is in relation to these two goals. If the two goals are located at the same spot (or at least close to one another), then the project team can head toward both goals simultaneously. If, however, the two goals are located far apart from each other and require radically different actions to be reached, then the project team has a major problem. The team has to decide which goal they want to reach—each step toward one goal takes them farther away from the other (see also Robinson and Redford this volume, Franks and Blomley this volume).

These two scenarios are relatively trivial cases. In most projects, however, the situation is probably one in which the goals are in the same general direction but are not in exactly the same place. To make things more complex, the goals are often not well defined and are measured against different metrics, making it much more difficult to compare whether progress toward one goal (such as biodiversity conservation) is a positive or negative contribution to progress toward another goal (such as economic development) (Robinson and Redford this volume).

Table 16.2 *A preliminary taxonomy of biodiversity conservation approaches and strategies*[a]

Protection and management	*Law and policy*
Protected Areas * *Reserves and parks:* IUCN Category I and II (Kenya Wildlife Service) *Private parks* (see Langholz et al. 2000)	**Legislation and Treaties** *Developing international treaties* (Convention on Biological Diversity) *Lobbying governments* (Sierra Club)
Managed Landscapes [b] * *Conservation easements* (see Gustanski 2000) *Community marine protected areas* (see Parks and Salafsky 2001)	**Compliance and Watchdog** *Developing legal standards* (CITES) *Monitoring compliance w/standards* (TRAFFIC)
Protected and Managed Species *Bans on killing specific species* (Convention for Regulation of Whaling) *Management of habitat for species* (Endangered Species Act)	**Litigation** *Criminal prosecution* (U.S. Fish and Wildlife Service) *Civil suits* (Sierra Club)
Species and Habitat Restoration * *Reintroducing predators* (U.S. Fish and Wildlife Service) *Recreating savannas and prairies* (see Stevens 1995, Dobson et al. 1997)	**Enforcement** * *Implementing sanctions* (U.S. Fish and Wildlife Service) *Military actions/nature keeping* (see Terborgh 1999)
Ex-Situ Protection *Captive breeding* (zoos, aquaria, and botanical gardens) *Gene banking* (Kew Gardens Millennium Seed Bank)	**Policy Development and Reform** *Research on policy options* (World Resources Institute) *Advocating devolution of control* (see Wyckoff-Baird et al. 2000)

Source: Adapted from Salafsky et al. 2002. This table categorizes the types of tools available to conservation practitioners. Columns contain broad categories of tools. Each cell contains a broad *approach* (bold font) and then two examples of more specific *strategies* (italic font) under this approach. Implementing each strategy involves using specific *conservation tools* (not shown). For each strategy, we also provide an example of an organization that is known for using this strategy and/or a reference describing and defining it.

[a]Approaches marked with an * are those most commonly used by ICDPs.

Education and awareness	*Changing incentives*
Formal Education [c]	**Conservation Enterprises** *
Developing school curricula (WWF Windows on the Wild)	*Linked:* e.g., ecotourism (see Salafsky and Wollenberg 2000)
Teaching graduate students (see Jacobson and Robinson 1990)	*Unlinked:* e.g., jobs for poachers (see Salafsky and Wollenberg 2000)
Nonformal Education [c]	**Using Market Pressure**
Media training for scientists (see Jacobson 1999)	*Certification:* positive incentives (Forest Stewardship Council)
Public outreach via museums (see Domroese and Sterling 1999)	*Boycotts:* negative incentives (Rainforest Action Network)
Informal Education [c] *	**Economic Alternatives** *
Media campaigns (Greenpeace)	*Sustainable agriculture/aquaculture* (see Margoluis et al. 2001)
Community awareness raising (Public Interest Research Groups)	*Promoting alternative products* (Viagra instead of rhino horn)
Moral Confrontation	**Conservation Payments** *
Civil disobedience (Greenpeace)	*Quid-pro-quo* performance payments (see Ferraro 2001)
Monkeywrenching/ecoterrorism (EarthFirst!)	*Debt-for-nature swaps* (Conservation International)
Communications	**Nonmonetary Values** *
Environmental publishing (Island Press)	*Spiritual, cultural, existence values* (see Ehrenfeld 1981)
Web-based networking (Forests.org)	*Links to human health* (see Meffe 1999)

[a] Citing specific organizations using a tool does not in any way imply that this is the only tool this organization uses, or that it is the only group using this tool.
[b] This category primarily includes taking conservation actions in lands managed for natural resource production that do not fall into IUCN categories I–V.
[c] These terms are following Fien, Scott, and Tilbury 1999.

Nonetheless, proponents of ICDPs often seem to assume that the multiple goals they are trying to reach are in more or less the same location—that one set of project actions will lead to win-win situations in which there are gains for both conservation of biodiversity and improvement of human welfare. In reality, however, although these two goals are clearly related, they are not the same thing. A model of the factors affecting conservation at a given site will clearly have human livelihood activities, institutional structures, and cultural values as factors positively and negatively affecting biodiversity. And a model of the factors affecting development at a given site will clearly have resources and services provided by biodiversity as factors affecting human welfare. But at best, these two goals are necessary but not sufficient conditions for each other. Thus, as shown in figure 16.3, human welfare depends at least in part on biodiversity to provide food, materials, and services. And conservation depends at least in part on human values for and the ability to protect biodiversity, which are related to human welfare (see also Shepherd this volume). But these relationships are not exclusive—more often than not, the goals are in at least partial opposition to each other. Achieving full conservation (almost by definition) demands restricting human economic development and vice versa.

Finally, having two (let alone three or four) widely divergent goals has a number of important effects on ICDPs in terms of making it difficult to conduct effective adaptive management. It is difficult to create a simple model of what is going on at the site. It is difficult to determine what actions the project should take. It is difficult to state assumptions about how proposed actions will affect the goals. And it is difficult to figure out what data to collect to test these assumptions.

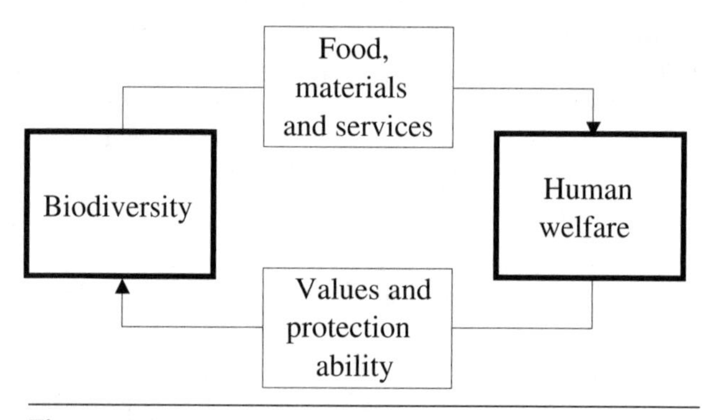

Figure 16.3.
Biodiversity and human welfare are related, but not the same. *Source:* Adapted from Locally Managed Marine Area Network 2001.

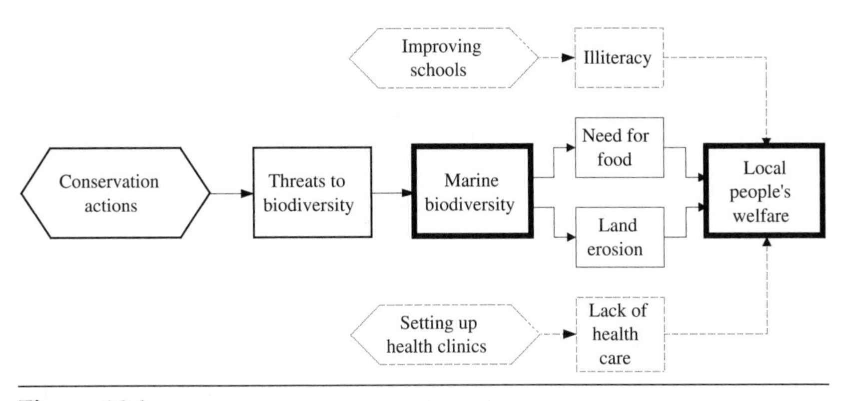

Figure 16.4.
Marine biodiversity as the focal goal in relation to an ultimate goal.
Source: Adapted from Locally Managed Marine Area Network 2001.

We believe that projects should pick one primary goal. If the project team insists on having two goals, then one solution is to ask the team to define these goals in relation to each other and then concentrate on one as the primary focus (see also Franks and Blomley this volume, Brown this volume). For example, consider a project that is focusing on the marine resources owned and managed by a local community. The project might state that its ultimate goal is to "promote the welfare of local people in the community." To this end, the team develops a conceptual model that has "local people's welfare" as their ultimate target, as shown in figure 16.4. The project recognizes that there are a number of factors that affect human welfare—in this case, erosion of the small island and the need for food, medical care, and basic education. Recognizing that they cannot do everything and that their strength is in environmental issues, the project team chooses to focus on those aspects of human welfare that are linked to maintaining marine biodiversity. To this end, they only take actions that are related to their primary goal of improving marine biodiversity. This does not mean that they are not interested in human welfare—as their model shows, conservation is directly linked to human welfare. But they do not undertake actions that are not related to their primary goal.

A second solution is to create separate models of the situation in relation to each goal that the project wants to achieve. If the project team goes this route, however, then they have to be prepared to decide on how to make allocations between the two goals—and, therefore, between what are effectively two separate projects. This means finding a way both to choose which actions to invest in and how to handle the inevitable trade-offs between the two goals (see also Brown this volume).

Both of these solutions essentially involve redefining traditional integrated conservation and development projects. Instead of a broad spectrum of stakeholder groups working on one large project, we are essentially proposing having each group define its own specific project with its own specific goal. The challenge then becomes for the various groups to figure out how to mesh their specific projects with one another so as to most effectively reach their desired ends.

UNDER WHAT CONDITIONS DOES EACH STRATEGY OR TOOL WORK?

Once we have clearly defined a tool to be used to achieve a specific conservation goal, we can begin to learn about its effectiveness and efficiency. As we discussed earlier, ICDPs almost always involve the simultaneous implementation of multiple interventions. Different tools have different levels of effectiveness in different situations. Similarly, the cost to implement various conservation tools differs widely. For example, a television-based media campaign in the United States would be much more expensive than working with teachers in a village in Guatemala to develop a curriculum for primary school children. Keeping in mind issues of effectiveness and efficiency, project managers need to know two things about specific tools when they are designing their projects. They need to know how a specific tool behaves independent of any other tool, and they need to know how the tool behaves or interacts with other tools.

Adaptive management can be used at both project and portfolio levels to determine the conditions under which specific conservation tools work, both on their own and in combination with other tools. This learning occurs in individual projects as managers determine what works best at their site, and it happens at a portfolio level in which practitioners can look across multiple projects—each with its own unique set of circumstances—to determine the conditions under which the tool works.

We illustrate this process using an example from Guatemala and Mexico in which project managers tested the effectiveness of sustainable agriculture at project and portfolio levels (Margoluis et al. 2001). For the purposes of this study, sustainable agriculture was defined as a strategy that promotes farmer-based subsistence agricultural technologies that intensify production and reduce deforestation. These programs typically incorporate a number of techniques such as cover crops, minimum tillage, barriers, integrated pest management, and terracing. Four organizations participated in this learning portfolio: Defensores de la Naturaleza, working in the Sierra de las Miñas Biosphere Reserve in

Guatemala; Línea Biósfera, working in the El Ocote Biosphere Reserve in Mexico; the Biodiversity Support Program; and the Center for International Forestry Research (CIFOR). Another similar example of project- and portfolio-level learning can be found in the results of the DGIS-WWF Tropical Forest Portfolio (McShane and Newby this volume).

Learning at the Project Level

As we discussed earlier, in project-level adaptive management, learning occurs by systematically testing assumptions. In practical terms, this involves monitoring various indicators associated with the factors that make up a causal chain that connects an intervention to a conservation goal. For example, the major causal chain that the Defensores de la Naturaleza project team wanted to analyze is illustrated in figure 16.5. Project managers believed that adoption of sustainable agriculture techniques would lead farmers to intensify their activities on existing plots, leading to increased yields of corn and beans. Furthermore, they believed that once farmers saw yields (kg/ha) rise, they would not perceive the need to expand their fields as their families grew and would therefore not cut down new forest areas to expand their fields. In this case, in order to test the association between sustainable agriculture and conservation success, Defensores needed to monitor the conservation goal (decreased rates of deforestation) and the key factors along the causal chain (adoption, labor/ha input, subsistence crop yield, and perceived need for more land). By monitoring these indicators, project managers could determine the extent to which assumptions held true.

When analyzing a causal chain and the factors that it comprises, there are four possible outcomes, as shown in figure 16.6. By testing the assumptions that are inherent along a causal chain and diagnosing why a specific tool leads to or does not lead to a specific conservation outcome, we can learn about what will lead to project success or failure at a given site. In some cases, we may find that the condition of a particular factor in the causal chain leads to project failure. Indeed, this was the case in the Sierra de las Miñas—farmers who used sustainable agri-

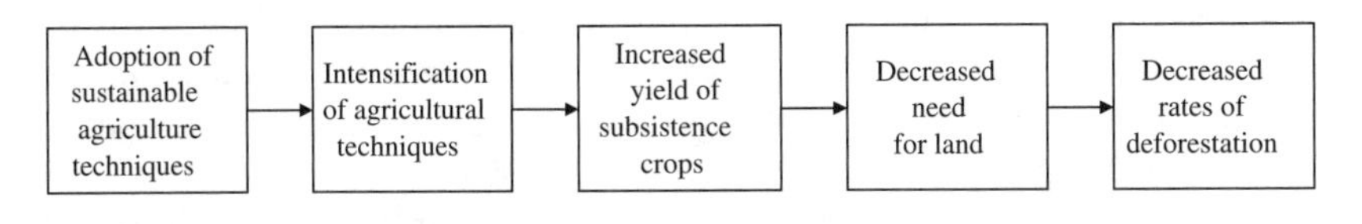

Figure 16.5.
Sustainable agriculture causal chain. *Source:* Margoluis et al. 2001.

Senario 1
Assumptions supported

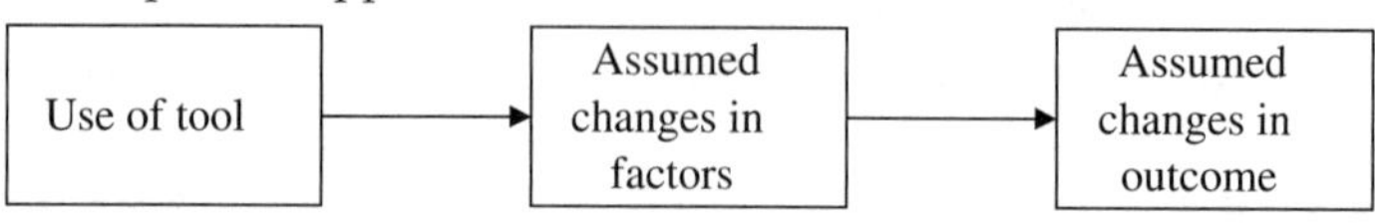

Senario 2
Assumptions not supported: terminal failure

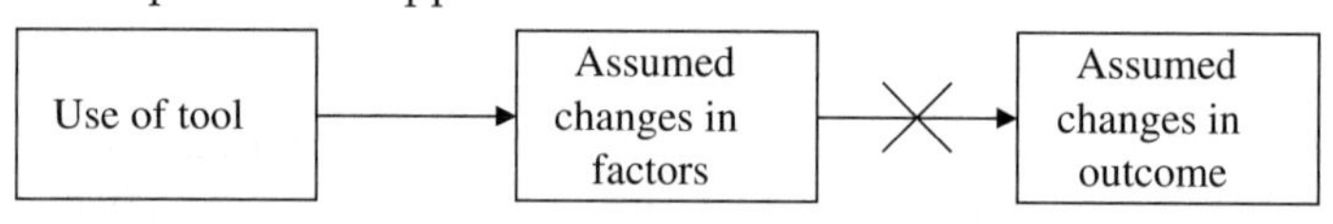

Senario 3
Assumptions not supported: Intermediate failure

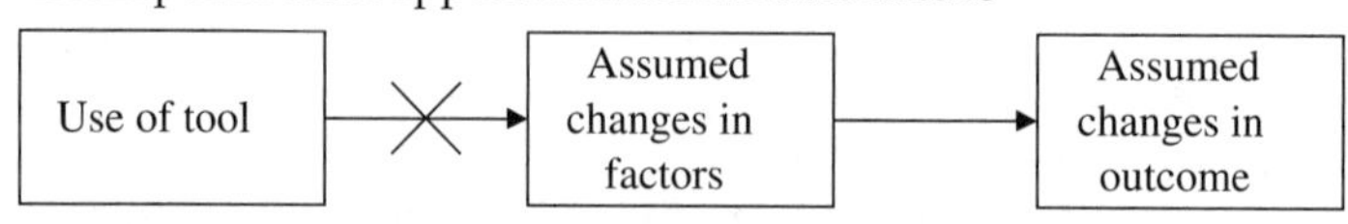

Senario 4
Assumptions not supported: complete failure!

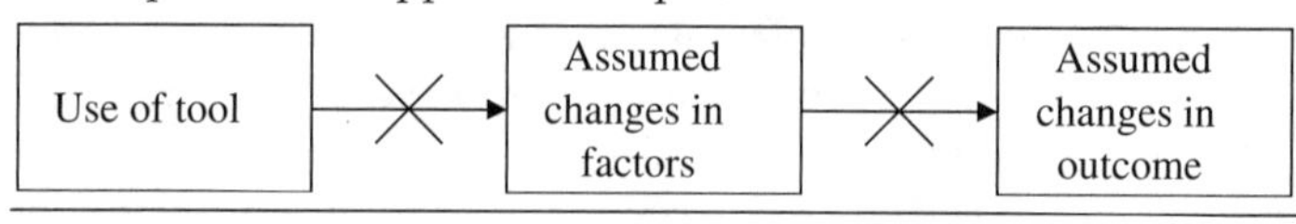

Figure 16.6.
Four scenarios describing possible cause-and-effect relationships between a tool and the desired outcome. *Source:* Adapted from Margoluis and Salafsky 1998.

culture techniques demonstrated no significant increase in crop yields over farmers who did not use the techniques.

In other cases, we may find that the logic of the entire causal chain is flawed. For example, in the Sierra de las Miñas, instead of using their saved labor (from more efficient techniques) for other less destructive activities, farmers who used sustainable agriculture actually invested their saved time in opening new agricultural lands to increase production. In still other cases, we may find that unanticipated factors have the greatest impact on outcome. Again, in the Sierra de las Miñas, the factors "access to land" and "ownership of land" were not included in the original causal chain and yet they proved to be the factors most related to deforestation, after controlling for family size.

We may also discover that the scale of the intervention does not match the scale of the desired outcome (see also Maginnis, Jackson, and Dudley this volume). While Defensores de la Naturaleza worked with only a few hundred families on sustainable agriculture, tens of thousands of subsistence families live in and around the Sierra de las Miñas. Finally, we may discover that the causal chain is totally unrelated to the outcome. In the Sierra de las Miñas, managers postulated that the need to open new lands for subsistence crops was the main driver of deforestation. In fact, subsistence crops had little bearing on deforestation. Whatever the case may be, project managers can learn about the conditions under which a tool works at a particular site by testing assumptions through monitoring during the life of the project.

In addition to analyzing a specific tool, project managers can test the effectiveness and efficiency of combinations of tools. This is especially important for ICDPs. Conceptually, how this is done is no different from the way we just described testing a single tool. In this case, however, additional tools are treated as additional factors in the causal chain and are measured in a similar fashion. How these tools interact with one another becomes additional information to determine the conditions under which they are effective. In Mexico, for example, Línea Biósfera discovered that the best combination of sustainable agriculture tools to decrease labor demands was cover crops, integrated pest management, and minimum tillage.

Operationally, testing the interaction of multiple tools is different from testing a single tool. In most cases, manipulating the factors in a causal chain is not possible or ethical. When analyzing how different tools interact with one another at a specific site, however, project managers usually have a greater opportunity to control the presence, intensity, and timing of the tools they use at their site. Often, this can be managed in controlled quasi-experimental ways (often referred to as "piloting" projects). By taking a somewhat experimental approach to learning about which tools or combination of tools work, do not work, and why, project managers can more effectively and efficiently determine the right combination of tools to optimize success at their site.

Learning at the Portfolio Level

Learning through adaptive management at a portfolio level is an even more powerful technique to determine the conditions under which strategies and tools work. As shown in figure 16.2, in a portfolio setting, individual projects can work together to test a single, mutually agreed on causal chain across multiple projects at multiple sites, thus dramatically increasing the sample size (see www.FOSonline.org for more information about learning portfolios and how to design and implement them). With greater sample size comes greater power to truly discern conditions under which an intervention works. This increase in power comes about because of the natural variation in each factor in the causal chain across the projects in the portfolio. Analyzing this variation helps determine the conditions under which the strategy or tool works, does not work, and why. Taking a portfolio approach to learning thus provides an opportunity to learn much more quickly and efficiently.

Fundamental to portfolio-level adaptive management is the need to develop a *learning framework* as described above. This framework clearly lays out a conceptual model, defines a specific strategy and set of tools, and outlines key factors, assumptions, data needs, and indicators that all projects in the portfolio agree to collect. For example, in our sustainable agriculture portfolio (Margoluis et al. 2001), the project partners focused on a sustainable agriculture strategy whose specific tools included planting velvet bean (a cover crop) and minimum tillage.

There is a great deal that can be learned by comparing already completed projects with one another to determine lessons about specific tools. Ultimately, however, we feel that these types of retrospective learning initiatives are severely handicapped because they often need data that were never collocated during the life of the project to adequately test assumptions. We thus feel that learning portfolios must be prospective—that is to say, they must lay out assumptions at the beginning and monitor information needed to test these assumptions throughout the life of the project. The ability to detect causal relationships between tools and outcomes is greatly enhanced if you can do it prospectively and see how slight adaptations of and modifications to the way tools are implemented affect conservation results. For example, the members of the sustainable agriculture learning portfolio first met to lay out their assumptions in 1997 and then worked to test these assumptions through the end of 2000. Learning portfolios cannot ignore previous work—indeed the first step is to gather existing information and knowledge. But they also must collect the specific information that they need to test key assumptions.

The "conditions under which a tool works" that we have described above are simply the variations that we see in a given factor across the projects in a learning portfolio. For example, in our example of the sustainable agriculture learning portfolio, for the factor "ownership of land," farmers in Mexico had legal title to their land, but farmers in Guatemala did not (see also Tongson and Dino this volume, Gartlan this volume). Similarly, for the factor "labor/ha inputs," we found that farmers in Mexico that used sustainable agriculture invested considerably more labor/ha than farmers that did not use sustainable agriculture. Meanwhile, we found that farmers in Guatemala that used sustainable agriculture invested considerably less labor/ha than farmers that did not use sustainable agriculture.

Analysis of these types of variations in the factors across the project sites gave us the opportunity to see how they interacted with the tool we were testing to influence the conservation outcome of interest. Some of the relationships between sustainable agriculture, conditions, and the conservation outcome are outlined below:

- Implementation of sustainable agriculture projects in areas where farmers have unlimited access to land does not act to decrease rates of deforestation.
- Implementation of sustainable agriculture projects in countries where government policies promote agricultural expansion does not act to decrease deforestation.
- Implementation of sustainable agriculture projects in areas where forest fires are a threat acts to reduce burning of agricultural lands and thus reduces deforestation.

One of the main goals of any learning portfolio is to develop general and yet nontrivial principles (Salafsky and Margoluis 1999). These principles are written as guidance for conservation practitioners inside and outside of the learning portfolio who may wish to attempt to use the specific conservation strategy or tools being tested by the portfolio. As shown on the right side of figure 16.7, at any given site there are *specific* principles that are of great use to people working at that site. However, these site-specific principles do not really help a person working at the next site over—let alone halfway around the world. On the left side of the diagram are *general* principles that apply to most or all sites. By nature, these principles are also *trivial*—they are true, but not very helpful to practitioners. The challenge, therefore, lies in discovering and defining the general and yet nontrivial principles represented by the box in the middle of the diagram.

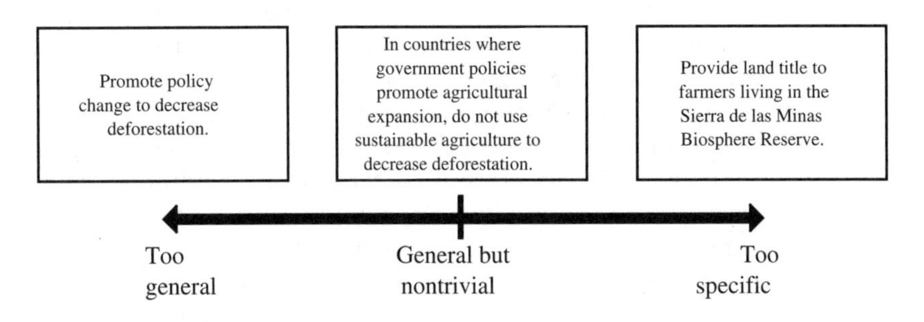

Figure 16.7.
Spectrum of possible principles related to sustainable agriculture. *Source:* Adapted from Salafsky and Margoluis 1999.

These principles are derived directly from the conditions analyzed in the learning portfolio. So, for example, our second statement of the relationships between the strategy or tool, conditions, and outcomes we discussed above can be represented by the general and yet nontrivial principle included in the middle box of figure 16.7.

Conclusions and Next Steps

Since the inception of ICDPs, there has been a heated debate about whether or not they work. As outlined in this paper, we believe that the conservation community cannot even begin to address this issue until we break it down into more fundamental questions and then use an adaptive management approach to answer them.

First, we must clearly define the strategies and tools that make up ICDPs. We believe that the conservation community should abandon its assertion that ICDPs are a specific strategy or tool. Instead, ICDPs should be viewed as a loose cluster of strategies and tools brought together to achieve both conservation and development goals. Letting go of the notion that ICDPs are a specific way to do conservation frees us to be able to look at the strategies and tools that constitute them in a much more discrete and useful way. By shifting our focus from ICDPs to their component strategies and tools, our unit of analysis and application becomes much more practical and much more powerful.

Second, we must clarify the goal of ICDPs—and reconcile possible incompatibilities between multiple goals. We believe that any project that integrates conservation and development must do so in a way that makes its goal unambiguous. While it might be more politically expedient to say that conservation and development will be equally served, this is usually unrealistic and impractical. One or the other—conservation or development—should be clearly articulated as the primary goal, and the relationship between the two should be clearly defined. This may also involve reorganizing a large project being implemented by many stakeholders into a set of related smaller projects.

Third, we must examine the conditions under which the strategies and tools that make up ICDPs work. To determine the conditions under which ICDPs work, we propose two approaches. Under one approach, the retrospective and cross-sectional evaluative work on ICDPs should be consolidated and summarized. This includes many of the case studies presented in this book. Under the other approach, we propose that prospective learning portfolios of ICDPs be explicitly designed and implemented so that we can evaluate over time the conditions under which specific tools—and combinations of tools—are most effective. The few examples of this approach to date (e.g., Margoluis et al. 2001; Salafsky et al. 2001; McShane and Newby this volume) have demonstrated a great ability to produce sound and practical principles for achieving success in conservation and development projects.

Just as we have to be clear about what hammer to use for what problem, we need to be clear about the strategies and tools that make up ICDPs and more accurately determine the conditions under which they work. These conditions will lead to our building a solid body of knowledge that practitioners can use to increase the efficiency and effectiveness of ICDPs around the world. Hopefully, practitioners will then have the right tools for the jobs they are facing.

References

Argyris, C. and D.A. Schön. 1978. *Organizational Learning: A Theory of Action Perspective*. Reading, Mass.: Addison-Wesley.

Dobson, A.P., A. D. Bradshaw, and A.J.M. Baker. 1997. Hopes for the future: Restoration ecology and conservation biology. *Science* 277:515–522.

Domroese, M. and E. J. Sterling. 1999. *Interpreting Biodiversity: A Manual for Environmental Educators in the Tropics*. New York: American Museum of Natural History.

Ehrenfeld, D. 1981. *The Arrogance of Humanism*. Oxford: Oxford University Press.

Ferraro, P. J. 2001. Global habitat protection: limitations of development interventions and a role for conservation performance payments. *Conservation Biology* 15:990–1000.

Ferraro, P.J. and R. D. Simpson. 2000. *The Cost-Effectiveness of Conservation Payments*. Discussion Paper 00–31. Washington, D.C.: Resources for the Future.

Fien, J., W. Scott, and D. Tilbury. 1999. *Education and Conservation: An Evaluation of the Contributions of Educational Programmes to Conservation Within the WWF Network*. Washington, D.C.: World Wildlife Fund.

Foundations of Success. 2001. *Improving the Practice of Conservation*. www.FOSonline.org (March 2001).

Gunderson, L., C. S. Holling, and S. S. Light, eds. 1995. *Barriers and Bridges in the Renewal of Ecosystems and Institutions*. New York: Columbia University Press.

Gustanski, J. A. 2000. Protecting the land: Conservation easements, voluntary actions, and private lands. In J. A. Gustanski and R. H. Squires, eds., *Protecting the Land: Conservation Easements Past, Present, and Future*, 9–25. Washington, D.C.: Island Press.

Holling, C. S., ed. 1978. *Adaptive Environmental Assessment and Management*. New York: John Wiley & Sons.

Jacobson, S. K. 1999. *Communication Skills for Conservation Professionals*. Washington, D.C.: Island Press.

Jacobson, S. K. and J. G. Robinson. 1990. Training the new conservationist: Cross-disciplinary education in the 1990s. *Environmental Conservation* 17:319–327.

Langholz, J., J. Lassoie, and J. Schelhas. 2000. Incentives for biological conservation: Costa Rica's private wildlife refuge program. *Conservation Biology* 14:1735–1743.

Lee, K. 1993. *Compass and Gyroscope: Integrating Science and Politics for the Environment*. Washington, D.C.: Island Press.

Locally Managed Marine Area Network. 2001. *Learning Framework for the Locally Managed Marine Area Network*. Bethesda, Md.: Foundations of Success.

Margoluis, R., V. Russell, M. Gonzalez, O. Rojas, J. Magdaleno, G. Madrid, and D. Kaimowitz. 2001. *Maximum Yield? Sustainable Agriculture as a Tool for Conservation*. Washington, D.C.: Biodiversity Support Program.

Margoluis, R. and N. Salafsky. 1998. *Measures of Success: Designing, Managing, and Monitoring Conservation and Development Projects*. Washington, D.C.: Island Press.

Meffe, G. 1999. Conservation medicine. *Conservation Biology* 13:953–954.

Parks, J. and N. Salafsky, eds. 2001. *Fish for the Future? A Collaborative Test of Locally Managed Marine Areas as a Biodiversity Conservation and Fisheries Management Tool in the Indo-Pacific Region*. Washington, D.C.: World Resources Institute (WRI).

Salafsky, N., H. Cauley, G. Balachander, B. Cordes, J. Parks, C. Margoluis, S. Bhatt, C. Encarnacion, D. Russell, and R. Margoluis. 2001. A systematic test of an enterprise strategy for community-based biodiversity conservation. *Conservation Biology* 15 (6):1585–1595.

Salafsky, N. and R. Margoluis. 1999. *Greater Than the Sum of Their Parts: Designing Conservation and Development Programs to Maximize Results and Learning*. Washington, D.C.: Biodiversity Support Program.

Salafsky, N., R. Margoluis, and K.H. Redford. 2001. *Adaptive Management: A Tool for Conservation Practitioners*. Washington, D.C.: Biodiversity Support Program.

Salafsky, N., R. Margoluis, K.H. Redford, and J.G. Robinson. 2002. Improving the practice of conservation: A conceptual framework and agenda for conservation science. *Conservation Biology* 16 (6):1469–1479.

Salafsky, N. and E. Wollenberg. 2000. Linking livelihoods and conservation: A conceptual framework and scale for assessing the integration of human needs and biodiversity. *World Development* 28:1421–1438.

Schön, D. 1984. *The Reflective Practitioner: How Professionals Think in Action*. New York: Basic Books.

Senge, P. M. 1994. *The Fifth Discipline: The Art and Practice of the Learning Organization*. New York: Currency Doubleday.

Stevens, W. K. 1995. *Miracle Under the Oaks*. New York: Pocket Books.

Terborgh, J. 1999. *Requiem for Nature*. Washington, D.C.: Island Press.

Wyckoff-Baird, B., A. Kaus, C.A. Christen, and M. Keck. 2000. *Shifting the Power: Decentralization and Biodiversity Conservation*. Washington, D.C.: Biodiversity Support Program.

PART THREE

Conclusion

17

The Future of Integrated Conservation and Development Projects: Building on What Works

Michael P. Wells, Thomas O. McShane, Holly T. Dublin, Sheila O'Connor, and Kent H. Redford

Introduction

Integrated conservation and development projects (ICDPs) moved from an untested concept in biodiversity conservation to conventional wisdom in a handful of years, despite a lack of convincing evidence that they were working effectively. Now, barely a decade after gathering momentum, ICDPs have been widely criticized for failing to meet expectations, and there are signs of their abandonment in favor of other, even less tested approaches.

Some conservation staff from the very same organizations that led the charge into ICDPs now claim to have turned away from this approach. While even a cursory examination of current field activities undermines these claims, the fact that they are made shows how far the pendulum has swung in a relatively short time. In practice, ICDP approaches are not about to disappear. While fewer conservation projects are now described explicitly as ICDPs, many current and planned conservation projects continue to be based on ICDP principles and objectives, with the word "integrated" remaining especially popular in project titles.

It seems that more than a decade of substantial investments in ICDPs has not delivered the anticipated biodiversity conservation benefits. Is it indeed time to consider and pursue alternatives? Or are there key elements of the ICDP experience that provide the foundation for more effective approaches? This paper addresses these questions.

Sorting Out the Jargon

Before discussing what ICDPs have or have not achieved, it is important to be clear about the terminology. The term *ICDP* was used originally to describe a project attempting "to ensure the conservation of biological diversity by reconciling the management of protected areas with the social and economic needs of local people" (Wells and Brandon 1992:ix). This original definition was developed as a first step toward describing and then analyzing a diverse set of independent project activities with shared conservation objectives. An ICDP is often, but not always, linked to one or more protected areas (PAs). Examples of ICDPs embraced by this definition include biosphere reserves, multiple-use areas, buffer zones and other initiatives on the boundaries of protected areas, regional land-use plans with biodiversity conservation components, and development projects incorporating biodiversity conservation.

Perhaps as a consequence of the rapid adoption of ICDPs by development agencies, the term became associated with large, "official" projects with an emphasis on livelihood opportunities that were often not clearly connected with biodiversity conservation (Sayer and Wells this volume). Perhaps to create some distance from these larger projects, the term *community-based conservation* (CBC) came into use in the early 1990s, referring to initiatives arising "from within the community—or at least at the community level—rather than nationally or internationally" (Western and Wright 1994:1). CBC also expanded the scope of ICDPs by being less tied to formal protected areas.

In some countries, including China, Nepal, Indonesia, and Vietnam, the ICDP label has "stuck" and become part of the standard conservation vocabulary. Comparable labels popular elsewhere are *ecodevelopment* in India (Singh and Sharma this volume) and *community-based natural resource management* (CBNRM) in southern Africa (Child and Dalal-Clayton this volume). Hulme and Murphree (2001) used the term *community conservation* (CC) to describe ICDP-like activities in Africa, while Robinson and Redford (this volume) call for a distinction between *development projects with conservation* and *conservation projects with development.*

Despite their subtle differences, all of these approaches originated from a shift in protected area management *away* from keeping people out by strict protection and *toward* more sympathetic treatment of local communities, including efforts to share benefits from the conservation of biodiversity. Adams and Hulme (2001:194) have characterized this as a profound change in the dominant "narrative" of conservation.

Alternative ICDP definitions have been proposed, including one recently developed for training purposes by CARE, the World Wildlife Fund (WWF), and the UN Development Program (UNDP). This one defines ICDPs as "an approach to the management and conservation of natural resources in areas of significant biodiversity value that aims to reconcile the biodiversity conservation and socio-economic development interests of multiple stakeholders at local, regional, national and international levels" (Franks and Blomley this volume). Compared to the original definition, this version takes ICDPs beyond protected areas while explicitly including a wider range of stakeholders and a range of spatial scales. This emphasis on stakeholders and spatial scale is consistent with the emerging ICDP model discussed here (Brown this volume; Maginnis, Jackson, and Dudley this volume).

Our review of the various terms in use suggests that ICDP continues to be a viable collective description for site-based conservation with social or economic development goals or components, encompassing CBC, CC, ecodevelopment, and perhaps some others. This is important, as recent experience shows that some initiatives professing not to be ICDPs—and thereby somehow immune from analyses of ICDP performance—do in fact have goals and activities that overlap significantly with an ICDP approach.

Notably, none of the various attempts to categorize or define ICDPs prescribe whether the objectives of ICDPs should be expressed in terms of biodiversity conservation or economic development or both. Project implementers and observers have continued to wrestle with the basic issue of whether ICDPs have had, can, or should have single or multiple objectives and how to define these (e.g., Robinson and Redford, Franks and Blomley, Salafsky and Margoluis, all this volume).

The Importance of Looking Back

Before examining ICDP performance, it is useful to briefly review the history of these projects, especially why and how ICDPs so quickly came to be regarded as *the* site-specific approach to biodiversity conservation in developing countries (for more detail, see Wells and Brandon 1992; Brandon and Wells 1992; Robinson 1993; Sayer 1995; Adams and McShane 1996; Larsen, Freudenberger, and Wyckoff-Baird 1998; Oates 1999; McShane and Wells this volume; Robinson and Redford this volume).

Conservation legislation was strengthened and many PAs established in newly independent developing countries during the 1960s. Most conservation efforts at this time were focused on particular species or sites. In 1972, the UN Environment Conference in Stockholm recognized growing human impacts on finite natural resources. Recommendations from the conference emphasized the need for careful planning of development projects to mitigate environmental damage, especially in developing countries. This helped initiate a new emphasis on the "rational" planning of both development and conservation programs.

In the late 1970s the conservation community argued strongly for ecological considerations to be reflected in economic development planning, most comprehensively in the *World Conservation Strategy* (IUCN 1980). This strategy called for strong linkages between conservation and development by affirming that the two goals were compatible and mutually supportive, needing only rational planning in order to coexist. Conservationists and protected area managers took these ideas further at the 1982 World Parks Congress in Bali, with a call for increased support for communities next to parks. Measures such as education, revenue sharing, participation in decision making, and appropriate development interventions near PAs were all discussed in relation to conservation objectives (McNeely and Miller 1984).

By the mid-1980s a new consensus had taken hold: that the successful long-term management of PAs depended on the cooperation and support of local people. Excluding poor people with limited resource access from PAs and their management arrangements without providing them with alternative means of livelihood was recognized as neither politically feasible nor ethically justifiable. The widespread adoption of this view led to increased efforts by PA managers and conservation organizations to obtain local cooperation, and consequently to a new wave of conservation projects subsequently referred to as ICDPs (Wells and Brandon 1992). By 1985, WWF had launched its Wildlands and Human Needs Program with more than twenty projects (McShane 1989; Adams and McShane 1996). Other conservation organizations followed suit, with the international conservation NGOs being extremely effective at promoting the ICDP approach as a key contribution to sustainable development.

The concept of "sustainable development" was popularized by the World Commission on Environment and Development (WCED 1987). The WCED's major thrust was for social and economic development under conditions of ecological sustainability, although there was little explicit mention of nature conservation. Undeterred, conservationists looked for synergies between the suddenly ubiquitous notion of sustainable economic development and their own priority of nature conservation—or, more precisely, of *biodiversity.* Biodiversity had emerged as the key international focus of nature conservation (cf. Wilson 1988). Leading conservationists realized the importance of identifying "sustainable forms of economic development to contribute positively to [the] conservation of biodiversity" (McNeely 1988:iv) and vice versa. *Caring for the Earth* (Munro 1991) attempted to link the WCED's concept of sustainable development with biodiversity conservation. There appeared to be a good fit between ICDPs and the priorities of this emerging international agenda.

By the end of the 1980s, public awareness and concern about accelerating environmental degradation had increased significantly. Pressure from NGOs and, eventually, the governments of richer countries led international development agencies—notably the World Bank and the bilateral agencies of the OECD countries—to look harder at the environmental impacts of their own projects. The World Bank adopted a Wildlands Policy in 1986 and launched its own Environment Department in 1987 as biodiversity conservation began to attract considerably more attention from both multilateral and bilateral development agencies.

The accelerating global interest in biodiversity produced landmarks in 1992, including the finalization of the Convention on Biological Diversity (CBD) at the UN Conference on Environment and Development (UNCED) in Rio de Janeiro and the launch of the Global Environment Facility (GEF). The CBD's emphasis on access, benefit sharing, and sustainable use reemphasized that most developing countries are not interested in biodiversity from anything other than a utilitarian perspective. The GEF, set up to support projects that address environmental problems of global significance in developing countries, immediately began to provide funding for biodiversity conservation on a much larger scale than all previous funding efforts combined (Kiss this volume; Singh and Sharma this volume). Virtually all GEF biodiversity projects have been channeled through the World Bank, UNDP, and UN Environment Program (UNEP), thereby leading to a substantial increase in biodiversity conservation activities within these agencies. The European and North American bilateral development agencies also made available substantially more funding for developing country biodiversity initiatives during the 1990s.

These factors combined in a potent mix. The growing interest of international development agencies in biodiversity, the sudden increase in financial resources for biodiversity, and the desire of the international community to help developing countries meet their obligations under the CBD all led to a search for viable project concepts. ICDPs based in and around PAs appeared to be a near-perfect fit to this converging set of interests. The recipient governments liked the economic development aspects of ICDPs, as these sounded more useful than biodiversity conservation on its own. For development agency staff, the ICDP approaches to address rural poverty in buffer zones sounded like rural development, something they were more familiar with than the rather intangible notion of biodiversity. Conservation organizations stood ready to promote and implement ICDPs, while popular authors promoted success stories to a new audience eager to hear about environmental achievements. ICDPs offered the attractive prospect of contributing to three of the most sought after goals on the sustainable development agenda at the same time: more effective biodiversity conservation, increased local community participation in conservation and development, and economic development for the rural poor. These features were irresistible to many NGOs, government departments, and development agencies. This set the stage, and ICDPs took off.

What Went Wrong?

The ICDP experience has now been documented and analyzed in some detail by a variety of authors (Wells and Brandon 1992; Western and Wright 1994; Barrett and Arcese 1995; Adams and McShane 1996; Blomley and Mundy 1997; Kramer, Van Schaik, and Johnson 1997; Sanjayan, Shen, and Jensen 1997; Brandon, Redford, and Sanderson 1998; Hart et al. 1998; Larsen, Freudenberger, and Wyckoff-Baird 1998; Wells et al. 1999; McShane 1999; Oates 1999; Roe et al. 2000; Hulme and Murphree 2001). Rather than reexamining all of these findings here, we will focus on those key issues that appear to offer the key to initiating a more convincing set of conservation projects in the future.

The basic idea of an ICDP is simple and intuitively appealing. Even better, in principle ICDPs offer something for everyone. As a consequence, the ICDP concept was easy to sell to a broad range of interests, including park managers, local communities, conservation organizations, development agencies, national governments, and local governments. Each of these stakeholders can see the prospect of their own interests being addressed and perhaps assume that competing interests will also be satisfied (McShane and Newby this volume; Brown this volume).

However, natural resource access and use are frequently complicated. Furthermore, the relevant interest groups vary in their capacity to exert power and to influence decisions. These difficulties are often compounded when a site is declared ecologically valuable and a protected area is imposed by a distant government using objectives or rules that may be alien to local practices. This context is challenging enough to understand, and extraordinarily difficult to influence through the short span of a project. Developing countries do not have a monopoly on this set of problems, however. While the ICDP approach is generally ascribed to developing countries, we see many of these same issues in the industrialized countries, including the Greater Yellowstone Ecosystem in the United States (Glick and Freese this volume).

We should not be too surprised that it has proven difficult to operationalize *single, site-based projects* that aim to (1) address the interests of everyone involved in land and resource access in and around PAs, (2) provide people with better livelihood opportunities and access to services, and (3) defuse the major threats to biodiversity. In fact, we now know that many ICDPs were based on naïve assumptions and were overambitious. Despite the best intentions and efforts of the conservation professionals that led many of these projects, too much was being expected in too short a time with inadequate tools (Sayer and Wells this volume).

Overoptimistic Goals and Weak Assumptions

Perhaps the most fundamental failing with ICDPs was that many of them were developed on a wave of enthusiasm for sustainable development based on two key assumptions: first, that rational planning and seed money were the main prerequisites for "win-win" solutions that would allow nature conservation to peacefully coexist with economic development; and second, that significant benefits from PAs could be generated and equitably distributed, thereby providing alternative livelihood opportunities for local people that would reduce pressure and threats from outside. Unfortunately, there is little evidence to show that either of these basic assumptions was sound in any more than a few rare cases (Robinson and Redford this volume; McShane and Newby this volume).

Both in principle and in practice, it has been difficult to decide if ICDP goals should be expressed in terms of conservation, development, or both (e.g., Robinson and Redford, Franks and Blomley, Singh and Sharma, Child and Dalal-Clayton, all this volume). Virtually all ICDPs have wrestled with this issue during the project design phase. While stakeholders have reached consensus in a few cases, typically this key point remains unresolved. As a result, project objectives have often been articulated in vague terms that are trivial, obscure, or impossible to measure (Salafsky and Margoluis this volume). This has perpetuated the unrealistic expectations of the different interest groups, each determined that its own objectives would be met, even when the incompatibility of the multiple objectives became clear. Establishing effective links between the conservation and the development activities of ICDPs has proven particularly hard, leading to frustration with the limited progress that ICDPs have made. The short time frames available to most projects and the lack of financial sustainability further hamper progress. These factors inhibit building on any progress made in the crucial areas of establishing relationships and trust with local stakeholders (Brown this volume).

It is surprising how many projects, not just ICDPs, are launched without their proponents having carefully thought through the assumptions they are implicitly making and testing (Brandon 2001; McShane and Newby this volume). For example, we believe poor people deserve support and opportunities to improve their own lives, and we think it is also clear that poverty can cause environmental degradation. However, this does not justify the near-ubiquitous ICDP assumption that making people in and around PAs better off will translate automatically into more effective biodiversity conservation. Not only is there is no hard evidence for this, but it is equally plausible that increasing local incomes adjacent to ecologically valuable areas will accelerate land clearing for agriculture (cf. Wunder 2001).

Closely linked to the preceding assumptions, early ICDPs assumed that stronger law enforcement for PAs was not a priority since the projects would deliver benefits at such a level that illegal park exploitation would no longer be attractive. This rather naïve idea, apparently based on a misunderstanding of the nature of incentives, has now been exposed as unrealistic in most cases (e.g., Singh and Sharma, Gartlan, Child and Dalal-Clayton, all this volume). Unfortunately, law enforcement related to biodiversity conservation in many countries—although often woefully inadequate—has often been carried out in a heavy-handed, unsympathetic way toward local people while turning a blind eye toward, or simply being unable to confront, illegal activities sponsored by the more rich and powerful. As a result, law enforcement has been too often discredited while its importance for PA management has been deeply underappreciated.

Unconvincing Local Participation

Local participation is synonymous with and an integral part of the ICDP approach. Almost no one now objects to consultations with local people in making decisions for biodiversity. In fact, most organizations expect and encourage it. However, while increased local participation in conservation and development seems a relatively simple and attractive idea, this general principle disguises a mass of contradictions and ambiguities that project planners and funders often overlook or are unable to come to grips with (Sayer and Wells, Brown, Tongson and Dino, Child and Dalal-Clayton, all this volume).

Although local people are usually one of the major intended beneficiaries of ICDPs, the original decision to launch an ICDP is rarely theirs. Many projects invest considerable effort in eliciting local people to participate in, or at least not to oppose, project activities. This can be extremely difficult, especially in the all-too-frequent case where there is no legal or institutional framework to encourage and perpetuate local participation. Projects seriously interested in fostering local participation may need to spend many years, if not a decade or more, helping to build the capacity of local institutions, even assuming that local and national laws, customs, and tenure arrangements permit and support such an approach (Wells and Brandon 1992; Brandon and O'Herron, Tongson and Dino, Gartlan, all this volume).

It can sometimes be difficult for projects to identify a discrete, recognizable local community. The idealized concept of the "local community" as used in project planning often bears little resemblance to any group of people in the real world because it makes little allowance for disparities in views, capacities, influence, cultures, or aspirations (cf. Agrawal 1997). Indeed, Roe et al. (2001) quote an observer defining a community as "a figment of the imagination of project managers and donors seeking quick fixes." Experience has emphasized the need to understand a much wider and more complex group of stakeholders with interests in PAs and their surrounding lands, and to acknowledge that it may not be possible to include *all* of these stakeholders in decision making.

The key and controversial question of how much local participation in biodiversity decisions is desirable or optimal has yet to be resolved. The principles of decentralization and local-level decision making for natural resource management are now so well entrenched that they are rarely challenged. However, having local interests manage ecologically valuable resources in legally protected areas may not be compatible with protecting or sustainably using the components of biodiversity that national governments are committed to conserving and that the international community is interested in supporting (Brandon and O'Herron this volume).

Given these complications, it is hardly surprising that the extent of genuine local participation in ICDPs has varied considerably. The practical issues for most ICDPs are *who* gets to participate, *what* do they get to participate in, *how* do they participate, and, finally, *who* decides this and *how* (Wells and Brandon 1992; Brown, Tongson and Dino, Gartlan, Singh and Sharma, Child and Dalal-Clayton, all this volume). Simply clarifying these issues would be a significant step forward for many projects.

Targeting the Wrong Threats

Many ICDPs did not focus on the major threats to biodiversity, and some ICDPs functioned on too small a scale to address even local threats. While these projects may have made some partially useful contributions to biodiversity conservation, it is now apparent that, as isolated efforts, they were a poor use of resources.

Most ICDPs began with the intention of focusing on the development needs of poorer people living in and around PAs, assuming that these people were the major threat to that area (McShane and Newby this volume). The tendency has been for PA managers and staff to focus aggressively on containing the "threat" of local villagers' small-scale farming and hunting activities. It is now becoming clear that the activities of local people are often less of a PA threat than activities such as mining, road building, dam construction, irrigation schemes, resettlement programs, plantations, logging, and hunting. However, the difficulties of confronting well-organized logging or poaching backed by rich, powerful, and often politically connected interests are numerous (cf. Tongson and Dino this volume; Child and Dalal-Clayton this volume). Such larger-scale threats—the root causes of biodiversity loss—are often well beyond the sphere of ICDPs that are limited in scope and scale over a short time period (Brandon, Redford, and Sanderson 1998; Wells et al. 1999; Wood, Stedman-Edwards, and Mang 2000; McShane and Newby this volume; Brandon and O'Herron this volume).

These problems illustrate the critical importance of analyzing and acting on threats to PAs at a variety of spatial scales (cf. Robinson and Redford, Maginnis, Jackson, and Dudley, Shepherd, all this volume). Conservation practitioners, including PA and ICDP managers, need to become informed and active in regional development and land-use planning, where many of the decisions potentially harmful to PAs are made (Wells et al. 1999).

Financial Sustainability

Many ICDPs were launched with the hope that they could become self-financing. In practice, most ICDPs required recurring funding after their initial term or faced the prospect of collapse. Environmental trust funds appear to provide one of the few reliable sustainable financing alternatives. There are three messages here for the next generation of projects. First, projects with little prospect of financial sustainability either should not be undertaken or should be scaled back to a more modest set of activities more consistent with local capacities. Second, every effort should be made for PAs supported by ICDPs to generate sustainable economic benefits without compromising their biodiversity conservation objectives, which may require involving the private sector (cf. Kiss, Glick and Freese, Child and Dalal-Clayton, all this volume). However, the ability of ICDPs to generate revenues sufficient to cover their operating costs and also benefit local populations is likely to be limited, and this reality needs to be reflected in project design. Finally, if developing countries are going to conserve their biodiversity, then richer countries will need to meet more of the cost than they have so far. However, there is so far little sign that more money leads to greater conservation successes, and many developing countries need to move further down the path of raising income levels and increasing political freedom, hopefully getting to the point where their own citizenry will start to demand effective conservation through legitimate political processes (see Child and Dalal-Clayton this volume).

Benefit Generation

The benefits generated as a result of ICDPs usually do not provide an adequate incentive to discourage activities that threaten the PA, such as hunting, logging, or the expansion of agriculture (cf. Norton-Griffiths and Southey 1995; Emerton 1998; Tongson and Dino, Gartlan, Singh and Sharma, all this volume). This is a major lesson of the ICDP experience. This does not mean that ICDPs will not work, but it does mean that PAs or ICDPs likely cannot generate sufficient alternative means of livelihood for local residents to assure the preservation of PAs. Generating and distributing benefits from conservation at the site level in a way that supports conservation has proven to be far more difficult in practice than was envisaged a decade or more ago (cf. Emerton 1998; Brandon and O'Herron this volume). Conservationists should therefore be cautious about committing protected areas to ambitious poverty reduction goals.

With hindsight, many of the components in ICDPs aimed at providing new livelihood opportunities and boosting local incomes now appear naïve and misdirected (Shepherd this volume). The selection of "development" or income-generating activities to be supported by ICDPs was based largely on the results of quick-and-dirty participatory rural appraisal (PRA) surveys. These identified community priorities and led to the implementation of "participatory" project interventions, even though most continued to be designed and then managed by outsiders. Based on PRA results, and sometimes because certain activities were donors' current "flavor of the month," many ICDPs enthusiastically launched local pilot projects in agriculture, agroforestry, aquaculture, apiculture, ecotourism, and so on. The hope was that new and sustainable income sources would be generated. These efforts focused almost entirely on production rather than financial viability and marketing. The results were usually disappointing, with relatively little impact on local incomes and little potential for scaling up or replication. While the use of PRA techniques did indeed give the impression of local participation, hindsight suggests that in practice many of these project components were prepackaged solutions (McShane and Newby this volume; Brown this volume). Some writers have gone as far as to dismiss the ICDP or community conservation approaches for these reasons (e.g., Agrawal 1997; Ghimire and Pimbert 1997).

Even where ICDP income-generation efforts did make a few people better off, it was rarely clear how this benefited biodiversity conservation (cf. Robinson and Redford this volume). A more thorough review of the rural development literature would have revealed the difficulties faced by outsiders trying to catalyze swift social or economic changes in communities (Lewis 1988; Wells and Brandon 1992; World Bank 1993). The introduction and adoption of new practices is an even greater challenge through the brief window of an ICDP. The inability to connect effectively with local development issues inevitably concerns donors, putting future project funding into question (Hart et al. 1998; see also Shepherd this volume).

The Culture of Success

ICDP experiences so far make it clear that single short-term projects alone cannot successfully address the major challenges facing the conservation of ecologically valuable areas (e.g., Robinson and Redford, McShane and Newby, Sayer and Wells, Kiss, Brandon and O'Herron, all this volume).

Why did so many ICDPs set out with such unattainable expectations? Donor pressure has had a powerful influence on the selection of ICDP objectives (Sayer and Wells this volume). No one wants to inform his or her donor or host institution of failure, or even of partial success. Reporting success has become the norm. Organizations enjoy hearing that their funds were well spent, so they can report positively to their own constituents. Renewal of funding naturally depends on success; so failures are rarely reported (Redford and Taber 2000). Within such a culture, project proposals and plans promise more and more in order to compete for funding. Accountability within this cycle is limited, as it can take several years for disappointing results to filter back to the organizations responsible. Even in the few cases where projects are explicit about the need for learning and experimentation, this does not seem to have increased the acceptability of unsuccessful outcomes. The prospect of learning from less-than-complete success is simply not an acceptable option in many institutions.

Looking Forward

Despite the rather demoralizing list of problems that ICDPs have experienced, the rationale for these types of projects has not disappeared. The notion that biodiversity can be conserved without considering local people's needs and aspirations is simply not viable. The need to address relations between PAs and their neighbors is now even more compelling and urgent. Illegal activity and land degradation (both inside and outside of PAs), decentralization of land management, the declining influence of central governments, persistent growth in the absolute numbers of poor people, and—at least in some countries—increasing participation of poor rural people in a democratic process all increase the pressure on biodiversity.

We therefore now turn to some emerging perspectives on ways of strengthening site-specific conservation initiatives, to help ICDPs and their equivalents make a sustained and more successful contribution to biodiversity conservation.

Root Causes and Policy Work

Our understanding of the root causes of biodiversity loss, and environmental degradation in general, has become more sophisticated (e.g., Perrings et al. 1995; Stedman-Edwards 1998; Wood, Stedman-Edwards, and Mang 2000). It is clear that many of the most important threats to biodiversity originate far from PA boundaries and involve issues and institutions well outside the traditional realm of conservationists (McShane and Newby, Kiss, Glick and Freese, Brandon and O'Herron, Tongson and Dino, Gartlan, Singh and Sharma, Child and Dalal-Clayton, all this volume).

What does this mean for project identification and design? Site-specific efforts will always be necessary. However, these need to be nested within broader-based strategies supportive of biodiversity conservation and more eco-friendly forms of economic development (Sayer and Wells this volume; Kiss this volume). Many conservation and development organizations are already engaged in crosscutting issues of land and resource access and tenure, agricultural and forestry policies and pricing, and international trade agreements. Unfortunately people who have little or no connection with conservation field projects often carry out these efforts. Compounding this separation, our experience is that it is initially difficult to engage field project managers in policy-level issues.

Site threats should be identified at all spatial scales (Robinson and Redford, Maginnis, Jackson, and Dudley, Shepherd, all this volume). More strategic analyses can be used to work out which challenges and threats lie within a project's sphere of influence. More effective protected area management requires local-scale interventions to be complemented by stronger law enforcement within protected areas, more effective environmental screening of nearby development projects, and more aggressive policy interventions in support of biodiversity conservation. It is often necessary to support or build partnerships to pursue these objectives, sometimes with those who are not traditional allies of conservation. The use of diverse field- and policy-oriented approaches must be *vertically integrated,* ensuring that site-based actions are directly supported by policy-level actions both nationally and internationally.

Adaptive Management

An adaptive management approach offers the opportunity to improve the implementation of ICDPs while avoiding some of the pitfalls experienced previously (Holling 1978; Bell and McShane-Caluzi 1984; Lee 1993). The approach has recently undergone a revival in popularity (cf. Gunderson et al. 1995). Salafsky, Margoluis, and Redford (2001:12) and Salfsky and Margoluis (this volume) describe recent applications to biodiversity conservation using this definition: "Adaptive management incorporates research into conservation action. Specifically, it is the integration of design, management and monitoring to systematically test assumptions in order to adapt and learn."

Many institutions and organizations desire to become "learning organizations" and thus profess to use or encourage the use of adaptive management approaches. However, in practice, very few have given the time, opportunity, or support to actively engage in designing, implementing, and monitoring their project, instead reverting to "learning by doing" as a poor substitute for adaptive management. Conservation organizations in particular are driven to have quick results to satisfy their varied constituents and allow them to compete for resources.

Adaptive management merges planning with both implementation and monitoring as part of a constantly rotating project cycle, not as three sequential and separate phases as is often the case now. In practice, this means devolving more responsibility to the field level, staffing projects with skilled managers who are able to exercise judgment, deploying resources in a flexible manner, and being able to draw on a toolbox of different actions (Child and Dalal-Clayton this volume; Salafsky and Margoluis this volume). A key change from many projects today, particularly those of the larger development agencies, would be to give much more attention to the *implementation* of projects rather than, as currently, to detailed *planning*.

Granted that developing-country PAs are almost all significantly underfunded, the adaptive management of ICDPs would be significantly enhanced by more flexible disbursement arrangements that are consistent with local absorptive capacities, since neither higher levels of funding nor faster project disbursement correlates with more successful community development. Reflecting the pace of the community, rather than attempting to meet externally imposed deadlines, contributes to a more efficient participatory planning process, genuine capacity development, and more profound community learning. Educating the many levels of people involved in an adaptive management approach about its value as well as its deficiencies is important if it is truly going to become an accepted practice and influence future ICDPs.

Identifying Targets and Addressing Trade-Offs

The *integration* of conservation and development goals through ICDPs has proven so difficult that we must question whether this remains a viable goal. The attractive idea of all parties achieving their objectives under win-win scenarios does not appear to be a realistic basis for planning conservation projects (Robinson and Redford, McShane and Newby, Brown, all this volume). In particular, experience shows that site-specific biodiversity conservation is rarely compatible with unfettered development, income generation, or livelihood interests.

The success of ICDPs, or any other site-based conservation projects, has been complicated by the difficulty conservationists have had in setting explicit conservation targets—those entities whose long-term persistence the conservation effort is attempting to ensure (Redford et al. 2003). The lack of such clearly stated targets has prevented all ICDP stakeholders from being able to participate in a transparent, explicit trade-off calculus (cf. Brown this volume). This lack of specificity has been made possible partially by the popularity and ambiguity of the term *biodiversity*. When defined as "the variety of life," this term allows many people to interpret the meaning in many ways, often contradictory. It is only when biodiversity is defined by specifying its site-specific components (genetic, population/species, and community/ecosystem) and attributes (structure, function, and composition) that it becomes a useful tool (Redford and Richter 1999). Each conservation area should attempt to identify both its conservation targets and the state or condition in which they should be conserved (e.g., number of individuals of a species, or area extent for an ecosystem). This specificity allows development of monitoring protocols and management interventions.

The same ambiguity and lack of explicit targets that pervade the conservation dimension of ICDPs are also manifested in the development side of the approach. Recognizing that not everyone's agenda can be satisfied or successfully integrated, ICDPs need to develop better techniques to identify and understand the goals and interests of the major stakeholders in and around PAs (Brown, Singh and Sharma, Maginnis, Jackson, and Dudley, Shepherd, all this volume). Once these different interests are identified, clarified, and understood, the opportunities for negotiation and trade-offs can be explored (e.g., Brown this volume).

While ICDPs can play key roles in helping bring together different stakeholder interest groups, it is important they be explicit and open about their own mission and objectives as outsiders. With the shift from win-win scenarios to the recognition of key stakeholder interests, it rapidly becomes clear that the ICDP itself also has a vested interest. Most ICDPs funded by global constituencies explicitly aim to conserve biodiversity while paying attention to development needs and priorities. Conversely, ICDPs led by development-oriented organizations tend to promote local social and economic interests ahead of conservation. In either case, project staff represent vested interests and should consider this carefully when presenting themselves as honest brokers seeking the "integration" of conservation and development (Franks and Blomley this volume).

Until recently, there have been few systematic attempts to help stakeholders identify and then make rational choices between competing scenarios in conservation or development, partly because of the persistence of the win-win myth. More recently, however, applied researchers have begun to develop and test tools that may prove extremely useful in helping diverse groups of stakeholders understand each other's viewpoints and make informed and appropriate choices (Brown this volume). One of the more exciting aspects of this work has been to dispel the conventional wisdom that outsiders can simplistically predict the outcome of such choices.

Using such approaches, ICDPs need to support the engagement of civil society and a broader variety of stakeholders in protected area planning and decision making. Projects should attempt to engage stakeholders more deeply in explicitly defining the objectives of project interventions, in monitoring progress, in learning from experience, and in systematically documenting and disseminating findings (see

Tongson and Dino, Singh and Sharma, Child and Dalal-Clayton, all this volume).

Scale Issues

The ecological and evolutionary processes that sustain biodiversity operate at large spatial and long temporal scales. These issues are rarely identified as of importance in discussing the challenges facing successful implementation of ICDPs. Yet scale questions are vital for successful implementation. Such questions include: At what scale should conservation and development trade-offs be considered? How do the heterogeneity of natural and human-dominated areas affect one another (Robinson and Redford this volume)? At what scales do differing human uses interact with ecological processes? At what scales do threats affect both human and natural systems that are critical to ICDPs? And, at what scales are conservation targets best set? Lack of explicit consideration of such scale questions has been possible partially because of the lack of specified targets for both conservation and development (see above). It is only when dealing with specific targets that scale considerations must be resolved.

Addressing spatial and temporal scale issues at the earliest stage of designing ICDPs offers advantages (Robinson and Redford, McShane and Newby, Franks and Blomley, Maginnis, Jackson, and Dudley, all this volume): (1) it identifies biodiversity-intense priority areas at scales that offer opportunities for diverse land and resource use; (2) it creates data and information sufficient to formulate robust conservation targets; (3) it provides a broader understanding of the social, political, economic, and historical fabric that will underlay conservation use and decisions; and (4) it looks to the long-term survival of biodiversity and of people associated with that biodiversity.

There is emerging interest in adopting "large-scale conservation approaches" to integrating conservation and development (see Redford et al. 2003 for a review of several of these approaches). Maginnis, Jackson, and Dudley (2001 and this volume) suggest that if properly chosen, a landscape approach can balance the ecological, social, and economic land uses necessary for sustainable development, including biodiversity conservation, through a process of land-use negotiations among a wide variety of stakeholders. Although still in their infancy, landscape approaches do offer the possibility of linking local initiatives with larger-scale regional and national policy processes (Wascher 2000). At the same time, while landscape approaches may indeed provide forums for conservation and development interests to negotiate, they cannot replace the need for effective on-the-ground local action to support biodiversity conservation in and around PAs, which leads directly back to ICDPs. For individual stakeholders, the idea of win-win will usually be no more viable at a landscape level than at a local level. The challenge for practitioners is not to decide the best scale at which to operate, but rather the combinations of actions required at different scales.

INCENTIVES FOR CONSERVATION

One of the assumptions underlying many early ICDPs was that providing park residents and neighbors with alternative opportunities to earn a living would provide a sufficient incentive for the beneficiaries to switch from environmentally threatening to environmentally friendly activities (cf. McShane and Newby this volume; Kiss this volume). As we have seen, this idea has proven extremely difficult to operationalize, and there are few persuasive examples (Brandon and O'Herron this volume). Are there alternative ways of providing adequate conservation incentives at the site level, particularly where conflicting stakeholder interests do not appear resolvable through a project intervention?

One possibility would simply be to pay cash in return for biodiversity protection (Ferraro and Kiss 2002; Kiss this volume; Brandon and O'Herron this volume). Selected local or national government entities or NGOs would receive payments, to use as they see fit, in exchange for PA management and conservation commitments. Payment schedules over extended periods would then be subject to independent performance reviews. The funding for such arrangements could originate from international sources or from government.

Governments could consider inviting tenders for the management of individual PAs: for example, a government would commit to taking whatever steps necessary to protect a particular PA, say for twenty-five years, while allowing independent monitoring. Interested parties (development agencies, NGOs, even private sector organizations) would then bid the amount they would be prepared to pay to secure this PA, payable over the full term of the agreement as long as the government continued to live up to its protection commitment. If adequate offers of international funds were not forthcoming, the government could then decide whether to finance conservation activities domestically (perhaps based on an assessment of watershed protection, tourism potential, or other national economic benefits) or to turn the PA over to other uses. Such an approach could also help sharpen the discussion concerning the level of financial resources that should be transferred to developing countries to support biodiversity conservation (Wells et al. 1999).

Ferraro (2001) has argued that paying individuals or communities directly for conservation performance may be simpler and more effective than the ICDP approach. This type of conservation contracting can simplify the achievement of conservation goals and strengthen the links between individual actions and habitat conservation, thus creating a local stake in ecosystem protection. While such approaches may prove an important addition to the conservation toolbox, their applicability may be limited within the countries where biodiversity is concentrated. Although conservation contracting does seem to offer considerable promise in North America and Europe, it depends on governance arrangements and an institutional framework that provides clarity in regard to land use and access rights as well as the consistent enforceability of legal contracts (cf. Glick and Freese this volume). These are still lacking in many developing countries.

Engaging with Stakeholders

Finding more effective ways to engage local stakeholders is a major challenge for both PAs and ICDPs. How to best achieve this has been extensively discussed in this book (e.g., Franks and Blomley, Tongson and Dino, Brown, Singh and Sharma, Maginnis, Jackson, and Dudley, Shepherd). As always, there is no one-size-fits-all approach.

An interesting set of institutional arrangements that may be replicable on a larger scale has recently been demonstrated in six countries by the Community Management of Protected Areas Conservation (COMPACT) Project of the GEF Small Grant Program (GEF/SGP). The COMPACT Project has made ten to twenty-five small grants (averaging U.S.$20,000) for community-level conservation and development activities in and around each PA. Interested groups, which often lacked experience in writing proposals or designing or managing projects, were then supported through the application process. The resulting portfolios of approved and funded projects are being carefully monitored.

This approach appears to be generating a number of positive results: (1) the key stakeholders are being brought together through the "local consultative bodies" (LCBs) to consider the future of high-priority PAs and are being provided with resources on a scale consistent with local absorptive capacities to address priorities they have identified themselves (see also Sayer and Wells this volume); (2) stakeholders who were previously unknown or even in opposition to each other are now working together and gaining an appreciation of each other's interests and priorities (see also Brown this volume); (3) conservation and development initiatives clearly linked to PA management are being tested and implemented by the local stakeholders themselves; (4) local resource user groups are being strengthened, in some cases preparing them for future roles in helping manage their natural resources in partnership with government agencies that lack the resources to adequately manage the PAs (see also Tongson and Dino this volume); (5) tangible benefits from the PAs are becoming apparent to local stakeholders; and (6) judicious selection of the members of the LCB is helping build bridges between local-level activities and the broader sets of policy and institutional constraints that also require attention if PA management is to be genuinely strengthened (Wells, Qayum, and Tavera in prep.).

Emerging Signs of Promise

Whether or not the ICDP label persists is of limited importance. However, the set of problems that ICDPs try to address is still present. Making PAs more effective remains one of the main challenges to biodiversity conservation, and PA relationships with their neighbors continue to be a high priority. Many PAs face formidable constraints as pressure from the sheer numbers of people surrounding them increases. Support from an external project is often the best option to help overcome these, even if the "project" is not a perfect delivery mechanism (Sayer and Wells this volume). However, it is essential to ensure that projects are designed with a very clear understanding of their objectives in terms of yielding benefits to local communities as well as mitigating threats to protected areas (Robinson and Redford, Shepherd, Salafsky and Margoluis, all this volume).

In practice, a small and expanding number of practitioners have recently carried out a quiet revolution by managing to work out some of the key elements of practical ways forward through a new generation of conservation projects that fit the ICDP definition, irrespective of their particular labels. Many of these resourceful and pragmatic project managers have adjusted their project objectives after becoming immersed in the practical reality of the local situation. Sometimes this has meant that detailed planning documents prepared by outside teams at great expense have been put aside, literally or figuratively, in favor of more modest and achievable goals. Aspects of the approach include the following:

- Clearly articulating the objectives of the PA and the external intervention
- Building alliances with and among local communities to help establish trust

- Building coalitions for conservation by engaging with stakeholders who can help address broader development-related issues and constraints beyond the scope of site-specific projects
- Supporting capacity building for independent local planning and action among emerging community-based organizations whose activities are linked to adjacent protected areas
- Increasing the capacities of PA and natural resource management agency staff, and facilitating better relations between these staff and local communities; helping reorient PA guards to be more sympathetic to local needs
- Opening lines of communication with local and sectoral government agencies who are in a position to deliver key services to PA residents and neighbors
- Supporting basic environmental education to broaden and deepen the constituency of support for biodiversity conservation
- Raising local awareness of the extraordinary values of local biodiversity and the importance of conservation
- Supporting carefully selected, tentative, small-scale pilot income-generating activities with genuine local support, real prospects of sustainability, and clear benefits for biodiversity conservation

Some "new" ICDPs have made commendable progress in these critically important areas, even though such achievements are extraordinarily difficult to measure in a convincing or cost-effective way and may appear relatively modest in comparison with the more ambitious objectives of earlier ICDPs (also see Tongson and Dino, Gartlan, Brown, Singh and Sharma, Child and Dalal-Clayton, Salafsky and Margoluis, all this volume).

A recent evaluation of the GEF's more than one hundred "medium-sized projects" (MSPs) provides some interesting examples of the next generation of more pragmatic ICDPs. This study examined a cluster of recently designed biodiversity projects being executed by national and international NGOs with support from the World Bank, UNDP, and UNEP that have incorporated various combinations of the approach listed above. Many of these projects appear to have made a good start and are held in high regard by their stakeholders. MSPs have an upper funding limit of $1 million, considerably less than many ICDPs, and there are strong indications that this cap on funding combined with more realistic assessments of threats at all spatial scales may well have resulted in more pragmatic assessments of problems and opportunities and, as a result, better project designs. Even though the length of these projects is typically only two to three years, many of them are building on work done by predecessor initiatives over several years (Wells, Ganapin, and Uitto 2001).

Future ICDPs

Future ICDPs will need to be designed on the basis of clearly stated objectives together with explicit and testable assumptions and tangible conservation targets. They need to be implemented using decentralized and adaptive management, and be able to draw on a toolbox of approaches. These projects should promote relatively simple and adaptive conservation and development initiatives that are consistent with an overall protected area strategy but based on specific site conditions and local community dynamics. ICDPs will need to play a more open and effective role in identifying and addressing diverse stakeholder interests, helping build PA management capacity, and supporting PA efforts to become more sensitive to and supportive of local needs. Finally, while local income-generating or other benefit-generating opportunities should be pursued energetically, these must make sense from cultural and conservation perspectives, as well as from economic and financial perspectives.

Perhaps most critical, ICDPs cannot act in isolation. They must seek effective partnerships to address larger-scale problems that defy local solutions. To effectively address these issues, approaches must be a "vertically integrated" mix of site-based programs, policy initiatives, and campaign action. The appropriate positioning of ICDPs relative to these other complementary conservation activities operating on a variety of spatial and temporal scales will be one of the major challenges of the emerging landscape- or ecoregion-scale conservation approaches. Although these methodologies are in their infancy, it already seems clear that their success will depend on the links between the constituent parts.

Readers may find that most of the ideas presented here are not new, and we recognize that many have been emphasized before by a variety of authors and practitioners, including ourselves. While new ideas and breakthrough solutions are always welcome, there is no substitute for carefully conceived, long-term commitments to high-priority field sites, complemented by relevant partnerships and actions to address larger-scale problems. That may be what the organizations supporting conservation consider themselves to be doing. Our simple message is that they need to be doing it better, in some cases a lot better.

References

Adams, J. S. and T. O. McShane. 1996. *The Myth of Wild Africa: Conservation Without Illusion*. Berkeley and Los Angeles: University of California Press.

Adams, W. M. and D. Hulme. 2001. If community conservation is the answer in Africa, what is the question? *Oryx* 35 (3):193–200.

Agrawal, A. 1997. *Community in Conservation: Beyond Enchantment and Disenchantment*. Gainesville, Fla.: Conservation and Development Forum.

Barrett, C. S. and P. Arcese. 1995. Are integrated conservation and development projects sustainable? On the conservation of large mammals in sub-Saharan Africa. *World Development* 23:1073–1084.

Bell, R. H. V. and E. McShane-Caluzi, eds. 1984. *Conservation and Wildlife Management in Africa*. Washington, D.C.: U.S. Peace Corps.

Blomley, T. and P. Mundy, eds. 1997. *Integrated Conservation and Development: A Review of Project Experiences from CARE*. Workshop Report. Copenhagen: CARE–Denmark.

Brandon, K. 2001. Moving beyond integrated conservation and development projects (ICDPs) to achieve biodiversity conservation. In D. R. Lee and C. B. Barrett, eds., *Tradeoffs or Synergies?* 417–432. Wallingford, Oxon, U.K.: CABI.

Brandon, K. and M. Wells. 1992. Planning for people and parks: Design dilemmas. *World Development* 20 (4):557–570.

Brandon, K., K. Redford, and S. Sanderson, eds. 1998. *Parks in Peril: People, Politics, and Protected Areas*. Washington, D.C.: Island Press.

Emerton, L. 1998. *The Nature of Benefits and the Benefits of Nature: Why Wildlife Conservation Has Not Economically Benefited Communities in Africa*. Community Conservation Research in Africa: Principles and Comparative Practice, paper no. 9. Manchester, U.K.: Institute for Development Policy and Management, University of Manchester.

Ferraro, P. J. 2001. Global habitat protection: Limitations of development interventions and a role for conservation performance payments. *Conservation Biology* 15 (4):1–12.

Ferraro, P. J. and A. Kiss. 2002. Direct payments to conserve biodiversity. *Science* 298:1718–1719.

Ghimire, K. B. and M. P. Pimbert, eds. 1997. *Social Change and Conservation: Environmental Politics and Impacts of National Parks and Protected Areas*. London: Earthscan.

Gunderson, L. H., C. S. Holling, and S. S. Light, eds. 1995. *Barriers and Bridges to Renewal of Ecosystems and Institutions*. New York: Columbia University Press.

Hart, T., C. Imboden, D. Ritchie, and F. Schwartzendruber. 1998. *Biodiversity Conservation Projects in Africa: Lessons Learned from the First Generation*. Environment Department Dissemination Note no. 62. Washington, D.C.: The World Bank.

Holling, C. S. 1978. *Adaptive Environmental Assessment and Management*. New York: John Wiley & Sons.

Hulme, D. and M. W. Murphree, eds. 2001. *African Wildlife and Livelihoods: The Promise and Performance of Community Conservation*. Oxford: James Currey.

IUCN (International Union for Conservation of Nature and Natural Resources). 1980. *World Conservation Strategy: Living Resource Conservation for Sustainable Development*. Gland, Switzerland: International Union for Conservation of Nature and Natural Resources.

Kramer, R., C. van Schaik, and J. Johnson, eds. 1997. *Last Stand: Protected Areas and the Defense of Tropical Biodiversity*. Oxford: Oxford University Press.

Larson, P. S., M. Freudenberger, and B. Wyckoff-Baird. 1998. *WWF Integrated Conservation and Development Projects: Ten Lessons from the Field 1985–1996*. Washington, D.C.: World Wildlife Fund.

Lee, K.N. 1993. *Compass and Gyroscope: Integrating Science and Politics for the Environment*. Washington, D.C.: Island Press.

Lewis, J. P., ed. 1988. *Strengthening the Poor: What Have We Learned?* U.S.–Third World Policy Perspectives, no. 10. Washington, D.C.: Overseas Development Council (ODC).

Maginnis, S., W. Jackson, and N. Dudley. 2001. *Guidelines to the Development of a Landscape Approach to Forest Conservation*. Gland, Switzerland: World Wildlife Fund and IUCN–The World Conservation Union.

McNeely, J. A. 1988. *Economics and Biological Diversity: Developing and Using Economic Incentives to Conserve Biological Resources*. Gland, Switzerland: International Union for Conservation of Nature and Natural Resources.

McNeely, J. A. and K. R. Miller, eds. 1984. *National Parks, Conservation, and Development: The Role of Protected Areas in Sustaining Society*. Washington, D.C.: Smithsonian Institution Press.

McShane, T. O. 1989. Wildlands and human needs: Resource use in an African protected area. *Landscape and Urban Planning* 19:145–158.

McShane, T. O. 1999. Voyages of discovery: Four lessons from the DGIS-WWF Tropical Forest Portfolio. *Arborvitae* suppl.: 1–6.

Munro, David, dir. 1991. *Caring for the Earth: A Strategy for Sustainable Living*. Gland, Switzerland: IUCN–The World Conservation Union, UN Environment Program, World Wildlife Fund.

Norton-Griffiths, M. and C. Southey. 1995. The opportunity costs of biodiversity conservation in Kenya. *Ecological Economics* 12:125–139.

Oates, J. F. 1999. *Myth and Reality in the Rain Forest: How Conservation Strategies Are Failing in West Africa*. Berkeley and Los Angeles: University of California Press.

Perrings, C., K.-G. Mäler, C. Folke, C. S. Holling, and B.-O. Jansson, eds. 1995. *Biodiversity Conservation: Problems and Policies*. Dordrecht, The Netherlands: Kluwer.

Redford, K.H. and B. Richter. 1999. Conservation of biodiversity in a world of use. *Conservation Biology* 13 (6):1246–1256.

Redford, K.H. and A. Taber. 2000. Writing the wrongs: Developing a safe-fail culture in conservation. *Conservation Biology* 14 (6):1567–1568.

Redford, K. H., P. Coppolillo, E. W. Sanderson, G. A. B. Fonseca, C. Groves, G. Mace, S. Maginnis, R. Mittermier, R. Noss, D. Olson, J. G. Robinson, A. Vedder, and M. Wright. 2003. Mapping the conservation landscape. *Conservation Biology* 17 (1):116–132.

Robinson, J. G. 1993. The limits to caring: Sustainable living and the loss of biodiversity. *Conservation Biology* 7:20–28.

Roe, D., J. Mayers, M. Grieg-Gran, A. Kothari, C. Fabricius, and R. Hughes. 2000. *Evaluating Eden: Exploring the Myths and Realities of Community-Based Wildlife Management*. Series no. 8: Series Overview. London: International Institute for Environment and Development (IIED).

Salafsky, N., R. Margoluis, and K. Redford. 2001. *Adaptive Management: A Tool for Conservation Practitioners*. Washington, D.C.: Biodiversity Support Program.
Sanjayan, M. A., S. Shen, and M. Jensen. 1997. *Experiences with Integrated Conservation-Development Projects in Asia.* Technical Paper no. 388. Washington, D.C.: The World Bank.
Sayer, J. 1995. *Science and International Nature Conservation.* Occasional Paper no. 4. Bogor, Indonesia: Center for International Forestry Research (CIFOR).
Stedman-Edwards, P. 1998. *Root Causes of Biodiversity Loss: An Analytical Approach*. Washington, D.C.: World Wildlife Fund.
Wascher, D., ed. 2000. *The Face of Europe: Policy Perspectives for European Landscapes.* Tilburg, The Netherlands: European Centre for Nature Conservation.
WCED (World Commission on Environment and Development). 1987. *Our Common Future*. Oxford: Oxford University Press.
Wells, M. 1997. *Economic Perspectives on Nature Tourism, Conservation, and Development.* Environment Department Paper no. 55 (Environmental Economics Series). Washington, D.C.: The World Bank.
Wells, M. and K. Brandon. 1992. *People and Parks: Linking Protected Area Management with Local Communities*. Washington, D.C.: The World Bank, U.S. Agency for International Development, and World Wildlife Fund.
Wells, M., S. Guggenheim, A. Khan, W. Wardojo, and P. Jepson. 1999. *Investing in Biodiversity: A Review of Indonesia's Integrated Conservation and Development Projects*. Washington, D.C.: The World Bank.
Wells, M. P., D. J. Ganapin, and J. I. Uitto. 2001. *Medium-Sized Projects Evaluation: Final Report.* Evaluation Report #2–02. Washington, D.C.: Global Environment Facility.
Wells, M. P., S. Qayum, and C. Tavera. In prep. Final Report on COMPACT—The Community Managment of Protected Areas Conservation Project. New York: UN Development Programme.
Western, D. and M. Wright, eds. 1994. *Natural Connections: Perspectives in Community-Based Conservation*. Washington, D.C.: Island Press.
Wilson, E. O., ed. 1988. *Biodiversity.* Washington, D.C.: National Academy of Science.
Wood, A., P. Stedman-Edwards, and J. Mang, eds. 2000. *The Root Causes of Biodiversity Loss*. London: Earthscan.
World Bank. 1993. *Area Development Projects: Lessons and Practices*. Washington, D.C.: Operations Evaluation Department, The World Bank.
Wunder, S. 2001. Poverty alleviation and tropical forests—What scope for synergies? *World Development* 29 (11):1817–1833.

Index